S0-AKJ-170

Chemometrics

Mathematics and Statistics in Chemistry

NATO ASI Series

Advanced Science Institutes Series

A series presenting the results of activities sponsored by the NATO Science Committee, which aims at the dissemination of advanced scientific and technological knowledge, with a view to strengthening links between scientific communities.

The series is published by an international board of publishers in conjunction with the NATO Scientific Affairs Division

A	Life Sciences	Plenum Publishing Corporation
B	Physics	London and New York
C	Mathematical and Physical Sciences	D. Reidel Publishing Company Dordrecht, Boston and Lancaster
D	Behavioural and Social Sciences	Martinus Nijhoff Publishers
E	Engineering and Materials Sciences	The Hague, Boston and Lancaster
F	Computer and Systems Sciences	Springer-Verlag
G	Ecological Sciences	Berlin, Heidelberg, New York and Tokyo

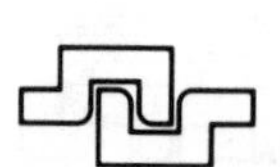

Series C: Mathematical and Physical Sciences Vol. 138

Chemometrics
Mathematics and Statistics in Chemistry

edited by

Bruce R. Kowalski
Department of Chemistry, University of Washington,
Seattle, Washington, U.S.A.

D. Reidel Publishing Company

Dordrecht / Boston / Lancaster

Published in cooperation with NATO Scientific Affairs Division

Proceedings of the NATO Advanced Study Institute on
Chemometrics - Mathematics and Statistics in Chemistry
Cosenza, Italy
September 12-23, 1983

Library of Congress Cataloging in Publication Data

NATO Advanced Study Institute on Chemometrics, Mathematics and Statistics in
Chemistry (1983 : Cosenza, Italy)
Chemometrics, mathematics and statistics in chemistry.

(NATO ASI series. Series C, Mathematical and physical sciences ; vol. 138)
Includes index.
1. Chemistry—Mathematics—Congresses. 2. Chemistry—Statistical methods—Congresses. I. Kowalski, Bruce R., 1942- . II. Title. III. Series: NATO ASI series. Series C, Mathematical and physical sciences ; vol. 138.
QD39.3.M3N38 1983 540'.1'51 84-16087
ISBN 90-277-1846-6

Published by D. Reidel Publishing Company
P.O. Box 17, 3300 AA Dordrecht, Holland

Sold and distributed in the U.S.A. and Canada
by Kluwer Academic Publishers,
190 Old Derby Street, Hingham, MA 02043, U.S.A.

In all other countries, sold and distributed
by Kluwer Academic Publishers Group,
P.O. Box 322, 3300 AH Dordrecht, Holland

D. Reidel Publishing Company is a member of the Kiuwer Academic Publishers Group

Printed in The Netherlands.

TABLE OF CONTENTS

PREFACE

At a time when computerized laboratory automation is producing a data explosion, chemists are turning to applied mathematics and statistics for the tools to extract useful chemical information from data. This rush to find applicable methods has lead to a somewhat confusing body of literature that represents a barrier to chemists wishing to learn more about chemometrics. The confusion results partly from the mixing of chemical notation and nomenclature with those of statistics, applied mathematics and engineering. Additionally, in the absence of collaboration with mathematicians, chemists have, at times, misused data analysis methodology and even reinvented methods that have seen years of service in other fields.

The Chemometrics Society has worked hard to solve this problem since it was founded in 1974 with the goal of improving communications between the chemical sciences and applied mathematics and statistics. The NATO Advanced Study Institute on Chemometrics is evidence of this fact as it was initiated in response to a call from its membership for advanced training in several areas of chemometrics. This Institute focused on current theory and application in the new field of Chemometrics: Use of mathematical and statistical methods, (a) to design or select optimal measurement procedures and experiments; and (b) to provide maximum chemical information by analyzing chemical data.

The Institute had two formal themes and two informal themes. First, lectures from statistics, mathematics and engineering presented the state of the art methods from their fields that have been found useful in chemistry. The emphasis for these lectures was on clarity, completeness, and attention to all of the important mathematical detail associated with the theory. Multivariate statistical methods such as principal components, multiple regression, canonical correlation and path modeling with latent variables received the lion's share of attention due to the focus they have drawn in recent years from chemists. Topics from spectrum and waveform analysis, control theory, experimental design, state and parameter estimation, cluster analysis, pattern recognition, computer graphics and optimization also were the topics of lectures that stressed theory and detail. These lectures had an enormous impact on the participants since for many of them, it was the first time they were offered the complete theory of methods on which they depend for research results.

The second theme of the Institute was a critical examination of the areas of chemical research where chemometrics has seen significant activity. These areas include chemical analysis, drug design, food chemistry, geochemistry, clinical medicine and environmental chemistry. For these lectures. chemists skilled in applying chemometrics to these areas of application briefly reviewed some of the successes in the area but spent more time presenting unsolved problems that chemometrics may address in the future. These lectures spawned many discussions and informal working groups that initiated research collaborations and technology exchange.

A primary problem that surfaced constantly was the lack of a commonly agreed upon notation and nomenclature for chemometrics. Therefore, one informal theme chosen was the development of a slate of chemometrics notation to be presented to the International Union of Pure and Applied Chemistry for eventual adoption.

A second informal theme was education in chemometrics. New courses are being taught at several universities around the world by chemists, many of whom are unfamiliar with the breadth of topics currently being addressed by chemometricians. Attendees had an opportunity to compare circular and course contents at a number of leading centers of chemometrics around the world.

While the present text is not offered as a course textbook, it will no doubt serve to make chemists and mathematicians aware of current problems and methods in use. The Institute could only accomodate 100 of the over 300 applicants so those that could not attend should benefit from this book. To those that did not apply, we invite you to learn more about chemometrics.

NATO's commitment to the international exchange of information by providing the most significant funding source for the Institute enabled the participants to gain a fuller perspective of Chemometrics. Because of the Institute, many participants have successfully collaborated on a variety of research projects. Truly, this is an important outcome. Other contributors include: Monsanto Company, the National Science Foundation (U.S.A.), and the Norwegian Food Research Institute.

LIST OF PARTICIPANTS

Dr. Guneri Akovali
Middle East Technical
University
Inonu Bulvari
Ankara
TURKEY

Christer Albano
Research Group of Chemometrics
Institute of Chemistry
University of Umea
S-901 87 Umea
SWEDEN

Thomas Blaffert
Philips Forschungslaboratorium
Hamburg
Vogt Kollnstrasse 30
D 2000 Hamburg 54
WEST GERMANY

Odd Borgen
Trondheim University
The Norwegian Institute of
Technology
Physical Chemistry Division
N-7034 Trondheim - NTH
NORWAY

Dr. Joseph J. Breen
OTS - Monitoring Team Leader
E.P.A. Office of Toxic
Substances
TS-798
401 M Street SW
Washington, D.C. 20460

Dr. Steve D. Brown
Professor
Department of Chemistry
Washington State University
Pullman, WA 99164-4630

Roy E. Bruns
Professor of Chemistry
Enderecamento Postal 13.100
Cidade Universitaria "Zeferino Vaz"
Barao Geraldo
Campinas, S.P.
BRAZIL

Kenneth William Burton
Science Department
The Polytechnic of Wales
Pontypridd
CF37 ADL
UNITED KINGDOM

Lutgarde Buydens
Vrje Universiteit Brussel
Laar Beeklaan 103
B-1090 Brussel
BELGIUM

Dr. Ing. Edip Buyukkoca
Chairman of EDP Center
Yildiz Universitesi
Kocaeli Muh Fak
Izmit-Kocaeli
TURKEY

M. Filomena Camoes
CECUL
Departmento de Quimica
Faculdade de Ciencuas
Rua da Escola Politecnica
1294 Lisboa Codex
PORTUGAL

Dr. Sergio Clementi
Dipartimento di Chimica
Universita di Perugia
Via Elce di Sotto 10
06100 Perugia
ITALY

Giovanni Consiglio
Cattedra di Chimica Organica
Facolta di Farmacia
Universita di Bologna
Via A. Zanolini 3
40126 Bologna
ITALY

Lloyd A. Currie
Research Chemist
Gas and Particulate Science Div.
Center for Analytical Chemistry
United States Dept. of Commerce
National Bureau of Standards
Washington, D.C. 20234

John C. Davis
Geologic Res. Section and
Professor Department of
Chemical & Petroleum Engineering
Kansas Geological Survey
1930 Avenue "A", Campus West
The University of Kansas
Lawrence, KS 66044-3896

Francesco Dondi
Istituto Chimico
Dell'Universita di Ferrara
Via L. Borsari, 46
Ferrara 44100
ITALY

Professor Domenico de Marco
Istituto Chimica Analitica
Universita di Messina
Via Dei Verdi
98100 Messina
ITALY

Stanley N. Deming
Department of Chemistry
University of Houston
Central Campus
Houston, TX 77004

Marc Demuynck
LISEC
B-3600 Genk
BELGIUM

Mrs. Marie-Paule Derde
Pharmaceutisch Instituut
Vrje Universiteit Brussel
Laarbeeklaan 103
B-1090 Brussel
BELGIUM

Dr. L. Domokos
Max-Planck-Institut fur
Kohlenforschung
Kaiser-Wilhelm-Platz 1
4330 Mulheim A.D. Ruhr
WEST GERMANY

A. M. Nunes Dos Santos
Faculdade de Ciencias e
Tecnologia
Seccao de Engenharia Quimica
Universidade Nova Lisboa
Quinta da Torre
2825 Monte Caparica
PORTUGAL

William J. Dunn
Univ. of Illinois at Chicago
Department of Medicinal
Chemistry and Pharmacognosy
833 S. Wood
Chicago, IL 60680

Kim H. Esbensen, Ph.D.
Chemometric Research Group
Dept. of Organic Chemistry
University of Umea
S-901 87 Umea
SWEDEN

John F. Feikema
3M Center Building 518 - ES&T
St. Paul, MN 55144

Max Feinberg
INRA Laboratoire de Chimie Analytique
16, Rue Claude Bernard
75231 Paris Cedex 05
FRANCE

Dr. M. H. Florencio
Centro de Espectrometria
de Massa
da Univ. Losboa
Complexo 1 do I.N.I.C. (1st)
Avenida Rovisco Pais
1096 Lisboa Codex
PORTUGAL

Michele Forina
Universita Degli Studi
Istituto de Scienze
Farmaceutiche
Viale Benedetto XV, 3
16132 Genova
ITALY

Ildiko Frank
Laboratory for Chemometrics
Department of Chemistry
University of Washington
Seattle, WA 98195

Stephen W. Gaarenstroom
Analytical Chemistry Department
General Motors Res.
Laboratories
Warren, MI 48090

Joseph A. Gardella, Jr.
Department of Chemistry
State Univ. of New York
- Buffalo
Buffalo, NY 14214

Dr. Paul Geladi
Avdeling Kemometri
Umea University
S 90187 Umea
SWEDEN

Stanley L. Grotch
Lawrence Livermore National Lab
P.O. Box 808 (L-329)
Livermore, CA 94550

Alice M. Harper
Biomaterials Profiling Center
391 South Chipeta Way
Research Park
Salt Lake City, UT 84108

Joel M. Harris
Associate Professor
Department of Chemistry
The University of Utah
Salt Lake City, UT 84112

Dr. Techn. Kaj Heydorn
Head of Isotope Division
Riso National Laboratory
DK 4000 Roskilde
DENMARK

Philip K. Hopke
Institute for Environmental
Studies
1005 West Western Avenue
Urbana, IL 61801

William G. Hunter
University of Wisconsin
Department of Statistics
1210 W. Dayton Street
Madison, WI 53706

J. Stuart Hunter
Statistics Consultant
100 Bayard Lane
Princeton, NJ 08540

Deborah L. Illman
Laboratory for Chemometrics
Department of Chemistry
University of Washington
Seattle, WA 98195

Tove Jacobsen
Brewing Industry Research
Laboratory
Forskningsveien 1
Oslo 3
NORWAY

Erik Johansson
Research Group of Chemometrics
Institute of Chemistry
University of Umea
S-901 87 Umea
SWEDEN

Dr. Ali Ersin Karagozler
Inonu University
Faculty of Science and Literature
Department of Chemistry
Malatya
TURKEY

G. Kateman
Dept. of Analytical Chemistry
Faculty of Science
Toernooiveld, Nijmegen
THE NETHERLANDS

Leonard Kaufman
Vrije Universeiteit Brussel
Fakulteit der Geneeskunde en der Farmacie
Farmaceutisch Instituut
Dienst Analytische Scheikunde en Bromatologie
Laarbeeklaan 103 B-1090 Brussel
BELGIUM

Guy W. Kornblum
Lab. voor Analytische Scheikunde
Delft University of Technology
P.O. Box 5029
2600 GA Delft
THE NETHERLANDS

Bruce R. Kowalski
Laboratory for Chemometrics
Department of Chemistry
University of Washington
Seattle, WA 98195

Dr. Kris Kristjansson
Science Institute
University of Iceland
Dunhaga 3, 107 Reykjavik
ICELAND

Olav Martin Kvalheim
Department of Chemistry
University of Bergen
Allegt. 41
N-5000 Bergen
NORWAY

Hans Laeven
Dept. of Analytical Chemistry
Van't Hoff Instituut
Universiteit van Amsterdam
Achtergracht 166
1017 WV Amsterdam
HOLLAND

Dr. Leonardo Lampugnani
Ist Chimica Analitica Strumentale C.N.R.
Via Risorgimento, 35
56100 - Pisa
ITALY

Dr. Silvia Lanteri
Istituto di Analisi e Tecnologie Farmaceutiche ed Alimentari
V. le Benedetto XV, 3
16132 Genova
ITALY

Giovanni Latorre
Dipartimento di Economia Politica
Universita dello Calabria
87036-Rende(es)
ITALY

Paul J. Lewi
Janssen Pharmaceutica
Research Laboratoria
2340 Beerse
BELGIUM

Professor Paolo Linda
Istituto Chimico
Universita di Trieste
Piazzale Europa 1
34127 Trieste
ITALY

Professor David Lunney
Chemistry Department
East Carolina University
Greenville, NC 27834

Roger Phan Tan Luu
IUT
Avenue Gaston Berger
Aix-en-Provence
FRANCE

H. J. H. MacFie
Meat Research Institute
Langford, Bristol BS18 7DY
UNITED KINGDOM

Edmund R. Malinowski
Professor of Chemistry
Department of Chemistry and
Chemical Engineering
Stevens Institute of Technology
Hoboken, NJ 07034

Dr. Eduard Marageter
Technical University Graz
Institute for Analytical
Chemistry
Micro- and Radiochemistry
Technikerstrasse 4/p
A-8010 Graz
AUSTRIA

Dr. Robert Marassi
Dipartimento di Scienze Chimiche
Universita di Camerino
62032 Camerino (MC)
ITALY

Harold Martens
Norwegian Food Research Inst.
P.O. Box 50
N-1432 Aas-NLH
NORWAY

Professor D. L. Massart
Vrije Universiteit Brussel
Farmaceutisch Instituut
Laarbeeklaan 103
B-1090 Brussel
BELGIUM

Robert R. Meglen, Ph.D.
Director, Analytical Lab
University of Colorado - Denver
1100 14th Street
Denver, CO 80202

Dr. Ivo Moret
Facolta' Chimica Industriale
Universita Venezia - Calle Larga
S. Marta 2137
30123 Venice
ITALY

Edward Morgan
Department of Science
The Polytechnic of Wales
Mid Glamorgan
South Wales, GREAT BRITAIN

Timothy J. Mulligan
Pacific Biological Station
Nanaimo, B.C. V9R 5K6
CANADA

Giuseppe Musumarra
Istituto Dipartimentale
di Chimica
Universita di Catania
viale A. Doria, 6
95125 Catania
ITALY

Tormod Naes
Norwegian Food Research Institute
P.O. Box 50
N-1432 Aas-NLH
NORWAY

Dr. G. Nickless
Department of Inorganic Chemistry
School of Chemistry
University of Bristol
Cantock's Close
Bristol BS8 1TS
UNITED KINGDOM

Dr. Tamerkan Ozgen
Department of Chemistry
Hacettepe University
Ankara
TURKEY

M. L. Parsons
Professor of Chemistry
Department of Chemistry
Arizona State University
Tempe, AZ 85287

Dr. Gregor Reich
Institute for Analytical Chemistry
University of Vienna
Waehringerstrasse 38
A-1090 Vienna
AUSTRIA

N. Lawrence Ricker
Assistant Professor
Department of Chemical Engineering
University of Washington
Seattle, WA 98195

Dr. Giuseppe Scarponi
Facolta' Chimica Industriale
Universita' Venezia
S. Marta 2137
J-30123 Venice
ITALY

Bruno Schippa
ENEA
C.N.S.-Cassaccia
TIB/CHIM Chiminal
C.S.N. Casaccia
Roma
ITALY

Hubert A. Scoble
Postdoctoral Associate
Dept. of Chemistry, Bldg. 56-028
Massachusetts Institute of Technology
Cambridge, MA 02139

Muhammad A. Sharaf
Box 293
Department of Chemistry
Cornell University
Ithaca, NY 14853

Dr. J. Smeyers-Verbeke
Vrje Universiteit Brussel
Farmaceutisch Instituut
Laarbeeklaan 103
B-1090 Brussells
BELGIUM

Dr. Ir. H. C. Smit
Laboratorium voor Analytische Scheikunde
Universiteit van Amsterdam
Nieuwe Achtergracht 166
1018 WV Amsterdam
THE NETHERLANDS

Dr. J. C. Smit
Radboud Ziekenhuis
Lab. voor Neurologie
Reinier Postlaan 4
65100 HB Nijmegen
THE NETHERLANDS

David L. Stalling, Ph.D.
Chief Chemist
United States Dept. of the Interior
Fish and Wildlife Service
Columbia National Fisheries
Route 1
Columbia, MO 65201

Scott W. Stieg
Harvey Mudd College
Department of Chemistry
Claremont, CA 91711

James Sneddon Swan
Head of Blending and Bottling Studies
Pentlands Scotch Whisky Research Ltd.
84 Slateford Road
Edinburgh EH11 1QU
UNITED KINGDOM

Nils Telnaes
Universitetet i Trondheim
NLHT - Kjemisk institutt, Rosenborg
7055 Dragvoll
NORWAY

Dr. Ing. Sorin Trestianu
Carlo Erba Strumentazione
Str. Rivoltana, Rodano
P.O. Box 10364
I 20110 Milano
ITALY

Antonio Trotta
Istituto Impianti Chimici
via Marzolo 9
35100 Padova
ITALY

Professor Nicola Uccella
Dipartimento di Chimica
Universita della Calabria
I-87100 Cosenza
ITALY

Professor Ivar Ugi
Organisch-Chemisches Institut
Technische Universitat Munchen
Lichtenbergstrasse 4
D-8046 Garching
WEST GERMANY

Pierre van Espen
Universitaire Instelling
Antwerpen
Department of Chemistry
Universiteitsplein 1
B-2610 Wilrijk
BELGIUM

Bernard Vandeginste
Laboratorium voor Analytische
Chemie
Tooernooiveld, Nijmegen
THE NETHERLANDS

Prof. Dr. H. A. Van't Klooster
Analyticäl Chemistry Lab RUU
Cruesenstraat 77A
3522 D Utrecht
THE NETHERLANDS

David J. Veltkamp
Laboratory for Chemometrics
Department of Chemistry
University of Washington
Seattle, WA 98195

Dr. Gabor E. Veress
Department for General and
Analytical Chemistry
Technical University
H-1521 Budapest
XI Gellert ter 4
HUNGARY

Lawrence E. Wangen
Los Alamos National Laboratory
Los Alamos, NM 87545

Charles L. Wilkins
Professor and Chairman
Department of Chemistry
University of California,
Riverside
Riverside, CA 92521

Anthony A. Williams
Food and Beverages Division
Long Ashton Research Station
University of Bristol
Long Ashton, Bristol, BS18 9AF
UNITED KINGDOM

W. Todd Wipke
Department of Chemistry
University of California
Santa Cruz, CA 95064

Dr. Svante Wold
Dept. of Organic Chemistry
University of Umea
S-901 87 Umea
SWEDEN

VIETATO FUMARE

EXPERIMENTAL DESIGNS: FRACTIONALS

J.S. Hunter

Professor Emeritus
Princeton University

Multivariate methods of data collection and analysis extend to experimental designs, most particularly to the fractional factorials. This paper is an exposition of the various forms of these designs and their analysis.

An outstanding characteristic of all Chemometricians is their interest in multivariate methods of data collection and analysis. This interest extends to multivariate experimental designs, most especially to the 2^k factorial and fractional designs. These designs can provide independent and simultaneous estimates of the main effects of k factors $x_1, x_2, \ldots, x_k$ upon a response η. Each effect is estimated with the same precision that would have existed if all the experiments had been performed only to estimate the effect of a single factor. These experimental strategies can also be designed to provide simultaneous estimates of all k(k-1)/2 two-factor interactions of the k factors and, if required, all other interactions up to order k. The simplest of these experimental designs are the full or complete 2^k factorials for $k \leq 5$. Their applications are described in many textbooks on experimental designs (1), (2), (3), (4). The concern of this paper is with fractional factorial designs, that is, strategies for obtaining estimates of the k main effects and possibly the k(k-1)/2 two factor

B. R. Kowalski (ed.), Chemometrics. Mathematics and Statistics in Chemistry, 1–15.

interactions in many fewer than the $n = 2^k$ experiments required by the full 2^k factorial. The economic advantages of a fractional factorial become obvious for values of $k>4$ since, for example, a full 2^7 factorial requires 128 experiments while the fractional 2^{7-4} design, the 1/16 th of the 2^7, requires only 8 experiments.

Consider the full 2^4 factorial design given on the left hand side of Table 1. Here two notations are employed to identify the four factors: x_1, x_2, x_3, x_4 or A, B, C, D. The "x" notation is convenient when mathematical models or geometry are discussed, the capital letter notation for the factors proves useful later when the "confounding pattern" or "alias structures" of the designs are discussed.

In this example the + and - notation for factor x_1 might represent two temperature settings, 100° C and 105°C; for factor x_2 two levels of pH, 4.5 and 4.8; for x_3, two types of catalyst A and B; for x_4 the presence (+) or absence (-) of some compound. We note the factors can be continuous (temperature and pH) or qualitative (catalysts and compound). The estimated effect of each factor is obtained by computing $\bar{y}_+ - \bar{y}_-$, the difference between the average yields for all runs carrying a + sign and all runs carrying a - sign for each factor. Thus, the estimated main effect of x_1 (or factor A) is (13.6 + 11.3 + ... + 22.1)/8 - (8.7 + 9.2 + ... + 16.4)/8 = 5.6. The other estimated main effects are x_2 = -0.3; x_3 = 7.7 and x_4 = 0.9. To estimate the 4(3)/2 = 6 two factor interactions additional columns of + and - signs are required. For example, to estimate the x_1x_2, (AB), interaction the column of signs given in Table 1 labelled x_1x_2 is produced by multiplying together row-wise the signs in columns x_1 with those in column x_2. The estimated x_1x_2 interaction effect is once again given by the difference between two averages $\bar{y}_+ - \bar{y}_-$, where the + and - signs are those given in the vector x_1x_2 as shown on the right hand side of Table 1. Thus, the x_1x_2 interaction estimate is given by (8.7 + 11.3 + ... + 22.1)/8 - (13.6 + 9.2 + ... + 16.4)/8 = -.04. Similar vectors of + and - signs can now be constructed for the x_1x_3, x_1x_4, ..., x_3x_4 interactions. To estimate a three factor interaction, say, $x_1x_2x_4$, the column of signs given in Table 1 is similaraly constructed from the original x_1, x_2 and x_4 columns. The estimate of the $x_1x_2x_4$ interaction equals -0.1. Finally the vector of signs for the four factor interaction $x_1x_2x_3x_4$ is similarly constructed. The estimated $x_1x_2x_3x_4$ effect equals -1.2. The complete list of estimates is given in Table 2. An algorithm, "Yates' Algorithm", exists that provides all the estimates without the need for constructing all $2^k - 1$ vectors of signs.

The magnitudes and signs of all 16 statistics are "clear" one of another, i.e., the estimates are all orthogonal. The variance of each effect $V(\bar{y}_+ - \bar{y}_-) = 4\sigma^2/n$ where $n = 2^k$, the total number of observations, and σ^2 the variance of the observations.

The 2^k factorials are often called "equal opportunity" designs since the magnitude and sign of each effect is determined from all the data as though it were the only effect to be estimated. Thus, before the experiments are run each factorial effect has an equal opportunity to demonstrate its influence upon the response to the experimenter. After the data are in hand, as always, some few effects (most often main effects) will be large, i.e., confidence limits about these effects will exclude zero, while most other estimated effects will be small. In this example the large effects are $x_3 = 7.7$ and $x_1 = 5.6$ and the next largest is the two factor interaction $x_1x_3 = 1.8$. Thus, the experimenter might conclude that only factors x_3 and x_1 have demonstrated any ability to influence the response while factors x_2 and x_4, although perhaps theoretically important, are ineffective over the ranges explored in these experiments. The experimenter has "screened" out the important subset of factors, and the 2^4 design now becomes a 2^2 factorial, in the important factors, replicated four times. These artificial replicates are sometimes used to provide an (inflated) estimate of σ^2.

Often the key question in the mind of the experimenter is, "Which of the many factors I can control have the greatest influence upon the response?" In the example above two of the four candidate factors proved important. Might the experimenter have identified these two factors with fewer runs? In this example we find the identification of the leading factors could have been accomplished with half the number of experiments, that is, with a 2^{4-1} <u>fractional</u> factorial design. The $2^{4-1} = 8$ runs that form this fractional are a carefully selected sub-set of the original 16 runs of the 2^4. The 2^{4-1} fractional design is displayed on the left-hand side of Table 3 along with the corresponding observations. The estimated effects are displayed in Table 4.

As with the full factorial, each estimated effect is a statistic of the form $\bar{y}_+ - \bar{y}_-$, the observations entering each average depending on the + and - signs appearing in the corresponding vectors.

Thus, the estimated x_1 effect is (15.1 + 11.3 + 22.3 + 22.1)/4 - (8.7 + 9.7 + 14.7 + 16.1)/4 = 5.4. Note however, the identical statistic is obtained when the experimenter attempts to obtain the estimated $x_2x_3x_4$ effect, as illustrated by the vector of signs for this effect also displayed in Table 4. The vector of signs for the x_1 effect is identical to those required for the $x_2x_3x_4$ effect. In statistical parlance, these two effects are thus "confounded". Viewed another way, the statistic $\bar{y}_+ - \bar{y}_- = 5.4$ carries two names, has an "alias" structure. The reader will note from the columns of signs displayed in Table 3 that the estimate of the x_1x_2 interaction effect will be identical to the estimate for the x_3x_4 interaction effect. Thus, x_1x_2 and x_3x_4 effects are confounded, or equivalently, the corresponding statistic -1.6 has aliases, that is, is named $x_1x_2 + x_3x_4$.

Confounding is a natural consequence of running fractional designs. Usually experimenters largely ignore names of statistics that involve high-order interactions. Thus, the statistic $5.4 = \underset{\sim}{1} + \underset{\sim}{234}$ (or A + BCD) would be simply identified as the main effect of factor $\underset{\sim}{1}$ (or A). The experimenter would observe on viewing Table 4 that factors $\underset{\sim}{1}$ and $\underset{\sim}{3}$ (A and C) were most influential, the same conclusion obtained after the 16 runs described in Tables 1 and 2. Of course, the precision of the estimates, $4\sigma^2/n$, is less since here n = 8. However, the larger estimates are already discernable despite the lower precision.

The theory underlying the construction of the $(1/2)^p$ fractions of the 2^k factorials i.e., the 2^{k-p} fractionals, is exposited in (5) and (6). Table 5 has been extracted from (4) and may be used to construct fractionals for $k \leq 8$. To illustrate the use of Table 5, let us construct the one-quarter replicate of the 2^6 factorial, that is, the 2^{6-2} fractional. The design requires $2^{6-2} = 16$ runs, and we begin to construct the design by first writing down the 16 runs of the 2^4 factorial as illustrated in Table 6. (Here letters are employed to identify the factors.) We note now in Table 5, in the section marked 2^{6-2}, the notations E = ABC and F = BCD. We therefore write down the two vectors of signs associated with the ABC, and the BCD interaction as illustrated in Table 6. The ABC columns of signs is now used to identify the + and - signs for factor E, and the BCD column the + and - signs of factor F. The six columns of signs provide a k = 6 factor design in n = 16 runs. The design is said to have two "generators", here the four letter "words" ABCE and BCDF, that is, the words simply obtained by combining the letters used to generate the design.

The experimenter would now run in some random sequence the 16 trials. Suppose, for this example, he obtained the same observations as given earlier in Table 1. To analyze the results he would at first imagine the data to be from a standard 2^4 factorial and hence get the table of estimates given earlier in Table 2. The key question is, "What are the full names of these fifteen estimated effects?" The answer is given by constructing a "defining relation" using the design generators. The defining relation is a sentence contining the identity $\underset{\sim}{I}$ (a symbol that acts as unity, or one), the generators, and all the words that can be constructed by multiplying the generators together. Thus we obtain initially the four word sentence: $\underset{\sim}{I}$ + ABCE + BCDF + AB^2C^2DEF. We now adopt the rule that any letter appearing an even number of times disappears and obtain the defining relation for this 2^{6-2}, i.e., $\underset{\sim}{I}$ + ABCE + BCDF + ADEF.

We note from Table 7 that the main effect of factor A = 5.6. Actually, the statistic 5.6 is here named A + BCE + ABCDF + DEF. This list of four names in the alias structure is obtained by multiplying through the defining relation by the symbol A and remembering the rule that any letter in a word appearing an even number of times disappears. Similarly, the four factor interaction in Table 2 equals -1.2. The names of this statistic are ABCD + DE + AF + BCEF. If we now drop from consideration all names with 3 or more symbols we obtain the estimates displayed in Table 7.

Reviewing Table 7, the experimenter would note the large estimated effects identified as A and C. The statistic 1.8 = AC + BE would likely be taken as evidence of an AC interaction. The experimenter has thus started out with 6 factors and in n = 16 runs he has identified the important few.

The example is illustrative of the usefulness of the 2^{k-p} fractionals. When k factors are studied simultaneously any one of the k may prove important, or any k(k-1)/2 pairs, or k(k-1)(k-2)/6 triplets, etc. The object is to screen for the leaders. The confounding, alias structure, can occasionally cause difficulty in interpretation, but this difficulty can be adjusted usually by two to four additional experiments. The 2^{k-p} fractionals are a powerful tool in the hands of the experimenter.

The "saturated" fractions (the Plackett and Burman designs (7)) allow the estimation of all k first order effects (confounded by two factor interactions which must be assumed equal to zero) in n = k + 1 runs.

The designs for k = 7 and k = 11 are displayed in Table 8. The k = 7 design can be obtained using Table 5.

Many other two level fractionals exist, for example an experimenter may want certain two-factor interactions not to be biased by other effects along with only the first order effects of still other factors. Methods also exist for changing two level factorials and fractionals into three or four level designs. The construction of such designs is described in the papers by Addelman (8), (9). Saturated factorials for n not a multiple of four have been provided by S. Webb (10). Although the Webb designs cannot provide clear estimates, they are the best designs possible. "D-optimal" designs are provided by T. J. Mitchell (11), and mixed level 2^k3^m fractionals by B. Margolin (12). Fractionals derived from the hyper-graeco-latin squares, called "orthogonal arrays" have recently generated much interest due to their application by Japanese statisticians, in particular T. J. Taguchi (13). The orthogonal array designs provide no estimates of interaction effects. The field of study in fractional designs continues to grow with recent (unpublished) work on "bits and pieces" of fractionals offering promise of further useful designs. Interesting texts written by industrial statisticians describing varied applications are (2), (4) and (14).

BIBLIOGRAPHY

1. Cochran, W.G., and G.M. Cox, Experimental Designs, 2nd ed., John Wiley & Sons, Inc., New York, 1957.

2. Davies, O.L., The Design and Analysis of Industrial Experiments, Oliver and Boyd, London, 1954.

3. John, P. W. M., Statistical Design and Analysis of Experiments, Macmillan, New York, 1971.

4. Box, G.E.P., W.G. Hunter, and J.S. Hunter, Statistics for Experimenters, John Wiley & Sons, Inc., New York, 1978.

5. Box, G.E.P., and J.S. Hunter, "The 2^{k-p} Fractional Factorial Designs I, II", Technometrics, vol. 3, pp. 311-351, 1961.

6. Raktoe, B.L., A. Hedayat, and W.T. Federer, Factorial Designs, John Wiley & Sons, Inc., New

7. Plackett, R.L. and J.P. Burman, "The Design of Optimal Multifactorial Experiments", Biometrika, No. 33, pp 305-325, 1946.

8. Addelman, S., "Orthogonal Main Effect Plans for Asymmetrical Factorial Experiments", Technometrics, no. 4, pp. 21-46, 1962.

9. Addelman, S., "Recent Developments in the Design of Factorial Experiments", The Journal of the American Statistical Association, no. 67, pp. 103-111, 1972.

10. Webb, S.R., "Non-orthogonal Designs of Even Resolution", Technometrics, no. 10, pp. 291-298, 1968.

11. Mitchell, T.J., "D-optimal 1st Order Designs", Technometrics, no. 20, pp. 369-380, 1978.

12. Margolin, B.H., "Orthogonal Main Effect $s^k 3^m$ Designs with Two Factor Interaction Aliasing", Technometrics, no. 10, pp. 559-573, 1968.

13. Taguchi, G. and Y. Wu, Introduction to Off-line Quality Control, Central Japan Quality Control Assoc., 1980.

14. Diamond, W.J., Practical Experiment Design, Lifetime Learning, Belmont, CA, 1981.

APPENDIX

Models

The analysis of the 2^k factorial, and fractionals, are based upon a unique class of mathematical models. To introduce the 'factorial' model we begin with a more general description of linear models.

Let η be some "true" response to be measured. Let $y = \eta + \varepsilon$ be an observed value of η with ε a "shock" or "error" assumed to be a random variable. The analyst is now faced with the "two model" problem, a model for η, and a model for ε. Usually ε is assumed to be a random variable having a Gaussian distribution with zero mean and constant variance σ^2. Initially, we assume that η is an unknown smooth function of a single factor x_1 under the control of the experimenter, i.e., $\eta = f(x_1)$. The 1st order and 2nd order Taylor's series approximations to $f(x_1)$ are given by:

1st order $\eta = \beta_o + \beta_1 x_1$

2nd order $\eta \;\; \beta_o + \beta_1 x_1 + \beta_{11} x_1^2$

In practice, the experimenter should attempt to keep the order of the approximating model to be as small as possible. Thus, before investing in a second order model various transformation in x_1, or η, should be attempted. Only rarely would third or higher order models be considered.

If η is a function of two factors, i.e., $\eta = f(x_1, x_2)$ the associated Taylor's series approximations are:

1st order $\eta = \beta_o + \beta_1 x_1 + \beta_2 x_2$

2nd order $\eta = \beta_o + \beta_1 x_1 + \beta_2 x_2 + \beta_{11} x_1^2 + \beta_{12} x_1 x_2$

For $\eta = f(x_1, x_2, \ldots, x_k)$ a function of k factors, then:

1st order $\eta = \beta_o + \sum_{i=1}^{k} \beta_i x_i$

2nd order $\eta = \beta_o + \sum_{i=1}^{k} \beta_i x_i + \sum_{i=1}^{k} \beta_{ii} x_i^2 + \sum_{i>j} \beta_{ij} x_i x_j$

Note there are k first order terms β_1, k quadratic coefficients β_{ii} and $k(k-1)/2$ crossproduct (sometimes termed two-factor interaction) coefficients in the second order model. For $k \geq 4$ there are more crossproduct coefficients than quadratic coefficients in a second order model.

The factorial model associated with the 2^k factorial and fractional designs does not contain any coefficient beyond those of the first order. Thus, for k = 2, the factorial model is:

$$\eta = \beta_o + \beta_1 x_1 + \beta_2 x_2 + \beta_{12} x_1 x_2$$

The factorial model for k = 3 is:

$$\eta = \beta_o + \beta_1 x_1 + \beta_2 x_2 + \beta_3 x_3 + \beta_{12} x_1 x_2 + \beta_{13} x_1 x_3 + \beta_{23} x_2 x_3 + \beta_{123} x_1 x_2 x_3.$$

Note especially the third order coefficient β_{123}, the three factor interaction contribution. In general, for p factors, the factorial model becomes:

$$\eta = \beta_o + \Sigma\beta_i x_i = \sum_{i \neq j} \beta_{ij} x_i x_j + \Sigma_{i>j>k} \beta_{ijk} x_i x_j x_k + \ldots + \beta_{ijk\ldots p} x_i x_j x_k \ldots x_p.$$

In practice, it is usual not to find coefficients higher than second order to be required in approximating unknown functions, due largely to the small regions of investigation, i.e., each x_i is varied over a narrow range and thus first or second order approximations prove effective.

A further considerations when the two level factorial design and model are used is that each x_i may represent some qualitative distinction (for example two types of catalyst, or the presence versus the absence of a catalyst). When the x_i are qualitative, no power terms can exist in the model.

Alias Structure

Let X be the matrix of derivatives (the matrix of 'independent' variables) associated with a 2^k factorial (or fractional) design and factorial model. Then in standard matrix notation, we write the model as:

$$\underset{\sim}{Y} = \underset{\sim}{X}\underset{\sim}{\beta} + e \qquad \begin{array}{l} \underset{\sim}{Y} \text{ is } n \times 1 \\ \underset{\sim}{X} \text{ is } n \times k \\ \underset{\sim}{\beta} \text{ is } k \times 1 \\ e \text{ is } n \times 1 \end{array} \qquad n + 2^k \text{ or } 2^{k-p}$$

The least squares estimate of $\underset{\sim}{\beta}$ is given by:

$$\underset{\sim}{B} = (\underset{\sim}{X}^T\underset{\sim}{X})^{-1}\underset{\sim}{X}^T\underset{\sim}{Y}.$$

The expected value of the extimates $\underset{\sim}{B}$ is given by:

$$\begin{aligned} E(\underset{\sim}{B}) &= E[(\underset{\sim}{X}^T\underset{\sim}{X})^{-1}\underset{\sim}{X}^T\underset{\sim}{Y}] = E[(\underset{\sim}{X}^T\underset{\sim}{X})^{-1}\underset{\sim}{X}^T(\underset{\sim}{X}\underset{\sim}{\beta} + G)] \\ &= E[(\underset{\sim}{X}^T\underset{\sim}{X})^{-1}\underset{\sim}{X}^T\underset{\sim}{X} + (\underset{\sim}{X}^T\underset{\sim}{X})^{-1}\underset{\sim}{X}^Te] = \beta, \end{aligned}$$

since $(\underset{\sim}{X}^T\underset{\sim}{X})^{-1}\underset{\sim}{X}^T\underset{\sim}{X} = \underset{\sim}{I}$ and $E[(\underset{\sim}{X}^T\underset{\sim}{X})^{-1}\underset{\sim}{X}^T\underset{\sim}{e}]$ = zero since $E(\beta) = 0$. Thus the estimates are unbiased estimates of the unknown parameters $\underset{\sim}{\beta}$.

Suppose now the correct model was $\underset{\sim}{Y} - \underset{\sim}{X}_1\underset{\sim}{\beta}_1 + X_2\underset{\sim}{\beta}_2 + \varepsilon$ and that the experimenter had fitted the simpler model $\underset{\sim}{Y} = \underset{\sim}{X}_1\underset{\sim}{\beta}_1 + \underset{\sim}{e}$. The expected value of $\underset{\sim}{B}$ is now:

$$\begin{aligned} E(\underset{\sim}{B}) &= (\underset{\sim}{X}_1{}^T\underset{\sim}{X}_1)^{-1}\underset{\sim}{X}_1{}^T\underset{\sim}{Y} \\ &= (\underset{\sim}{X}_1{}^T\underset{\sim}{X}_1)^{-1}\underset{\sim}{X}_1{}^T(\underset{\sim}{X}_1\underset{\sim}{\beta}_1 + \underset{\sim}{X}_2\underset{\sim}{\beta}_2) \end{aligned}$$

which gives

$$\begin{aligned} E(\underset{\sim}{B}) &= \underset{\sim}{\beta}_1 + (\underset{\sim}{X}_1{}^T\underset{\sim}{X}_1)^{-1}\underset{\sim}{X}_1{}^T\underset{\sim}{X}_2\underset{\sim}{\beta}_2 \\ &= \beta_1 + \underset{\sim}{A}\underset{\sim}{\beta}_2 \end{aligned}$$

where $\underset{\sim}{A} = (\underset{\sim}{X}_1^T \underset{\sim}{X}_1)^{-1} \underset{\sim}{X}_1^T \underset{\sim}{X}_2$ is called the "alias" matrix. The estimates $\underset{\sim}{B}_1$ are biased by the unestimated coefficients $\underset{\sim}{\beta}_2$.

The use of the defining relation in the analysis of the 2^{k-p} fractional represents a convenient algorithm for determining the alias structure (confounding pattern) accompanying each estimated effect.

Consider the 2^{3-1} design with generator ABC and defining relation I + ABC. The design matrix, factorial model, and the matrix of derivatives $\underset{\sim}{X}$ are given below. (We adjust notation letting X_A, X_B, and X_C define the factors.)

The 2^{3-1} design is initially analyzed as though it were a 2^2 factorial in factors X_A and X_B using the simpler model

$$\eta = \beta_0 + \beta_A X_A + \beta_B X_B + \beta_{AB} X_A X_B .$$

Thus the matrix of derivatives X is partitioned with $\underset{\sim}{X}_1$ and $\underset{\sim}{X}_2$ equalling

$$\begin{matrix} & X_0 & X_A & X_B & X_{AB} \end{matrix} \qquad\qquad \begin{matrix} & X_C & X_{AC} & X_{BC} & X_{ABC} \end{matrix}$$

$$\underset{\sim}{X}_1 = \begin{bmatrix} + & - & - & + \\ + & + & - & - \\ + & - & + & - \\ + & + & + & + \end{bmatrix} ; \qquad \underset{\sim}{X}_2 = \begin{bmatrix} + & - & - & + \\ - & - & + & + \\ - & + & - & + \\ + & + & + & + \end{bmatrix}$$

The estimates b_i equal one half the corresponding effects which were earlier defined to equal $\bar{y}_+ - \bar{y}_-$. The question is the names of these estimated effects, or equivalently, coefficients.

$$\underset{\sim}{B}_1 = \begin{bmatrix} b_0 \\ b_1 \\ b_2 \\ b_3 \end{bmatrix} = \begin{bmatrix} 45.5 \\ 1.5 \\ 2.5 \\ 0.5 \end{bmatrix}$$

The expected value of these estimates is given by:

$$E(\underset{\sim}{B}_1) = \underset{\sim}{\beta}_1 + \underset{\sim}{A}\underset{\sim}{\beta}_2$$

where

$$\underset{\sim}{A} = (\underset{\sim}{X}_1^T \underset{\sim}{X}_1)^{-1} \underset{\sim}{X}_1^T \underset{\sim}{X}_2$$

$$E\begin{bmatrix} b_0 \\ b_A \\ b_B \\ b_{AB} \end{bmatrix} = \begin{bmatrix} \beta_0 \\ \beta_A \\ \beta_B \\ \beta_{AB} \end{bmatrix} + 1/4 \begin{bmatrix} 0 & 0 & 0 & 4 \\ 0 & 0 & 4 & 0 \\ 0 & 4 & 0 & 0 \\ 4 & 0 & 0 & 0 \end{bmatrix} \begin{bmatrix} \beta_C \\ \beta_{AC} \\ \beta_{BC} \\ \beta_{ABC} \end{bmatrix} = \begin{bmatrix} \beta_0 + \beta_{ABC} \\ \beta_A + \beta_{BC} \\ \beta_B + \beta_{AC} \\ \beta_{AB} + \beta_C \end{bmatrix}$$

The defining relation I + ABC similarly tells us that the identical alias structure (confounding pattern) exists, i.e.

1.5 estimates A + BC
2.5 estimates B + AC
0.5 estimates C + AB

TABLE 1

The 2^4 Factorial Design

Factors Estimation					Interaction		
A	B	C	D	Response	AB	ABD	ABCD
x_1	x_2	x_3	x_4	y	x_1x_2	$x_1x_2x_4$	$x_1x_2x_3x_4$
–	–	–	–	8.7	+	–	+
+	–	–	–	13.6	–	+	–
–	+	–	–	9.2	–	+	–
+	+	–	–	11.3	+	–	+
–	–	+	–	15.8	+	–	–
+	–	+	–	22.3	–	+	+
–	+	+	–	16.1	–	+	+
+	+	+	–	24.2	+	–	–
–	–	–	+	11.8	+	+	–
+	–	–	+	15.1	–	–	+
–	+	–	+	9.7	–	–	+
+	+	–	+	14.6	+	+	–
–	–	+	+	14.7	+	+	+
+	–	+	+	24.0	–	–	–
–	+	+	+	16.4	–	–	–
+	+	+	+	22.1	+	+	+

TABLE 2

Estimated Effects from a 2^4

Grand Average	Main Effects	Two Factor Interactions
y = 15.6	1 or A = 5.6	12 or AB = -0.4
	2 or B = -0.3	13 or AC = 1.8
	3 or C = 7.7	14 or AD = 0.2
	4 or D = 0.9	23 or BC = 0.8
		24 or BD = -0.4
		34 or CD = -1.2

Three Factor Interactions	Four Factor Interactions
123 or ABC = -0.1	1234 or ABCD = -1.2
124 or ABD = -0.1	
134 or ACD = -0.1	
234 or BCD = -0.2	

TABLE 3

The 2^{4-1} Fractional Factorial Design

A x_1	B x_2	C x_3	D x_4	Response y	B C D $x_2x_3x_4$	A B x_1x_2	C D x_3x_4
-	-	-	-	8.7	-	+	+
+	-	-	+	15.1	+	-	-
-	+	-	+	9.7	-	-	-
+	+	-	-	11.3	+	+	+
-	-	+	+	14.7	-	+	+
+	-	+	-	22.3	+	-	-
-	+	+	-	16.1	-	-	-
+	+	+	+	22.1	+	+	+

TABLE 4

Estimated Effects from the 2^{4-1} Fractional

1 + 234 or A + BCD = 5.4	12 + 34 or AB + CD = -1.6
2 + 134 or B + ACD = -0.4	13 + 24 or AC + BD = 1.4
3 + 124 or C + ABD = 7.6	14 + 23 or AD + BC = 1.0
4 + 123 or D + ABC = 0.8	

TABLE 5

Generators for the 2^{k-p} Fractionals

k = 4	k = 5	k = 6	k = 7	k = 8
2^{4-1} n=8 D = ABC	2^{5-1} n=16 E = ABCD	2^{6-1} n=32 F = ABCDE	2^{7-1} n=64 G = ABCDEF	2^{8-1} n=128 H = ABCDEFG
	2^{5-2} n=8 D = AB E = AC	2^{6-2} n=16 E = ABC F = BCD	2^{7-2} n=32 F = ABCD G = ABDE	2^{8-2} n=64 G = ABCD H = ABEF
		2^{6-3} n=8 D = AB E = AC F = BC	2^{7-3} n=16 E = ABC F = BCD G = ACD	2^{8-3} n=32 F = ABC G = ABD H = BCDE
			2^{7-4} n=8 D = ABC E = AB F = AC G = BC	2^{8-4} n=16 E = ABC F = ABD G = ACD H = BCD

Table adapted from a more extensive table given in Box, Hunter and Hunter (4).

TABLE 6

Constructing a 2^{6-2} Fractional

A	B	C	D	E ABC	F BCD	Response y
−	−	−	−	−	−	8.7
+	−	−	−	+	−	13.6
−	+	−	−	+	+	9.2
+	+	−	−	−	+	11.3
−	−	+	−	+	+	15.8
+	−	+	−	−	+	22.3
−	+	+	−	−	−	16.1
+	+	+	−	+	−	24.2
−	−	−	+	−	+	11.8
+	−	−	+	+	+	15.1
−	+	−	+	+	−	9.7
+	+	−	+	−	−	14.6
−	−	+	+	+	−	14.7
+	−	+	+	−	−	24.0
−	+	+	+	−	+	16.4
+	+	+	+	+	+	22.1

TABLE 7

Estimated Effects for 2^{6-2}

5.6 = A	−0.4 = AB + CE	−0.1 = E	−1.2 = DE + AF
−0.3 = B	1.8 = AC + BE	−0.1 = *	
7.7 = C	0.2 = AD + EF	−0.1 = **	
0.9 = D	0.8 = BC + DF	−0.2 = F	
	−0.4 = BD + CF		
	−1.2 = CD + BF		

* −0.1 = ABD + CDE + ACF + BEF

** −0.1 = ACD + BDE + ABF + CEF

TABLE 8

The Plackett and Burman Designs for $k = 7, 11$

A	B	C	D	E	F	G
−	−	−	−	+	+	+
+	−	−	+	−	−	+
−	+	−	+	−	+	−
+	+	−	−	+	−	−
−	−	+	+	+	−	−
+	−	+	−	−	+	−
−	+	+	−	−	−	+
+	+	+	+	+	+	+

A	B	C	D	E	F	G	H	I	J	K
+	−	+	−	−	−	+	+	+	−	+
+	+	−	+	−	−	−	+	+	+	−
−	+	+	−	+	−	−	−	+	+	+
+	−	+	+	−	+	−	−	−	+	+
+	+	−	+	+	−	+	−	−	−	+
+	+	+	−	+	+	−	+	−	−	−
−	+	+	+	−	+	+	−	+	−	−
−	−	+	+	+	−	+	+	−	+	−
−	−	−	+	+	+	−	+	+	−	+
+	−	−	−	+	+	+	−	+	+	−
−	+	−	−	−	+	+	+	−	+	+
−	−	−	−	−	−	−	−	−	−	−

TABLE 9

Matrices: 2^{3-1} Factorial

Design Matrix			Matrix Derivatives $\underset{\sim}{X}$				Interaction Vectors				OBS $\underset{\sim}{Y}$
X_A	X_B	X_C	X_0	X_A	X_B	X_C	X_AX_B	X_AX_C	X_BX_C	$X_AX_BX_C$	
−	−	+	+	−	−	+	+	−	−	+	42
+	−	−	+	+	−	−	−	−	+	+	44
−	+	−	+	−	+	−	−	+	−	+	46
+	+	+	+	+	+	+	+	+	+	+	50

Factorial Model

$$y = \beta_0 + \beta_A X_A + \beta_B X_B + \beta_C X_C + \beta_{AB} X_A X_B + \beta_{AC} X_A X_C$$
$$+ \beta_{BC} X_B X_C + \beta_{ABC} X_A X_B X_C + \epsilon$$

MULTIVARIATE DATA ANALYSIS IN CHEMISTRY

Svante Wold, C. Albano, W. J. Dunn III, U. Edlund, K. Esbensen, P. Geladi, S. Hellberg, E. Johansson, W. Lindberg and M. Sjöström

Inst. of Chemistry, Umeå Univ., 901 87 Umeå, Sweden

Part 1: Principal Components and Factor Analysis and Multidimensional Scaling. Part 2: Pattern Recognition. Part 3: Multiple regression (MR) and Partial least squares modelling with latent variables (PLS).

Abstract: Any data table produced in a chemical investigation can be analysed by bilinear projection methods, i.e. principal components and factor analysis and their extensions. Representing the table rows (objects) as points in a p-dimensional space, these methods project the point swarm of the data set or parts of it down on a F-dimensional subspace (plane or hyperplane). Different questions put to the data table correspond to different projections.

This provides an efficient way to convert a data table to a few informative pictures showing the relations between objects (table rows) and variables (table columns).

The methods are presented geometrically and mathematically in parallell with chemical illustrations.

B. R. Kowalski (ed.), Chemometrics. Mathematics and Statistics in Chemistry, 17–95.

The chemical data explosion.

Chemistry has changed dramatically in the last 25 years. Before 1960, most chemical problems were solved by "wet chemistry", i.e. titration, precipitation, weighing and specific reagents (fig.1). Today the situation is different. Chemical samples are now inserted into big machines called instruments, e.g. NMR, IR, UV, INAA, XRF, SSMS, ICP and ESCA spectrometers and gas and liquid chromatographs, etc. (figure 2).

The problem with these instruments is that they do not give the direct information of orange precipitates or the amount of added 0.1 M NaOH. Rather they pour out large amounts of signals, curve forms, numbers, ..., in short, data. The need to convert these data to information and the need to design experiments and measurements so that the resulting data indeed contain information has led to the development of a new field in chemistry, chemometrics. Methods and approaches are fetched from statistics and applied mathematics, while the problems dealt with remain those of chemistry.

Data tables

We shall now discuss a special but common form of data analysis, namely the analysis of data tables. Chemical data can often in a natural way be arranged as a table, a data matrix (figure 3). We assume that we have measured p variables on n objects. Typical chemical objects are analytical samples, chemical compounds and chemical processes, but the objects can be anything we like to think of as conceptually separate units. The variables are often derived from amounts of chemical constituents in the objects as for instance inferred from the size of gas or liquid chromatographic peaks. Other typical variables are those of spectroscopy, e.g. NMR, IR, UV, INAA, XRF and MS.

We use x to denote data, the index i for objects and the index k for variables. Hence, the element x_{ik} denotes the value of variable k on object i. The matrix (table) containing the data is called X.

Figure 1. In classical chemistry, information was directly obtained by "wet" methods.

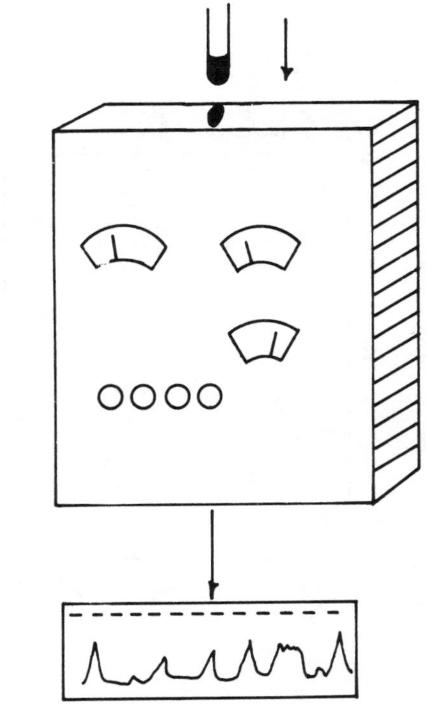

Figure 2. The instruments of today spew out data, which must be analysed to give information.

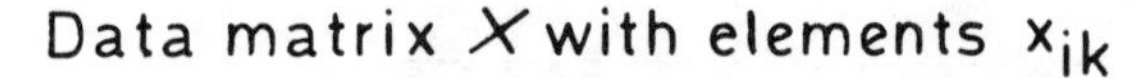

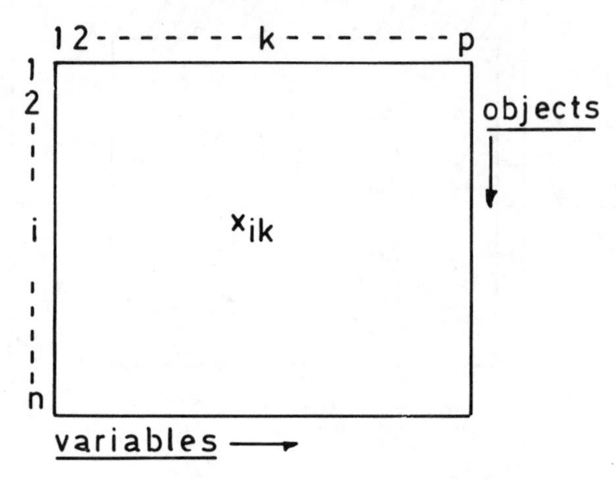

Figure 3. A data table, X, of n objects times p variables.

Part 1: Principal components and factor analysis

A basic method in the analysis of data tables is to calculate a representation of the n x p table X as a product of two matrices T (n x F) and P (F x p) plus a table of deviations, residuals, E. This corresponds to projecting X down on a few dimensional space (line, plane or hyperplane) as will be discussed below. When the number of components or factors, F, is small compared to p and n, this modelling provides a considerable simplification and reduction of the data set X.

This analysis has many names. In chemistry factor analysis, principal components analysis, eigen vector projections, singular value decomposition and Karhunen-Loeve expansions are commonly used. The way these variants of the same analysis are applied in chemistry, they correspond to principal components analysis in the statistical nomenclature. We shall henceforth use the name principal components analysis, briefly PCA.

Example 1, TLC of 54 drugs in 8 solvent systems.

Musumarra et.al. 1983 run 54 drugs in thin layer chromatography (TLC) with 8 different solvent systems. We can see each drug as one objects and the measured R_f values as eight variables measured on each object. Thus the data are contained in a 54 x 8 data matrix, X.

The scope is to develop an identification of drugs on the basis of its TLC R_f values in a battery of different solvent systems. One also wishes to use the given data to find out about noise, amount of information, relevant eluents, possible simplifications. And a graphical display of the results is preferred.

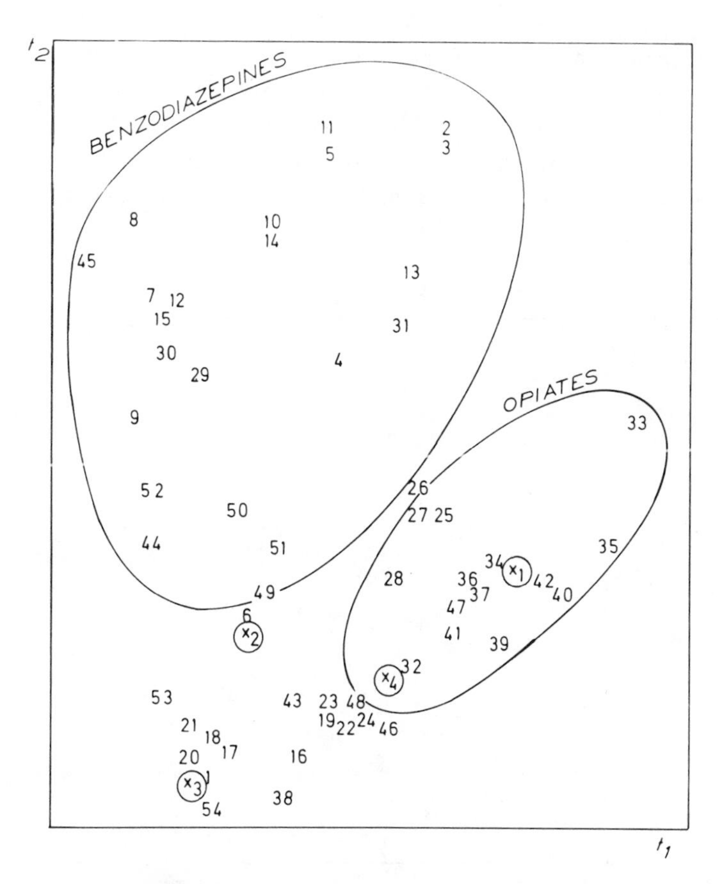

<u>Figure 4.</u> The first PC (t_1) plotted against the second (t_2) for the TLC data of Musumarra et.al.(1983). The circles are repeated observations of compounds 1, 6, 32 and 34, which indicates the small range of candidates if these were unknown drugs.

The data matrix was scaled to unit variance over the variables and analysed in terms of principal components (PC). Two significant PC.s were obtained which together describe 82 % of the variance of X.

A plot of these two PC.s is shown in figure 4. We see each object as a point. Chemically similar compounds are grouped together. Moreover, circles corresponding to repeated TLC of compounds 1 (cocaine), 6 (flurazepame), 32 (diacetylmorphine) and 34 (codeine) are seen to be close to the corresponding points of the "training set". Hence, the number of possible candidates for these "unknowns" is reduced to very few. The R_f values indeed have information which is useful for drug identification.

A plot of the PC loadings, i.e. the coefficients p_k for each of the 8 variables (R_f values) in the PC model, is shown in figure 5. The variables are clearly grouped. This shows that three R_f values -- e.g. 2, 4 and 7 -- are sufficient to give the

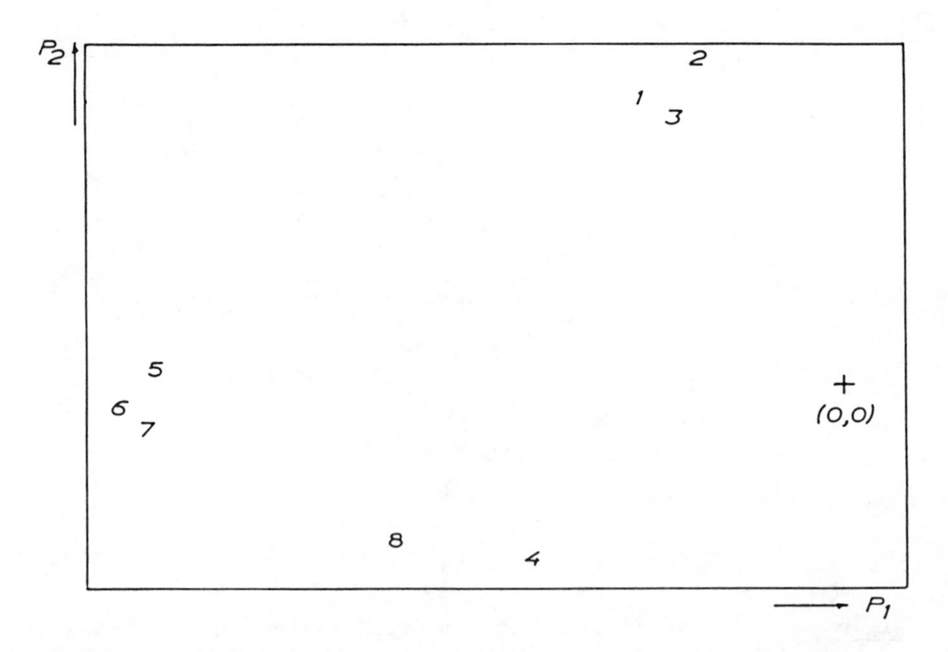

Figure 5. The loading plot of the first two PC.s of the Musumarra et.al.(1983) TLC data. The point (0,0) is indicated by a +.

same information as the original eight, albeit with a somewhat decreased precision.

To summarize: The analysis of the 54 x 8 data table shows that the systematic part of the data can be described by two "factors", principal components. These can be used to visualize the place of the 54 drugs with respect to each other and to indicate the likely candidates for "unknown" drug samples. The loading plot shows a strong grouping of the variables into three clusters.

Variables.

The variables used to characterize chemical objects usually are of either of two types.

(I) Those derived from spectra (NMR, IR, UV, MS, ESCA, Raman, X-ray,...). Raw data, i.e. spectra, can either be converted to variables by directly digitizing the spectra at given frequencies, or by specifying the position of specified signals, say the given carbons in C-13 NMR or the C=C and the C=O peaks in IR. Later we shall discuss this important question of translating raw data to variables.

(II) Those related to the amounts of chemical constituents in the samples. Such variables can be peak heights or integrals of reoccuring peaks in chromatographic profiles. GC, HPLC, electrophoresis, isoelectric focusing, ion chromatography, gel filtration, ..., are common chromatographic techniques used in various chemical fields.

Amino acid concentrations and trace and main element concentrations are often used.

(III) Other. Curve forms from kinetic runs or electrochemical measurements can be transformed to variables. So can rate constants, calorimetrically measured enthalpies, quantum mechanical indices, distorsion energies calculated by molecular mechanics, melting points and refractive indices.

The important thing is that the variables mean the same thing in all objects in the data table. Thus if variable 12 is the size of a GC peak of ethylhexanoate in object one, it must be so also in objects two, three, ..., n. Hence peak identification is crucial when chromatograms are converted to variables. The same with NMR and IR spectra.

However a variable may well be an aggregate of several "constituents", as long as it is the same aggregate for all objects. It is better to have fewer well assigned variables than many of uncertain assignment.

Objects

In chemical applications of PCA, the most common objects are analytical samples or chemical compounds (as in example 1). The former are often characterized by chromatographic profiles and the latter by spectra, but nothing hinders the characterization of mixtures by spectroscopic profiles. See e.g. the chapter by Martens.

Other typical chemical objects are given by repeated observations of a chemical process or reaction. Several different things are measured, often with the hope to find patterns in these data that can be used to control the process or reveal the mechanism of the reaction.

Clinically and biologically oriented chemists often make multitudes of measurements on individuals (rats, frogs, patients). Each such individual then is an object in the present jargon. If repeated observations are made, each multivariate observation may constitute an object.

Thus, by an object we henceforth mean a vector of p numbers, p variable values, used to characterize a conceptually recognizable system. In short, one row in the data table is one object.

Scope of Principal Components Analysis (PCA)

The primary scope with PCA is to get an overview of the dominant "patterns" in a data table. These patterns are of two kinds: Relationships between objects and relationships between variables. These are visualized by pictures and graphs. Such phenomena as strong groups of objects or variables are seen. So are abnormal, anomalous objects or variables, outliers. These are often scientifically and technically interesting. For instance, anomalies in geochemical data may indicate oil or gold ore and anomalies in spectral data a chemical compound of a new type or a new "effect" that you can name after yourself!

The dimensionality reduction accomplished by PCA is often of great value. In example 1 it was found that the data contain two independent dimensions and not eight. Two is much easier to discuss and think about and display than eight. And these dimensions, factors, principal components, can often be given chemical names, e.g. inductive and resonance effects in organic chemistry, polarity and hydrogen bonding in gas chromatography.

When analysing data on mixtures, the number of factors is directly interpretable as the minimum number of chemical constituents in the samples. In solution chemistry this may be a nice way to find the number of "complexes". In pyrolysis-GC or MS, the number of factors may be related to the number of different macromolecules in the pyrolized samples, etc. The resolution of curveforms into components, e.g. spectral bands or chromatographic peaks, is a typical PC problem (Shurvell and Dunham, 1978 and Sharaf and Kowalski, 1981).

PCA is sometimes used for predictions. One can leave holes in the data matrix and use PCA to predict these missing values (see e.g. Malinowski, 1980). This has been successfully used in NMR signal assignment and the prediction of the biological effect of chemical compounds. It is emphasized here, however, that there are more efficient ways to predict single or multiple values by multivariate data analysis, the so called partial least squares (PLS) methods. See part 3 in this chapter and Johnels et al. (1983). The comparison of chemical analytical methods is another area where PCA has been used (Carey et.al., 1975), but where PLS methods are more efficient.

The number of components (factors), F.

An important result of PCA is how many significant factors are seen in the data. This number of factors is often directly interpretable as the minimum number of molecular species that can be postulated in a investigated samples or the number of "chemical effects" that influence an ensemble of reactions.

When discussing significance, it is important to remember the difference between statistical significance and chemical significance. Thus, in the present context, we have a number, say F, of statistically significant factors. This is the number seen in the data. But some of these factors may be very small, contribute very little to the variation in the data. We must decide how big a factor must be to be <u>chemically</u> significant. It may well be so that only one or two of the F factors are of this size and that the minor factors, while statistically significant, are chemically totally uninteresting.

If this difference between statistical and chemical significance is not understood, one gets into big trouble when the data set one investigates is increased in size, either by measuring more variables or by studying more objects or both. Often then the number of statistically significant factors increase, because the larger data set can reveal the existance of minute regularities. But these new factors often are so small and chemically unimportant that they don't change any conclusions or improve any results. But <u>you</u> must have the courage to decide that they are so small as to be uninteresting. Otherwize the numbers take command instead of yourself.

It must be remembered, however, that to be at all interpretable, a component must be statistically significant. The inclusion of statistically insignificant components into the rotation, for instance, will lead to a gross overinterpretation of the chemical results. In chemical practice this seems to be a more common error than the reverse.

<u>Geometrical structures in p-space.</u>

The essential ideas of PCA are easiest demonstrated in terms of geometry. This because we have good abilities to understand structures in two and three dimensional spaces, an understanding that can be generalized to higher dimensional spaces with p dimensions, p-spaces. These spaces are obtained by letting each of the p variables, columns in the data matrix X (figure 3), define one orthogonal coordinate axis. Though such a space cannot be seen directly or touched upon, it has a kind of real existance as discussed below.

In p-space, geometrical constructs such as points, lines, planes, angles, volumes and distances have mathematical definitions analogous to the definitions in 2- and 3-space. Hence we can use 2- and 3-spaces as conceptual models for p-spaces, remembering that their properties are analogous.

The most fundamental and practical property of p-space in the present context is that the p variable values of a single object, i.e. one row in X, are represented by a single point in this space as shown in figure 6. The data of the n objects are hence represented by a swarm of n points (figure 7).

Multivariate data analysis -- where PCA forms the basis -- can be seen as methods to describe the location and spread of this point swarm by means of pictures and statistical parameters such as averages and variances along different directions in p-space.

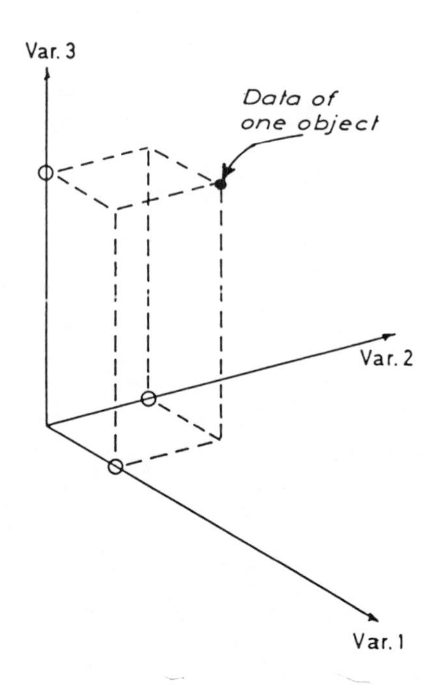

Figure 6. Three dimensional data spaces are henceforth used to illustrate p-dimensional spaces. In p-space the data of one object are represented by the position of a single point.

The PC model in p-space and in matrix form.

We shall now, in parallel, develop the PCA model of a given data matrix geometrically and in matrix notation. We assume that the data matrix X is appropriately transformed and scaled (this is discussed in a separate section below). The n objects are represented as the point swarm in p-space in fig. 7 and, parallelly, as n rows in the data table, X (right part of fig.7).

1. Represent the data swarm by its "mid point". The coordinates of this point are the averages of the variables, $\bar{x}_k$, which together form the row vector $\bar{x}$ (figure 8).

2. Subtract these averages from the data to get the residuals e_{ik}, elements in the matrix E. This corresponds to moving the coordinate system to now be centered in the point $\bar{x}$.

3. Rename the residuals E to X. We now have scaled and centered data, X.

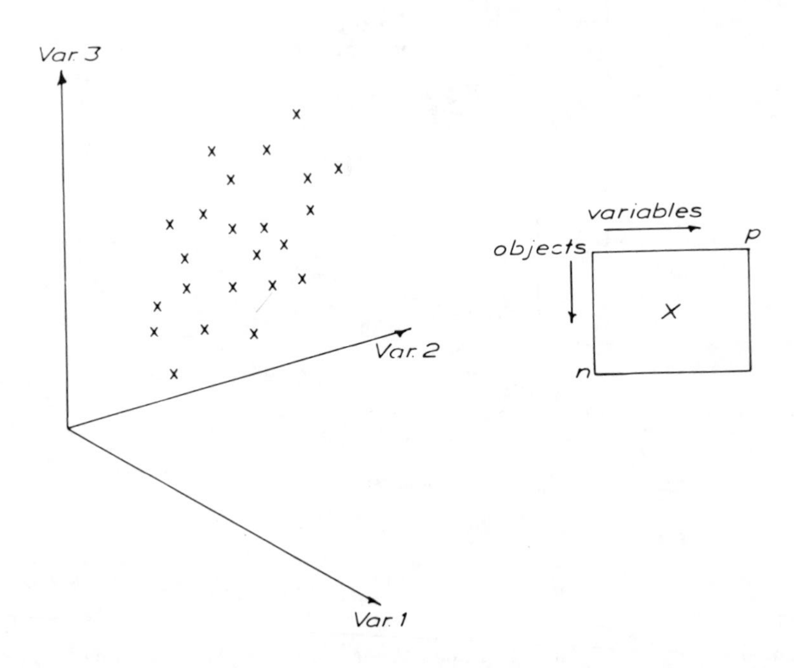

Figure 7. The n objects in the data set constitute a swarm of points in p-space.

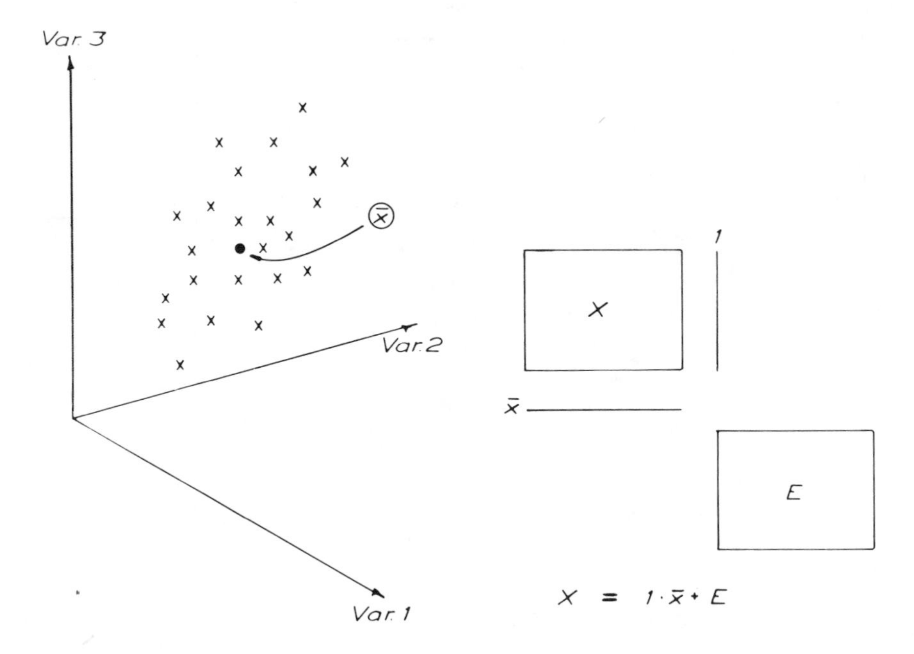

Figure 8. The simplest model is to represent the data set by its middle point, $\bar{x}$. This has the coordinates defined by the averages of the variables. In matrix form: $X = 1 \cdot \bar{x} + E$. The matrix E contains the residuals e_{ik}.

---

4. Fit a straight line to the n points in p-space so that the deviations are as small as possible in the least squares sense (figure 9). The direction coefficients of this line are called the "loadings" -- one for each variable k -- and denoted by p_{1k} forming the row vector p_1 (the first loading vector). When each point is projected down on this line, we get the scores t_{i1}, i.e. the coordinate of point i along the axis p_1 (figure 9). Subtract $t_i p_k$ from x_{ik} to get the residuals e_{ik}.

5. Rename the residuals E to X. This corresponds to removing the direction p_1 from the data. When now the new X is used to fit another straight line to its "objects", this corresponds to fitting a second line through $\bar{x}$, orthogogonal to the first line, and again making the residuals as small as possible (least

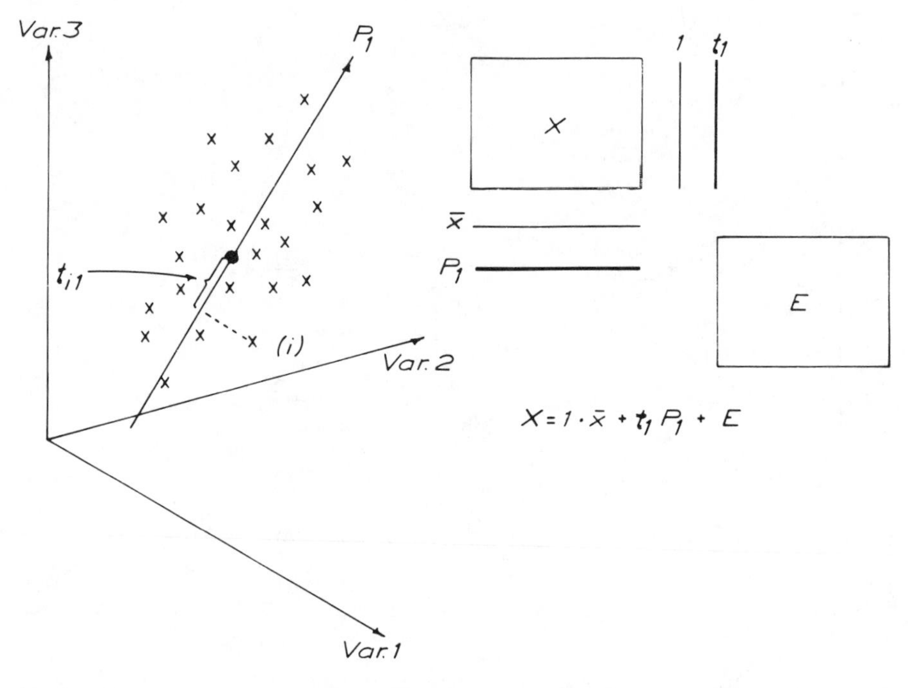

Figure 9. The data are least squares modelled by a line through the average $\bar{x}$. This line is the first principal component (PC) and its equation is defined by the loading vector p_1. The projection of point i on the line lies on the distance t_{i1} from $\bar{x}$

squares). This is shown in figure 10.

6. This can now be repeated over again until, after p or n factors, whichever comes first, the residuals are identically zero. We have then constructed a new coordinate system with H axes (H is the smaller of n and p) and represent the n objects as points in this H-space.

7. The useful thing with PCA, however, is that we can use only the first F components to represent X. This, we see, is a projection of X down on an F-dimensional space, represented in figure 11 with F=2. The plane in figure 11 actually has the character of a window in p-space. We can lift out this plane and look at the projections of the n points as a picture of what goes on in p-space, just like we look at the real 3-dimensional world through 2-dimensional windows (our eyes, house windows, etc.).

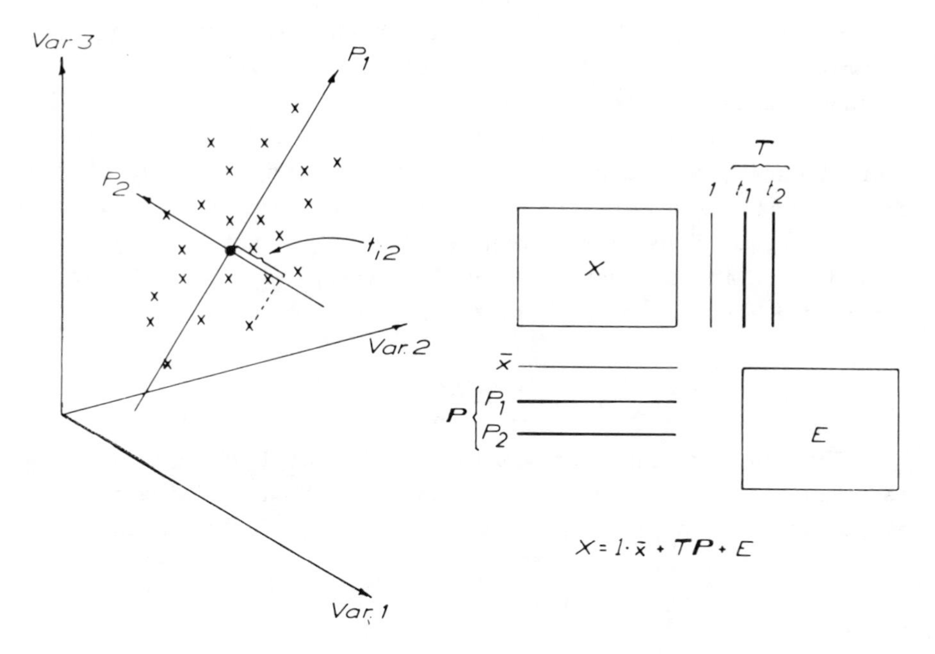

Figure 10. The second PC is a line (p_2) through $\bar{x}$, perpendicular to p_1.

The principal components model

With PCA we have decomposed X as

$$X = 1 \cdot \bar{x} + T \cdot P + E$$

where the n x F score matrix T describes the projection of the n object points down on the F dimensional hyperplane defined by the F x p loading matrix P. The residual matrix E contains what is left over. When these residuals are small compared to the variation in X, the PC model is a good representation of X itself. We realize that a crucial problem is to determine the statistically significant number of components, F. This is discussed in a separate section below.

The residuals can be used to construct a tolerance level around the PC hyper plane (figure 12). This interval which contains most of the object points and which also on a given level of probability contains future object points which are "similar" to those used to develop the model.

When F is two or three, the columns in T can be plotted against each other to get a few two dimensional pictures of the objects and their relations in p-space. Similarly, the loading vectors -- the rows in P -- can be plotted against each other to show the relationships between the variables. Figures 4 and 5 above were examples of such plots.

Rotations

The new coordinate system defined by the loading vectors p_f is calculated so that each successive direction explains as much as possible of the remaining variance in X or E. Often, however,

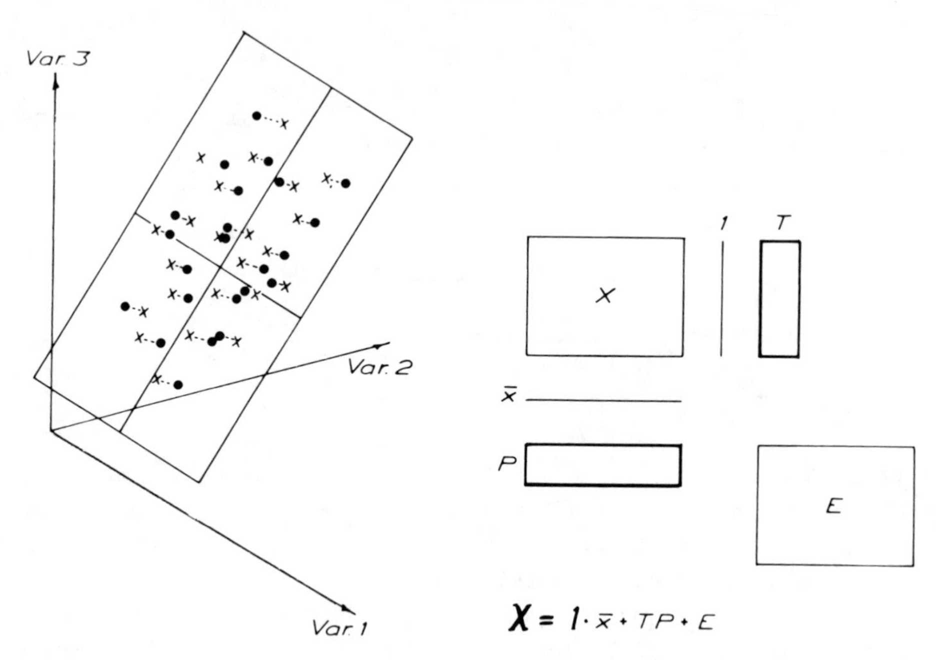

Figure 11. Together, the first and second PC define a plane. This plane with the projections of the n object points can be lifted out and displayed on a computer screen or a piece of paper. Thus we have a window letting us look into p-space and see a two-dimensional picture of the object point configuration.

one is also interested if certain external vectors, here denoted by z, are valid coordinate axes in the factor solution. One may then rotate the first F loadings to see if their linear combination is highly correlated with z. This indicates that indeed the "chemical effect" described by z is present in the data. This is called target rotation by Malinowski and Howery (1980).

Numerically, this rotation is accomplished by using z as the "dependent variable" and the loading vectors as predictor variables in a multiple regression

$$z = P'b + e$$

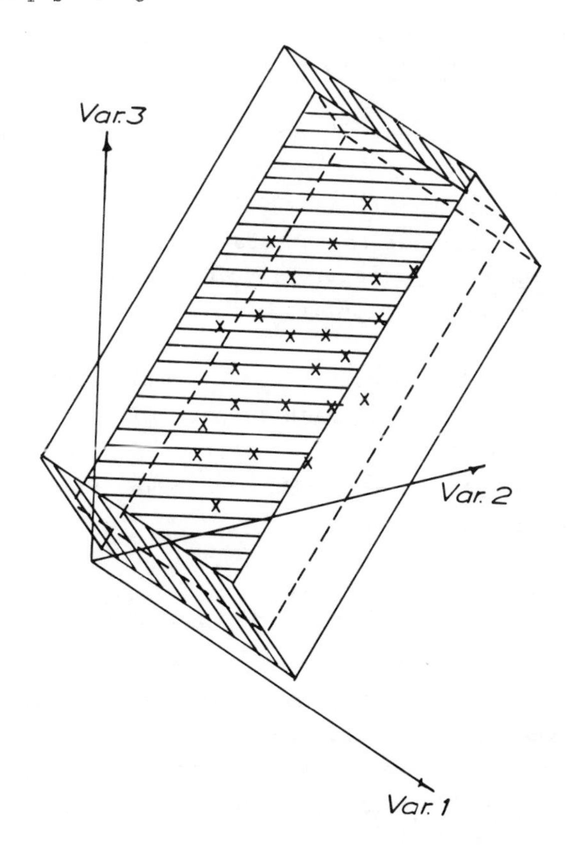

Figure 12. From the dispersion of the object points around the PC plane and from the distribution of the t-values on the plane, one can calculate a tolerance interval containing, say, 95 percent of future object points similar to those of the given data set.

The vector of regression coefficients, b, then describes how the p-vectors should be combined to best fit z. The residual vector e contains the part of z not "explained" by the rotation. In this way target rotation is a special case of principal components regression (Draper and Smith, 1981).

In cmemical applications other types of rotations are sometimes of interest. Using PCA for the deconvolution of twodimensional chromatograms gives initially a solution where parts of the components are negative. This solution must be rotated to give a solution with positive "chromatographic peaks" and constituent spectra.

The residuals, e_{ik}

The residuals e_{ik} after the F significant components can be used to calculate measures of fit to the PC model for each object and each variable. Thus the standard deviation (SD) of e_{ik} for object i, s_i, is proportional to the distance between the object and the PC hyper plane in p-space. This SD can be compared with the total SD, s_0, by an approximate F-test to see whether object i is an outlier.

In the same way, the SD of e_{ik} for variable k measures how much of the variable is "noise" with respect to the model. Dividing this SD with the corresponding SD of x_{ik} gives the ratio unexplained to total SD. One minus this ratio is often called the "modelling power" of the variable.

Outliers

As discussed in previous section, moderately outlying objects can be found by their too large residuals. However, because of the least squares properties of the PC model, objects very far from all the others, i.e. gross outliers, pull the model so that it fits them closely. This is analogous to fitting a straight line to bivariate data (y,x). One point far from the others will tilt the line so that it passes very close to the far away point.

Gross outliers are immediately seen in PC score plots, i.e. plots of t_1 against t_2, etc. We see that two criteria are used to find outlying objects, the object positions in PC score plots and the size of the object residual SD.s.

Predictions with the PC model for new objects and variables

When a new object with the data row vector x_j can be projected down on the PC model the resulting scores t_j describe the place of the new object with respect to the ones used earlier to develop the PC model. The residual SD s_j measures the degree of fit. If s_j is "large" compared to the total residual SD, s_0, object j is inferred to be dissimilar to the earlier objects. Since the loading matrix P is fixed, this projection corresponds to the linear multiple regression (after scaling):

$$x_j - \bar{x} = t_j P + e_j$$

Thus $t_j = P'(x_j - \bar{x})$

Analogously, a new "variable", z can be related to the PC model by the regression discussed above under rotation.

Computational aspects

The calculation of the parameter vectors $\bar{x}$, p_f and t_f (f=1,..,F) can be done in many ways. Historically, this was first done by calculating p_f as the f.th eigenvector to the covariance matrix $(X - 1\cdot\bar{x})'(X - 1\cdot\bar{x})$. When X is scaled to unit variance, this corresponds to the correlation matrix. Nowadays, much more efficient algorithms utilizing the raw data matrix have been developed in numerical analysis; the conjugate gradient methods presently being the fastest and most stable. Hence the name eigen vector analysis is no longer warranted.

In case only the first few components are wanted, a variant of the conjugate gradient method or the NIPALS method developed by H.Wold (see Jöreskog and H.Wold, Ed.s, 1982) is efficient. The latter applies also to incomplete data matrices(missing data).

Another example. The nonclassical carbonium ion controversy.

In 1949 Winstein proposed the existance of so called nonclassical carbonium ions to explain the abnormally fast solvolysis of exo 2 norbornyl compounds (no 20,25 in figure 13) in comparison with endo compounds (21,26). H.C.Brown proposed an

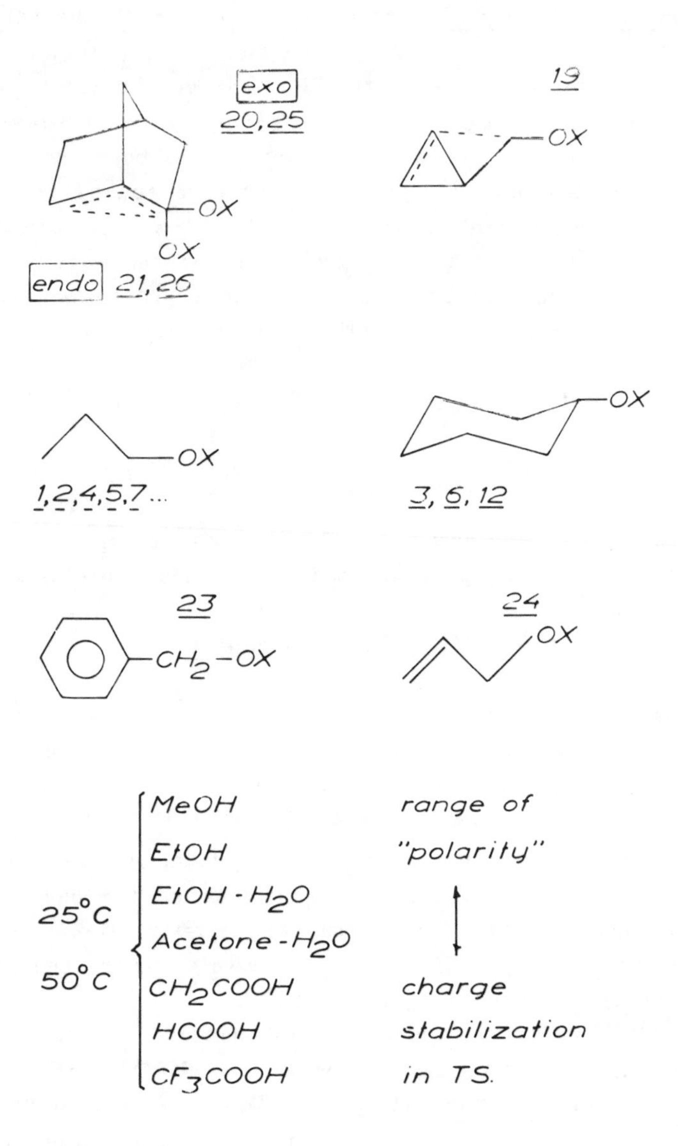

Figure 13. The data set contains data for (I) molecules representative for solvolysis without charge delocalization in transition state (CDTS) such as 1-7 and 12, (II) those with CDTS (23,24) and (III) the interesting compounds 19-21 and 25,26. The rate constants of their solvolyses were found in the literature for seven solvents of different polarity at two different temperatures.

alternative explanation in terms of steric strain release and rapid equilibria between classical ions. For references, see Albano and Wold (1980).

To investigate whether published but non-analysed kinetic data could provide information about the controversy, Albano and Wold (1980) structured the problem as one of empirical similarities which might be approached by multivariate analysis.

Thus there are compounds which all agree react via classical ionic transition states (TS) with delocalized charge, i.e. benzyl and allyl compounds (23 and 24 in figure 13). Other compounds examplified by no 1-7 and 12 (fig.13) react via classical ions with localized charge in TS.

Hence if appropriate data were measured on the reactions of these compounds and on the controversial ones (19,20,21,25,26), and these data were subjected to PCA, the emerging similarities might indicate which view was more likely to be correct.

The logaritmic reaction rate in a number of solvents ranging in polarity from methanol to trifluoro acetic acid at two different temperatures might be such data. A reaction with localized charge in TS should be more affected by a change in polarity that a reation with delocalized charge in TS. Consequently, these data were collected from literature sources and arranged in a 26 x 14 table, X (26 compounds and 7 solvents times 2 temperatures). The data matrix was incomplete; around 15 % of the elements were missing.

A PC analysis of the part of this matrix corresponding to secondary "classical" compounds (3,6,8,9,12,14,18) gave two significant components according to cross validation (see below). When the other objects are projected down on this plane, figure 14 was obtained. We see a clear grouping according to the degree of charge delocalization in TS. Interestingly, the two exo 2 norbornyl compounds fall among the ones with "known" charge delocalization in TS (23,24), while the two endo compounds (21,26) fall right in the middle of the "classical" secondary compounds.

These results support the Winstein interpretation and are difficult to interpret with Brown´s formalism. Thus, PCA of data selected to bear information on a given question, reveal informative patterns.

We emphasize the importance of "model systems" with "known behaviour", in this case ones with and without charge delocalization in TS and the need for good data that are believed to be related to the given problem. However, there is no need for a fundamental model in the present case, i.e. an a priori model relating the degree of charge delocalization to the measured rate constants. Such a model would today be very difficult to construct.

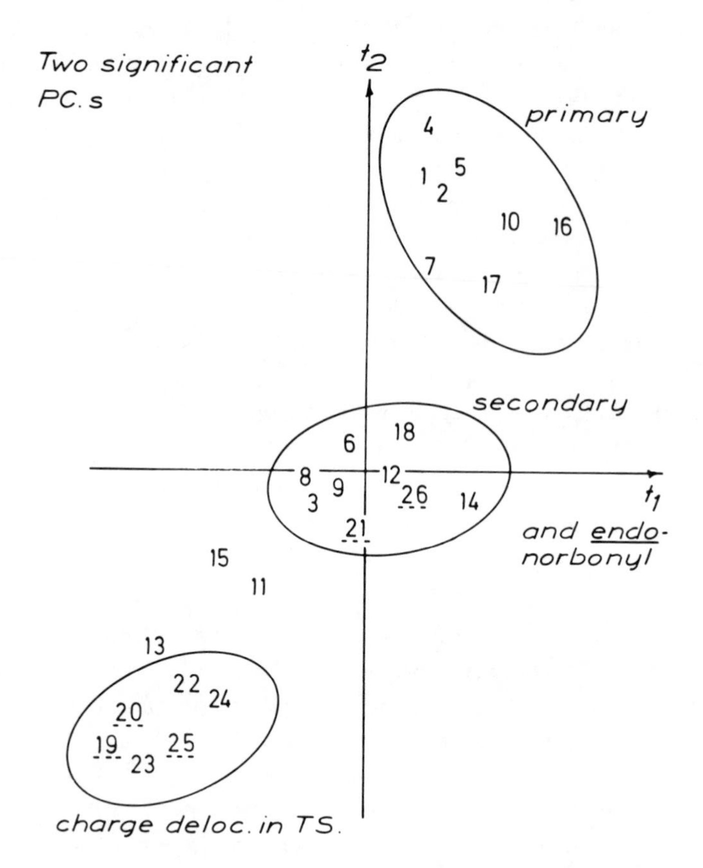

Figure 14. The data projected on the plane of the two first PC.s of the matrix of the seven secondary compounds without CDTS. The two endo compounds fall in the middle of these seven points indicating their lack of CDTS while the two exo points and the cyclopropenylmethyl (19) fall among those with known CDTS supporting the Winstein conjecture of nonclassical charge delocalisation.

Missing data

In practice, chemical data tables are usually incomplete. Hence, computational methods have been developed for estimating the coefficients in the PC model also for these cases. The NIPALS algorithm used by Wold et al in the SIMCA package works well with moderate amounts of holes in X. The basic idea with all methods handling missing data is to fill in values so that the model coefficients are unaffected by these values. These values usually are estimated from an initial round of PCA with missing data filled in by row or column means. See Wold et al. (1983).

Interpretation of the PCA model

The traditional interpretation of the PC model is that it is the result of a linear combination of "factors". Thus, when F chemical compounds are mixed in different proportions, the resulting samples give spectra which can be modelled by a PC expansion with F components, factors.

However, it can be shown that provided that the data table contains objects which are in some way similar, the data measured on these objects can always be PC modelled. The PC model has the same properties for data tables as polynomials have for bivariate data (y,x), see Wold (1976). Then, of course, in a practical case, the PC model may be a "mixture" of these two alternative interpretations.

Finally, for non-homogeneous data, the PC model can be seen as just a numerical projection as discussed above.

Information content of the PC model

Data tables are frequently "analysed" one variable at a time, giving, say, averages and standard deviations (SD.s). For a table with p variables, this gives 2p parameters. When applying PC models, the covariation between the variables are also utilised, i.e. an additional $p(p-1)/2$ parameters. For $p=10$, the latter are 45 in number, while the averages and SD.s are 20. For $p=50$, these numbers are 1225 and 100. This shows that much information may be contained in the covariation. This is not seen by looking at the data one variable at a time.

Model complexity, the number of components, F

Like in all least squares models, the PC residuals e_{ik} decrease when the number of parameters increases, i.e. when the number of components F increase. This gives the model a better fit, a better power, but beyond some limit the validity of the results decreases (figure 15). Hence, F must be carefully determined so that all information is extracted from the data, but no spurious components are included in the model.

If the precision of measurement is known, and if one is certain that the model must fit the data to this precision, the problem is simply solved by including so many components, F, that the residual SD corresponds to this precision. This may be the case in some applications, notably the modelling of spectra of dilute mixtures.

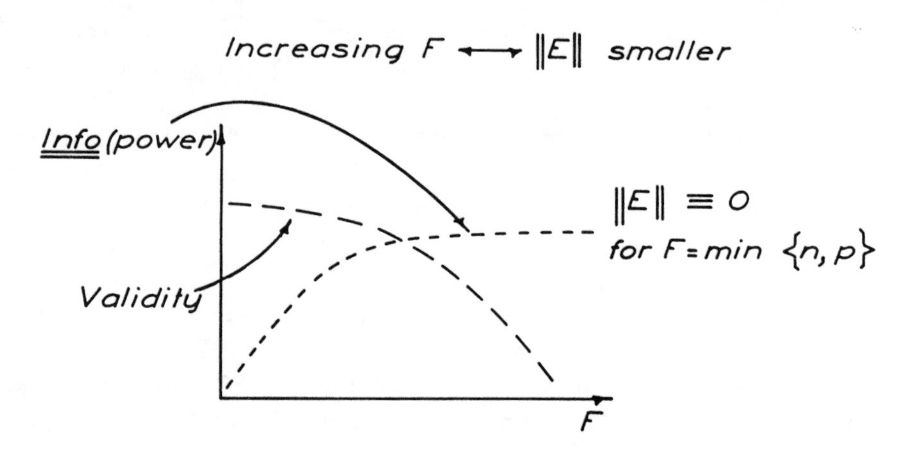

Figure 15. In analogy with all other empirical models, the residuals of the PCF models decrease with the increased number of model parameters (here the number of components or factors, F). The residuals are identically zero when F equals the smaller of n and p. Hence the power of the model -- the amount of data explained -- increases with F while the validity of the results decreases with F. The crucial problem is to find the right number of components, F_{optim}, where the model has both high power and high validity.

In general, however, there is an unknown model error involved. One cannot at all be certain that the model should fit within the precision of measurement. Then F must be determined by other criteria. The criterion favored by us is the one of cross-validation. This corresponds to the scientific objective to make the model be an optimal predictor of future events.

With the computer -- even a microcomputer -- it is easy and fast to recalculate the model several times for slightly different data sets. Thereby an estimate of the predictive power of the PC model with different number of components can be obtained as follows (Wold, 1978, Eastment and Krzanowski, 1982):

A few data elements are kept out from the data matrix X each round, and the PC model with different number of factors (F) is fitted to the remaining data. So many rounds are made as needed to keep each data element out once and only once. The values of the kept out elements are calculated from the resulting models with different F and the deviations between calculated and predicted values are formed. When the squares of these deviations from the separate "rounds" are summed up, a predictive sum of squares, PRESS, is obtained for each F. The F-value corresponding to the minimum PRESS is chosen to be the one used in the model.

Two objectives of the data analysis

In a given data set, X, there is only a certain amount of information about a given problem, say I. The data analysis should extract as much as possible of I. If successful, the data analysis has high power. No false negatives are obtained, no errors of the second kind committed.

However, when the power of a data analytic method is maximized, there is a risk that too much of the data is judged to be information. I is overestimated, the model overfitted, the results have low validity, one commits errors of the first kind.

The difficult balance is to find what is there (high power), but not find too much, i.e. not be fooled (high validity). Cross-validation is a good tool for finding this balance.

Scaling

The PC model is fitted to the data using the criterion of least squares. This makes the results depending on the scaling of the data. Thus, if a variable is multiplied by ten, its variance will increase by hundred. The residual sum of squares will then depend very much on this variable, and very much of it will hence be modelled. Thus, the initial variance of a variable partly determines its importance in the model.

In factor analysis, it is customary to scale the variables to unit variance before the analysis. Hence the results are scaling independent. This is by many considered as an asset of factor analysis.

However, if one is uncertain about the importance of the variables, one can always scale the data to unit variance -- autoscaling. In the case of prior information this can be utilized . Hence, the scaling dependence of PCA is an asset if correctly used.

No prior information: Compute scaling weights w_k as $1/s_k$ (SD of variable k).

Prior information: Use scaling weights w_k proportional to this information.

Scaling: Transform x_{ik} to $w_k \cdot x_{ik}$.

Choice of representation

PCA projects the data matrix X (properly transformed and scaled) down on the hyperplane defined by the loading matrix P, the factor space. The resulting coordinates of the objects on this hyperplane, the score matrix T, are usually interpreted visually. This corresponds to making inference about similarities between objects from their closeness in the factor space. This, in turn, is related to their closeness in p-space if the PC model fits the data well, i.e. describes a large proportion of its variance, say more than 75-80 percent.

If this inference shall be valid, closeness in p-space must correspond to similarity between objects. Hence, the raw data must be translated to the data matrix X so that indeed X has the property of revealing object similarites in terms of a small distance i p-space. This condition should be fulfilled not only for PCA but for all methods of multivariate data analysis. They are all based on relating similarity to closeness in p-space.

Let us look at an example where common chemical practice does not accomplish this, namely mass spectroscopy. In the multivariate analysis of mass spectra, these are usually represented as data vectors where the intensity of each mass number is one variable. Let us think of a chemical compound and the resulting single point in p-space. Now, we make a minute change in the chemical structure of the compound by substituting one hydrogen to a deuterium in a single place far out in the structure. This gives a compound that in all practical respects is chemically identical to the first. But parts of the mass spectrum are shifted one unit up because some fragments are one unit heavier than before. Hence, for some variables m, the intensity of variable m+1 corresponds to the earlier m.th variable. In p-space this corresponds to a grand rotation, variable m is shifted to m+1 for several m. Two almost identical compounds lie very far from each other (figure 16).

We realize that for PCA, this representation of mass spectra is unsuitable. What to do? Well, mass spectroscopists have always

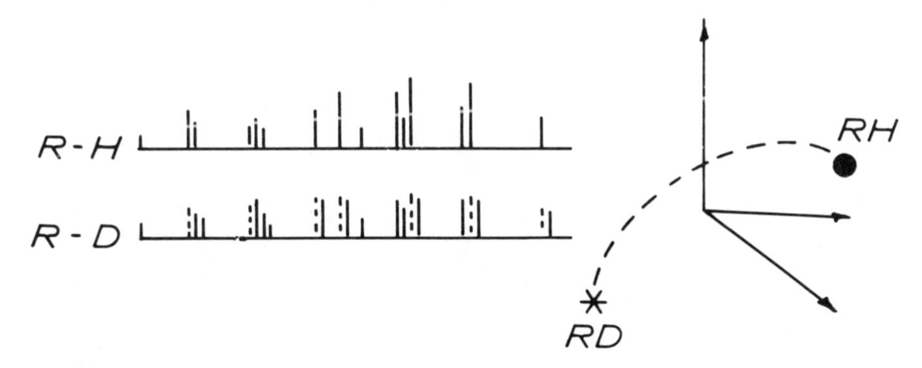

Figure 16. The mass spectra (MS) of a compound RH and its deutero isomer (RD) are partly shifted one unit with respect to each other. Representing the MS by the intensities at each mass number makes the points of RH and RD lie very far from each other in p-space because the variables (intensities at given mass numbers) do not all correspond to similar fragments for the two compounds.

looked for periodicities in their spectra as being informative. Several transformations of a mass spectrum -- Fourier, Hadamard, autocorrelation -- reveal such periodicities, and have indeed been used in PARC-MS applications. See e.g. Kowalski, 1974 and Varmuza, 1980. Unfortunately very inefficient PARC methods were used which masked the advantage of such transformations. The simplest transformation which is independent of translation of the spectrum, is the autocorrelation transform.

The autocorrelation transform is simply defined as the result of the vector multiplication of the spectrum and itself

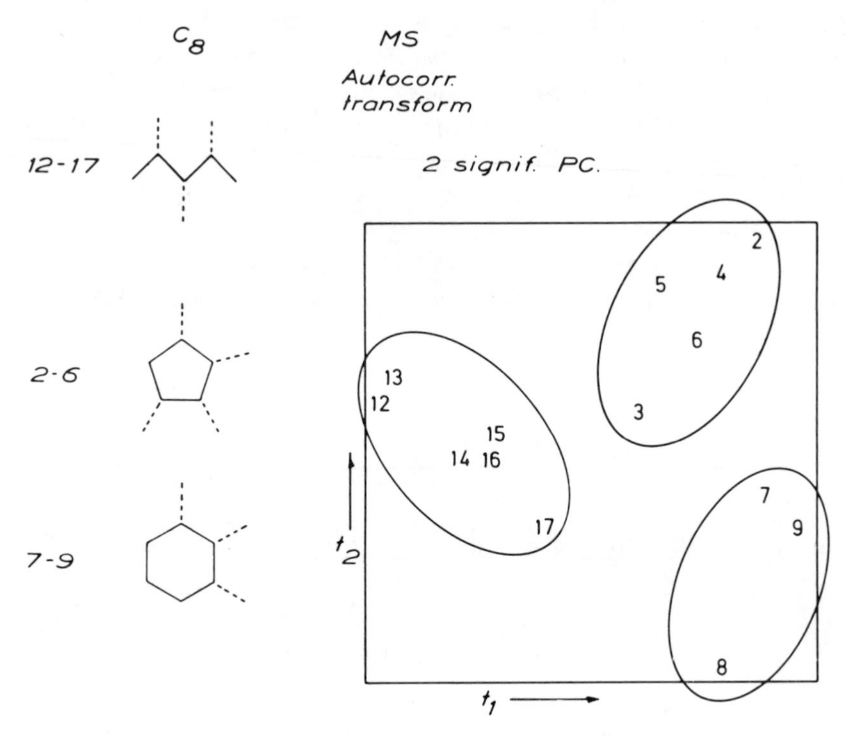

Figure 17. By instead representing the MS as their autocorrelation transforms, variables with good properties for PCFA are obtained. The picture to the right is the PC projection of the autocorrelation transformed MS of the 16 compounds shown to the left. Compounds with similar skeletal structure cluster in p-space.

No 2 - 6 are substituted in three of the four indicated positions and no 7 - 9 in two.

shifted r mass numbers. The transform is calculated for r=1,2,...,R with R being approximately the highest mass number divided by 4. The new R variables have the right properties with respect to small changes in chemical structure and can thus be subjected to PCA. Figure 17 shows an example, where the mass spectra of C_8 alkanes are analysed in this way.

Fortunately, chemical signals are usually tranformed to data X in a way suitable for PCA and other multivariate analysis. The exceptions are mass spectroscopy (MS) and infra red (IR) spectroscopy, where the spectra often just are mechanically digitized at regular mass or frequency numbers, but where the rapidly varying curve form makes the resulting data have the wrong properties. In both these cases, autocorrelation or Fourier or similar transforms have better properties than the raw data. In IR, we recommend the identification of characteristic peaks -- carbonyl peaks, C=C peaks and the like --and then making their position and intensity in the spectrum be the x-variables.

Peak identification is essential when translating chromatographic (GC, HPLC, ...) or spectral (NMR, IR, UV, XRF,...) profiles to the data X. It is essential that a given peak always appears as the same x-variable. Otherwise grand rotations in p-space occur and the data analysis is botched.

Direct digitization can be done only with slowly varying curve forms such as those of kinetics and ultraviolet and near infrared spectroscopy (NIR). Rapidly changing curveforms must be either transformed or handled by peak identification.

Closure

An interesting difficulty arises from a certain way of transforming the raw data, namely the normalization of the variables for each object to a constant sum, often 100 or 1000. This is common in chromatography and mass spectroscopy to avoid the influence of the amount of sample injected into the instrument. But this normalization introduces a correlation between the variables. If the values of p-1 of the variables are known, the value of the last one can be calculated from their sum and the normalization constant.

If the variables are numerous and all of similar size, this closure of the data does not matter very much. However, it is rather common in chromatography that one or two variables are much larger than the rest. Then the closure introduces serious problems. This because when the large peak(s) goes up and down, the closure, the normalization, forces the sum of remaining peaks to go down and up. Hence, the closure introduces a strong but spurious correlation between the large peak(s) and the remaining ones. This spurious correlation turns up as an early and strong factor in PCA.

Moreover, negative correlations between the large peak(s) and the smaller ones in the original data are masked and disappear in the closed data. Much of the information is distorted or lost.

Closure was recognized long ago in geochemistry where the data (amounts of elements or oxides) usually are expressed in percent and therefore add up to hundred. But in the remaining parts of chemistry, this problem is unrecognized.

The closure problem can crudely be handled by normalizing the sum of all moderate variables to, say, hundred, keeping large and small variables out of the computation of the normalization constants. The latter are then, of course, also transformed by the multiplication of the calculated normalization constant (Johansson et al., 1983).

The best way to avoid the closure problem is <u>not to normalize the data</u>. The size factor can, if desired, be removed in other ways, see e.g. Lewi in this volume. Or it can be kept and interpreted in the PCA, it usually does not cause any trouble if recognized (Söderström et.al., 1982).

Multidimensional scaling

Sometimes the data table is not of the kind discussed above with objects times variables but is instead symmetrical objects times objects (dimension n x n). The elements d_{ij} in this table D then usually express either the distance between objects i and j or the similarity between these objects.

Such symmetrical data matrices are rare in chemistry except in food and fragrant chemistry where subjective measures of taste, smell, preference, are obtained as the distance or similarity between pairs of objects (wines, sausages, carrots, sweeteners, perfumes) by panels of judges.

Just as with rectangular data matrices X, the interpretation of the matrix D in terms of pictures and plots is desirable. This is accomplished in two steps. First, an n-dimensional space is constructed where each object is a point. Such a space can always be constructed so that the distances (or similarities) in the matrix D directly and exactly correspond to the interpoint distances in this n-space. Thereafter, hyper planes of various dimensions are inserted in to the n-space and the points projected down on these planes.

If the n-space can be considered to be Euclidean, the ordinary principal components expansion of D gives good projections, so called principal coordinates. Sensory data often do not conform with Euclidean metrics, however, and then non-metric projections must be made. Instead of planes one then uses smooth surfaces in n-space and a local metric for the projection. The resulting pictures are more difficult to interpret than the linear principal coordinate projections, but still reveal useful patterns. The interested reader is referred to Mardia et al. (1979) and Schiffman et al. (1981).

Summary of PCA

A data table X can be decomposed into an average, a score matrix T times a loading matrix P and a residual matrix E. The number of rows in P and columns in T, the dimensionality F, can be determined by cross validation to give the PC model good predictive properties.

The columns in T can be plotted against each other to picture the relations between the objects. In a connected picture of the rows in P, the variables are obtained as points. Thus PCA can be seen as a transformation of a data table X to a few informative pictures. Outliers strongly influence the PC model.

Provided that the data matrix X is observed on n similar objects, it can always be modelled by a PC model with few components. The same if X is observed on mixtures of a few chemical constituents.

The parameters P and T are scaling dependent. When no prior information is available, X should therefore be scaled to unit variance over each variable.

The transformation of raw data to X is essential. In mass and infra red spectroscopy, the intuitive data representation is usually highly impractical.

Part 2, pattern recognition (PARC)

We shall now continue with a special case of the analysis of data tables. Above we saw that data tables can be projected down on hyper planes to give informative pictures of X. Provided that the objects in the table are similar, a PC model with few components, dimensions, gives a good approximation of X. A necessary condition for this to be valid is that the data, the variables, indeed describe this similarity.

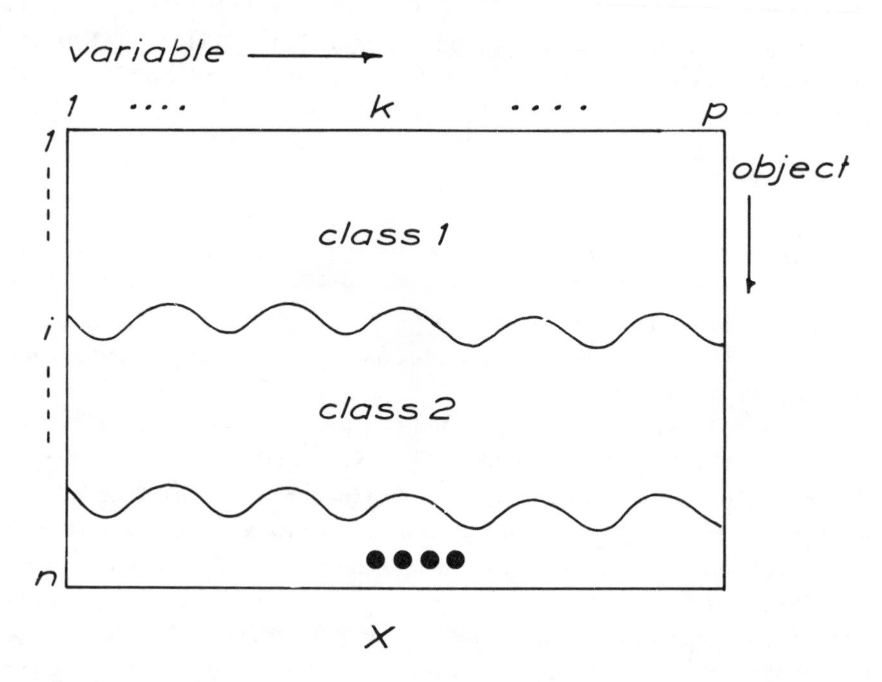

Figure 18. Frequently the data matrix X is divided into "classes" and the scope of the data analysis is to find "patterns" related to this classification. We then call the data analysis one of pattern recognition (PARC).

We shall now exploit this approximation property of PC models to analyse data tables divided into classes (figure 18). This analysis is often called pattern recognition (PARC), discriminant analysis and classification. We here use PARC.

The brain tissue example

An interesting way to characterize biological samples is to measure "chemical fingerprints" in terms of chromatographic profiles of various forms of preprocessed such samples. Often chromatographic methods especially developed for biological samples are used such as electrophoresis, isoelectric focusing and affinity chromatography, but we here use gas chromatographic (GC) data to illustrate PARC. There is no difference in principle between the analysis of GC data and any other type of chromatographic data.

Jellum et al. (1981) used this strategy to see if they could distinguish between samples coming from brain tumor tissue and from "normal" brain tissue (figure 19). The translation of the gas chromatographic profiles resulted in 156 variables (peak heights of recognizable peaks) for 16 objects in two classes, 6 normals and 10 tumors. Thus we have a 16 x 156 data matrix X divided into G=2 classes.

Figure 20 shows the hopeless task to see anything at all in the data table. In contrast, figure 21 reveals that in p-space (p=156) the two classes are well separated and that indeeed this type of data contains useful information. The complementing picture of the variables (fig.22) shows which variables discriminate between the classes.

Many variables and few objects

This example is shown to illustrate the following point. The classical methods of multivariate analysis, namely multiple regression, linear discriminant analysis and analysis of variance were all developed around 1930. At that time measurements were expensive and therefore one made few on each object. Typically one compensated this by having many objects. Hence the data tables were long (n large) and lean (p small).

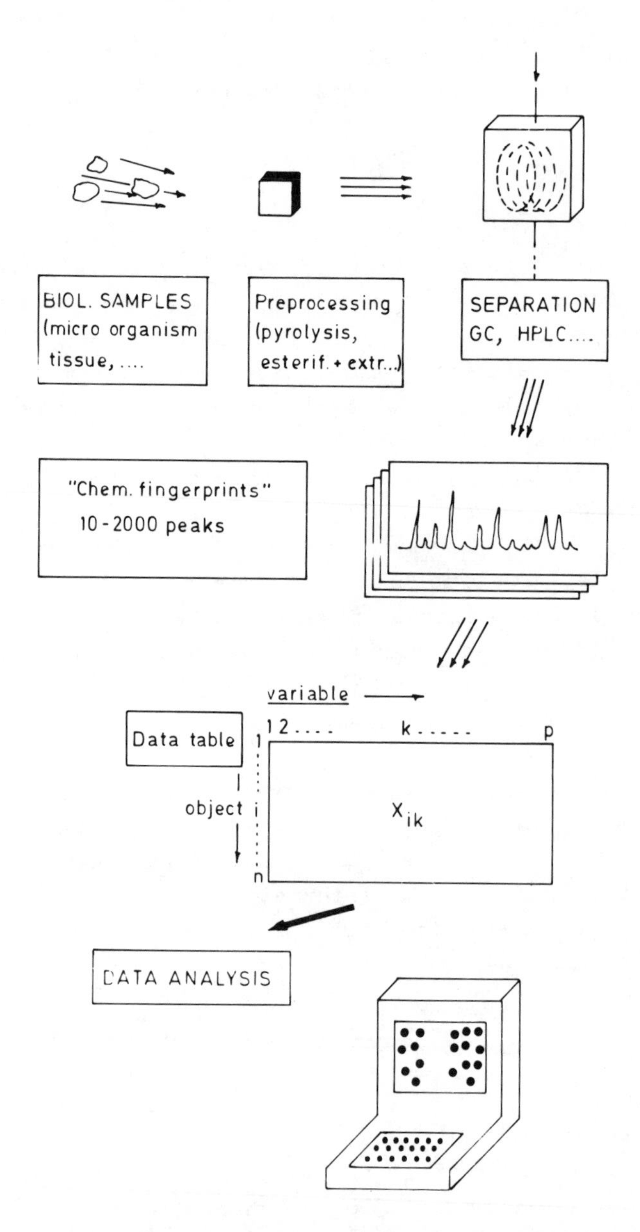

Figure 19. Biological samples can be characterised by various chromatographic profiles -- chemical "fingerprints". One interesting problem is to find out how samples of different kinds differ in these "fingerprints".

Consequently, the data analytical methods at the time were developed for the situation with n>>p. After a while everybody was taught that one has to have many more objects than variables and this is today almost a dogm in statistics and data analysis.

The typical situation 1983 in chemistry, biology, geology and medicine is, however, different. With the modern instruments and separation methods many measurements are made on each object. Each measurement is cheap and once one has an object in hand, one might as well measure many variables. The objects are now usually much more expensive than the variables (think of one brain operation compared with one GC peak).

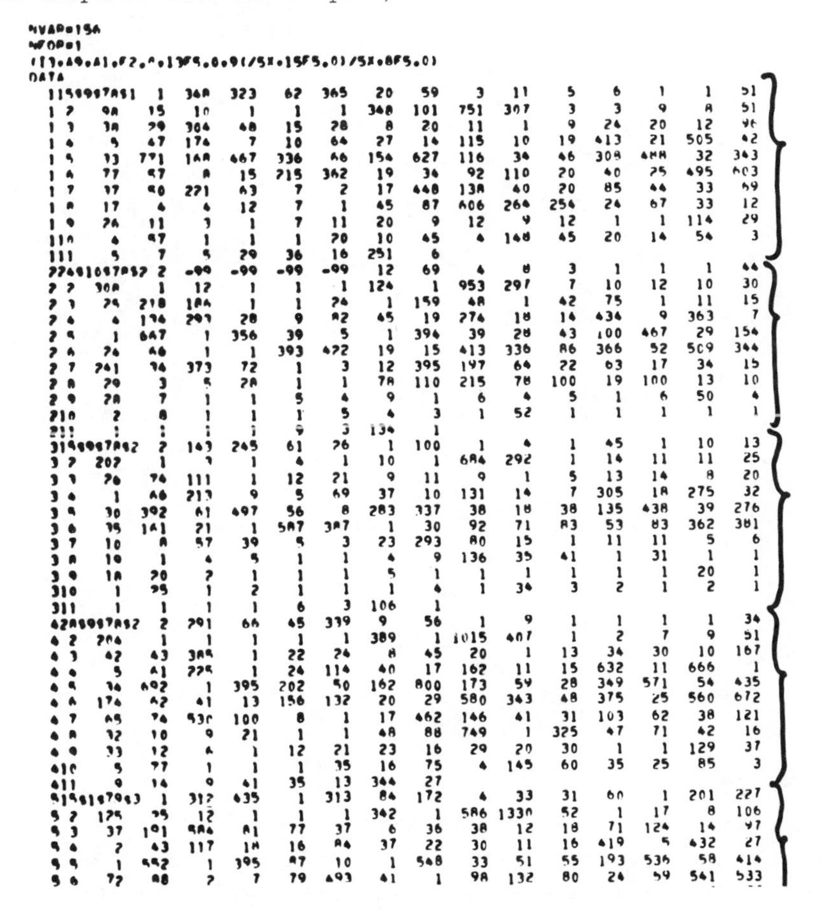

```
NVAR=156
NFOR=1
(I3,A9,A1,F2.0,13F5.0,9(/5X,15F5.0)/5X,8F5.0)
DATA
 1159997851  1  348  323   62  365   20   59    3   11    5    6    1    1   51
 1 2   98   15   10    1    1    1  348  101  751  307    3    3    9    8   51
 1 3   38   29  304   48   15   28    8   20   11    1    9   24   20   12   96
 1 4    5   47  174    7   10   64   27   14  115   10   19  413   21  505   42
 1 5   13  771  148  467  336   86  154  627  116   34   46  308  488   32  343
 1 6   77   57    8   15  215  362   19   34   92  110   20   40   25  495  603
 1 7   17   50  271   63    7    2   17  448  138   40   20   85   44   33   59
 1 8   17    4    4   12    7    1   45   87  606  264  254   24   67   33   12
 1 9   26   11    3    1    7   11   20    9   12    9   12    1    1  114   29
 110    4   97    1    1    1   20   10   45    4  148   45   20   14   54    3
 111    5    7    5   29   36   16  251    6
 2249108782 2  -99  -99  -99  -99   12   69    4    8    3    1    1    1   44
 2 2  308    1   12    1    1    1  124    1  953  297    7   10   12   10   30
 2 3   25  218  186    1    1   26    1  159   48    1   42   75    1   11   15
 2 4    4  134  293   28    9   82   45   19  274   18   14  434    9  363    7
 2 5    1  647    1  356   39    5    1  394   39   28   43  100  467   29  154
 2 6   26   46    1    1  393  422   19   15  413  336   86  366   52  509  344
 2 7  241   34  373   72    1    3   12  395  197   64   22   63   17   34   15
 2 8   29    3    5   28    1    1   78  110  215   78  100   19  100   13   10
 2 9   28    7    1    1    5    4    9    1    6    4    5    1    6   50    4
 210    2    8    1    1    1    5    4    3    1   52    1    1    1    1    1
 211    1    1    1    1    9    3  134    1
 3159987852    2  143  245   61   26    1  100    1    4    1   45    1   10   13
 3 2  202    1    3    1    4    1   10    1  684  292    1   14   11   11   25
 3 3   26   34  111    1   12   21    9   11    9    1    5   13   14    8   20
 3 4    1   86  213    9    5   69   37   10  131   14    7  305   18  275   32
 3 5   30  392   61  497   56    8  283  337   38   18   38  135  438   39  276
 3 6   35  141   21    1  587  387    1   30   92   71   83   53   83  362  381
 3 7   10    8   57   39    5    3   23  293   80   15    1   11   11    5    6
 3 8   10    1    4    5    1    1    4    9  136   35   41    1   31    1    1
 3 9   18   20    2    1    1    1    5    1    1    1    1    1    1   20    1
 310    1   95    1    2    1    1    1    4    1   34    3    2    1    2    1
 311    1    1    1    1    6    3  106    1
 4285987852    2  291   66   45  339    9   56    1    9    1    1    1    1   34
 4 2  204    1    1    1    1    1  389    1 1015  407    1    2    7    9   51
 4 3   42   43  385    1   22   24    8   45   20    1   13   34   30   10  167
 4 4    5   41  225    1   24  114   40   17  162   11   15  632   11  666    1
 4 5   34  602    1  395  202   50  162  800  173   59   28  349  571   54  435
 4 6  174   62   41   13  156  132   20   29  580  343   48  375   25  560  672
 4 7   85   74  530  100    8    1   17  462  146   41   31  103   62   38  121
 4 8   32   10    9   21    1    1   48   88  749    1  325   47   71   42   16
 4 9   33   12    6    1   12   21   23   16   29   20   30    1    1  129   37
 410    5   77    1    1    1   35   16   75    4  145   60   35   25   85    3
 411    9   14    9   41   35   13  344   27
 5158187983    1  312  635    1  313   84  172    4   33   31   60    1  201  227
 5 2  125   95   12    1    1    1  342    1  586 1330   52    1   17    8  106
 5 3   37  191  586   81   77   37    6   36   38   12   18   71  124   14   97
 5 4    2   43  117   18   16   84   37   22   30   11   16  419    5  432   27
 5 5    1  552    1  395   87   10    1  548   33   51   55  193  536   58  414
 5 6   72   88    2    7   79  493   41    1   98  132   80   24   59  541  533
```

Figure 20. Part of the Jellum et.al. data table containing the heights of 156 gas chromatographic peaks of 16 brain tissue samples (6 "normal" and 10 "tumor" of two types).

Hence, the data matrices in chemistry are today often short (n small) and fat (p large). The classical methods of statistics break down and chemists have to use other methods.

Projections work well also with many variables and few objects

Fortunately, the projection methods such as PCA discussed above give good results even when the number of variables far exceeds the number of objects. In fact, the object scores (matrix T) are better estimated the larger the number of relevant variab-

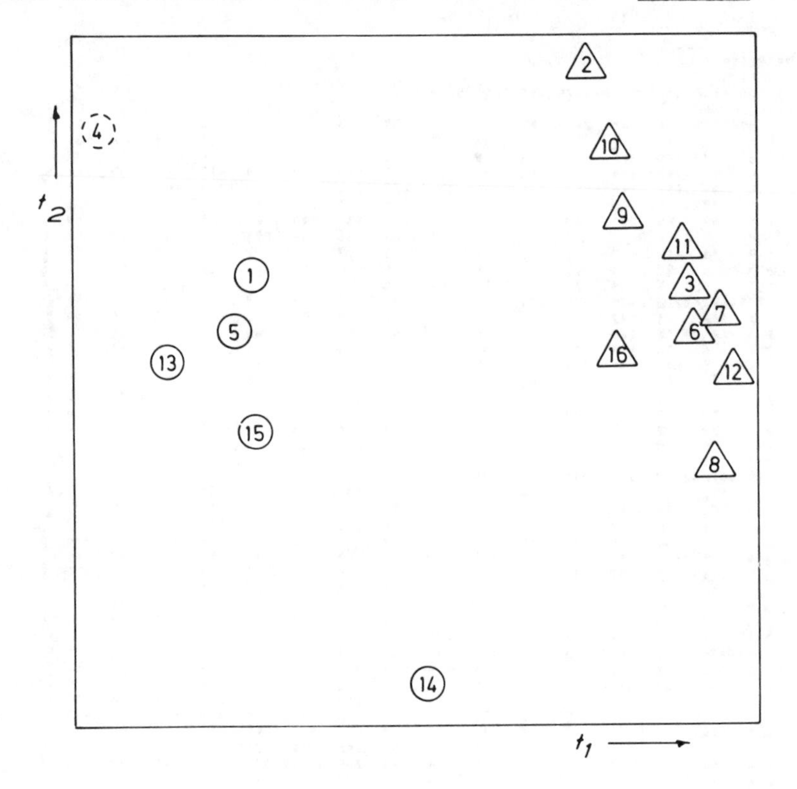

Figure 21. A PLS projection (see below) of the brain tissue data. Tumor samples (triangles) and "normals" (circles) are well separated. Sample 4 was initially mislabeled as a tumor sample but the data analysis consistently indicated it to be "normal" and the mislabeling was detected.

les (for a given number of objects). This because the t-values are linear combinations of all the variables and thus have the character of weighted averages. And averages are more precise the larger the number of relevant elements on which they are based.

We shall see that projection methods are useful also for the pattern recognition and regression problems. Hence there are methods which not only are useful when n >> p, but which indeed give better results the larger the number of relevant variables.

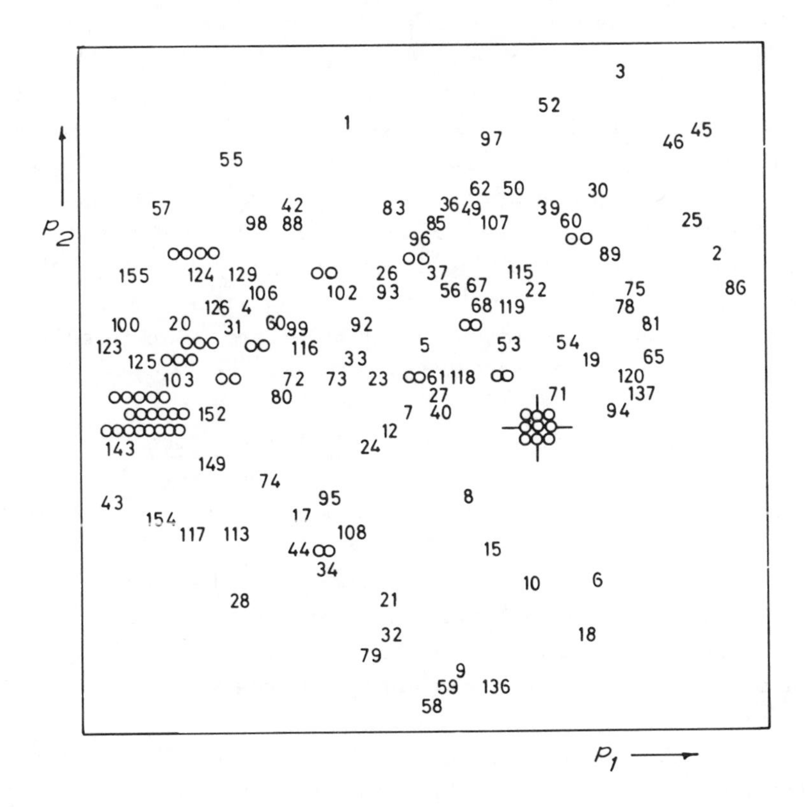

Figure 22. Loading map of the tumor PLS projection (fig.21) showing each variable as a point in the p_{1k} - p_{2k} plane. The directions in this plane correspond to those in fig. 21 so that variables 2,3,45,46 and 86 (upper right) are high for tumors compared to normals and vice versa. The cross at right center is (0,0). Rings indicate multiple points.

Why is pattern recognition (PARC) essential

All statistical models such as regression models, principal components and factor models, discriminant models, time series models, are derived as linearizations -- Taylor expansions -- of complicated unknown functions in many variables. Linearizations are valid only in limited intervals, as long as higher terms in the Taylor expansions don´t become large.

Therefore, whenever a problem is investigated for a wide range of objects, a single model cannot be applied, but the objects must be dividied into groups, classes. Separate empirical models (linearizations) can then be constructed for each class.

This has long been realized by man, who has since the dawn of history divided his systems and objects into conceptual classes. We must remember, however, that classes are man made constructs, abstract artifacts which may be or may not be practical for the problem at hand. The objects, in contrast, have a real existance and one relevant problem therefore is to investigate whether the proposed classification is a good one or not.

Two phase of PARC, training and prediction

PARC is based on the existance of a training set of objects, each "known" to belong to one of the given classes. In the brain tissue example, all 16 objects belonged to the training set having two classes, "normal" with n_1=6 and "tumor" with n_2=10.

This knowledge needs not be perfect, the PARC methods tolerate a minority of erroneously assigned objects, outliers, in each class.

In the first phase of PARC, the training set is used to develop mathematical rules for assigning the objects to the respective classes on the basis of their data, x_i.

In the second phase, that of prediction, these rules are used to predict the class of the objects in the test set. In the brain tissue example no test set was given.

Classes

The classes used in chemical problem formulations are usually one of two kinds. The first is related to the type of chemical objects (see part one above). This can be the type of reactivity of a compound, say a nucleophile or an electrophile, a catlyst in a certain process or not. Compounds are often divided into classes according to their type of biological activity: sweet or not, beta-adrenergic agonist, antagonist or neither, carcinogenic or not, etc. The determination of the chemical structure can be seen as the assignment of the compound to a union of classes: aromatic, aliphatic, alcohol, amine, hydrocarbon, cis, trans, saturated, unsaturated, straight chain, branched, ...

The problem of "diagnosis" in the general sense is one of class assignment and the "objects" to be diagnosed can be patients or reactions (class="mechanism") or processes or any kind of systems. The "diagnosis" of an object as forged or not by chemical means is done with pieces of art, stamps, coins, samples of orange juice, perfume, whisky and wine, just to mention a few.

The second kind of classes is related to the origin of the objects. This can be the origin

in the geographical sense -- classification of archeological artifacts, particulates in acid rain --

in time -- the dating of chinese porcelain or ancient coins or minerals or bones --

in terms of individual candidates in forensic chemistry -- does a blood stain come from this or that person, does an oil spill come from this or that or that tanker, does the PCB in a dead fish come from this or that manufacturer --

in terms of association to valuable resources -- is a sediment similar to those previously collected above oil fields or not, is a mineral sample similar to those previously collected in the neighborhood of gold ore bodies or not.

Similarity revealed by measured data

The idea of PARC is to use the similarities, hopefully existing in data measured on similar objects in such a way that new objects can be classified as similar to this class or that class in the training set. We note that in the beginning we have no idea if these similarities exist and how they look. It might seem a hopeless task to develop methods to reveal these similarities among sets so diverse as tumor tissue samples, oil spills, wines and trans alfa-beta unsaturated carbonyl compounds. It is a strange but encouraging fact, however, that data measured on any kind of chemical "objects" or processes (dynamic objects) can be approximated and analysed in the same way, provided only two conditions are fulfilled.

(1) The objects in each class must be in some way similar and (2) the majority of the data measured on the objects must in some way be related to this similarity.

We need not know _how_ these conditions are fulfilled or how the similarity relations look, we need just tentatively assume _that_ they are fulfilled. The data analysis then shows, among other things, if this is the case or not.

Selection of classes

We note again that the classes are man made. Hence they can be constructed in ways which are consistent with the data or in ways that are in conflict with the data. Only in the former case does PARC work. We can only recognize something that really is there. Hence the matter of validity is of utmost importance in PARC. The analysis should reveal what is in the data but not give results which are unsupported and which therefore are misleading.

The selection of classes is part of the problem formulation. It must be done to correspond to a real observable group similarity. The within class similarity between objects must be greater than the between class similarity. Again, one task of the data analysis is to find out if this has been accomplished or not.

A-variables expressing analogy

From part one we remember that the measured raw data could be translated to the table X so that similarity corresponded to closeness in p-space or not. PARC is based on the same assumption that closeness in p-space and similarity -- in the present case class membership -- are related. Hence the matter of how to represent the raw data is as essential here as for PCA. We call variables having the right properties for _A-variables_. These are such that closely similar objects are likely to have closer values of these variables than less similar objects. The variables describe the degree of analogy between the objects.

The PARC data set

We now assume that our chemical problem has been structured and data been measured so to give a data matrix in figure 23. We have as before n objects and p variables, but these data now comprise a training set divided into G classes.

PARC, levels 1 and 2

In the early days of PARC, the methods used were based on the assumption that all training set objects were correctly assigned to the given classes and that these classes were all that needed to be considered.

In chemistry the last assumption is rarely fulfilled, however. Objects in the test set may well belong to a new class which was unrepresented in the training set due to the lack of

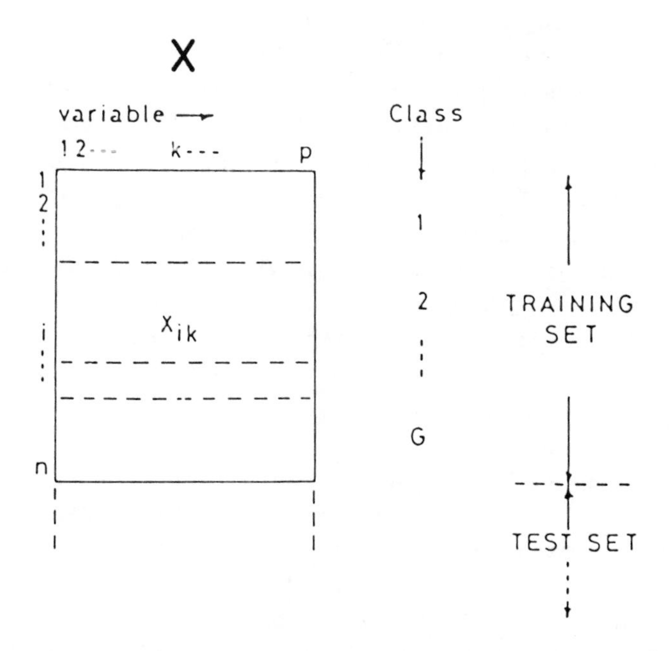

Figure 23. The data table of an ordinary pattern recognition (PARC) problem, i.e. levels 1 and 2.

knowledge or foresight. In structure determination, for instance, compounds of new structural types often appear in the test set. In forensic applications one can never be sure that all of the candidates for the given crime have been thought of, one must take into the account the possibility that the blood stain or the oil spill emanates from somebody who was clever enough to avoid suspicion in the first phase of the investigation.

In medical applications like the brain tissue example, outliers in the training set and unseen classes in the test set are common. A test set object may emanate from a rare type of special tumors not included in the training set. Or a patient may have taken some medicine unlike the previous patients or may have a different age or sex or have a methabolic disease or may be a heavy alcohol consumer or .... .

Therefore Albano et.al. (1978) proposed the use of PARC on a higher level in chemical applications. On level 2 a method must be able to recognise the presence of unseen classes in the test set, i.e. outliers with respect to the training set classes. The method should also recognize incorrectly assigned training set objects, i.e. outliers in the training set.

We shall henceforth discuss PARC only on levels 2 and higher, since the demands are easy to meet. Some older methods such as linear discriminant analysis (LDA) and the linear learning machine (LLM) don´t meet these demands, however. Since they are less useful anyway because they cannot handle asymmetric cases (see below) or cases with more variables than objects and since level 2 methods give better results than LDA and LLM whenever compared (see e.g. Sjöström and Kowalski, 1979), we shall not worry much about this slight limitation in our choice of methods.

The ketone example

As a second example we shall use the data of Mecke and Noack (1960) collected to get information about the conformation of alfa-beta unsaturated carbonyl compounds (figure 24) in solution. The compounds with small R_1 and R_3, i.e. hydrogen, were then as now assumed to have a planar trans conformation (fig.24). Mecke and Noack were interested to see what happened when large groups, i.e. methyl and larger, were put in positions 1 and 3. There is then, according to stick and ball models and Van der Waal´s radii, no longer room for a planar trans conformation. One could envision a twisting over to a planar cis-conformation (fig.24) or may be a non-planar skewed conformation.

To investigate this problem, Mecke and Noack prepared compounds locked in the trans and cis conformations by C-C bridges, compounds with hydrogens in positions 1 and 3 and compounds with "large" groups in these positions (fig.24). On all of these compounds, infrared (IR) and ultraviolet (UV) spectra were measured in solution. Seven variables were extracted from these spectra for each compound (object).

We divide part of these data into a training set with 13

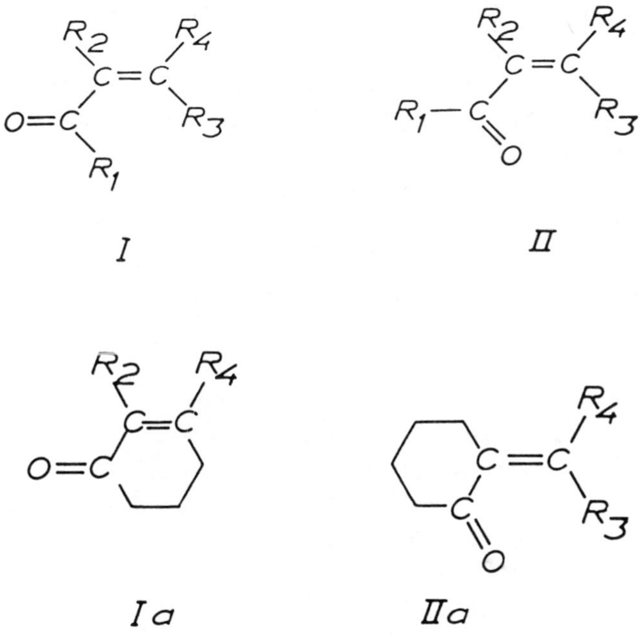

Figure 24. The compounds in the ketone data set have planar trans (I) or cis (II) conformations or nonplanar skew configurations (not shown). Some compounds in the trans training set are locked in planar conformation (Ia). Three compounds in the test set (14-16) are locked into planar cis conformation (IIa).

objects in two classes, class 1 = six trans compounds and class 2 = seven sterically hindered compounds (large R_1 and R_3). The cis compounds are too few (3) to constitute a separate class and are put in the test set together with four compounds of assumed trans conformation. This data table has been published several times in connection with its use as SIMCA test data and also been used in comparisons between PARC methods (Sjöström and Kowalski, 1978).

Do not use one variable at a time

The traditional way to look at chemical data tables is one variable at a time. One starts with variable one, computes average and mean for each class and then makes some kind of test of the class separation in that variable. Then variable two, followed by variable three, etc.

Figure 25 shows the results of this approach with two of the ketone variables. Neither seems to contain any information about the given problem.

When the same two variables are plotted simultaneously, however, the two classes are resolved and the cis compounds are seen to be close to the sterically hindered compounds (figure 26). This constitutes a geometrical proof that two variables sometimes show infinitely more than the same data seen one variable at a time.

The one variable at a time (OVAT) approach thus has a serious drawback: The information in multivariate data is often not seen. There is a second even greater problem with the OVAT "method". That of spurious results. The "significance" of the difference in the average of a variable between two classes is usually evaluated on the 5 % probability level. Thus there is less than 5 % risk that the difference between the two class averages is due just to chance. But with two variables this risk is about twice as large and with p variables this risk is $1-(0.95)^p$. For $p=7$ as in the ketone example this risk is about 30 % and for $p=50$ it is about 90%.

Therefore it is very difficult to judge whether the differences in single variables between classes are real or not. The only way to handle these two problems is to analyse all variables simultaneously.

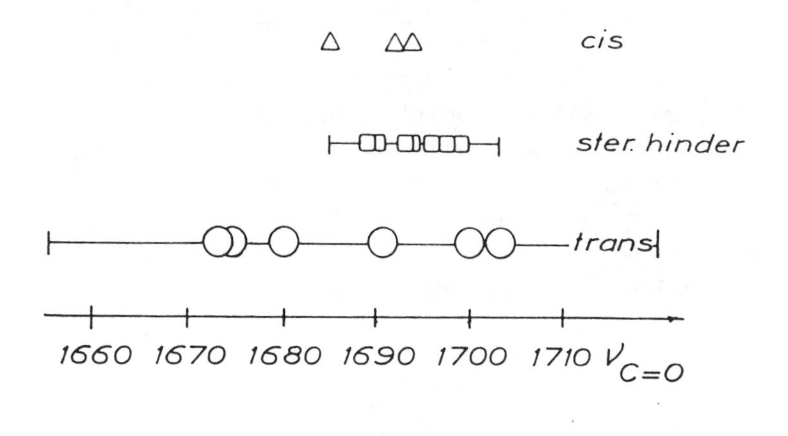

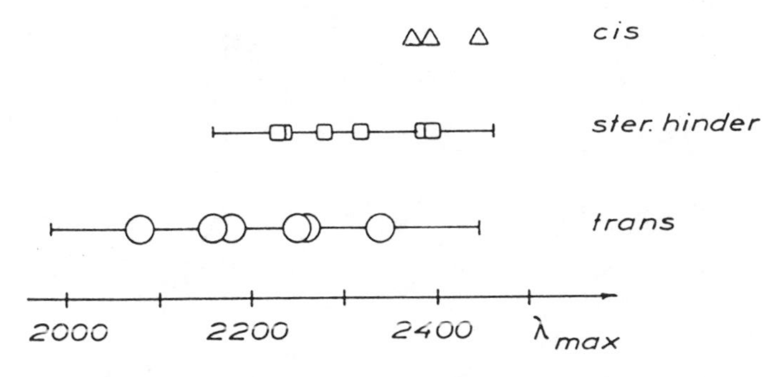

Figure 25. There is not sufficient class differentiating information in the variables 1 (carbonyl IR frequency) or 6 (wave length for max UV absorption) when they are looked at one at a time.

PARC as geometrical structures in p-space.

The p-space discussed in part one is highly relevant also for the PARC problem. The question now is if the point swarms corresponding to the training set classes are in some way separately situated in p-space (figure 27). In that case, PARC level 2 is simply to describe the domain of each class in p-space. If the class domains are at least partly separated, the data contain some information about the given problem. New test set objects can then be classified simply by seeing in which class domain in p-space they fall.

Pictorial overview of the PARC data set

A linear projection of the data down on a plane provides a two-dimensional window into p-space as discussed in part one. For some reason the least squares projection is usually called eigenvector projection when done in connection with PARC. We shall here continue to call it the PC projection (fig.28). Both the PC and the PLS projections work well when p exceeds n as exemplified in the brain tissue plot (figure 21).

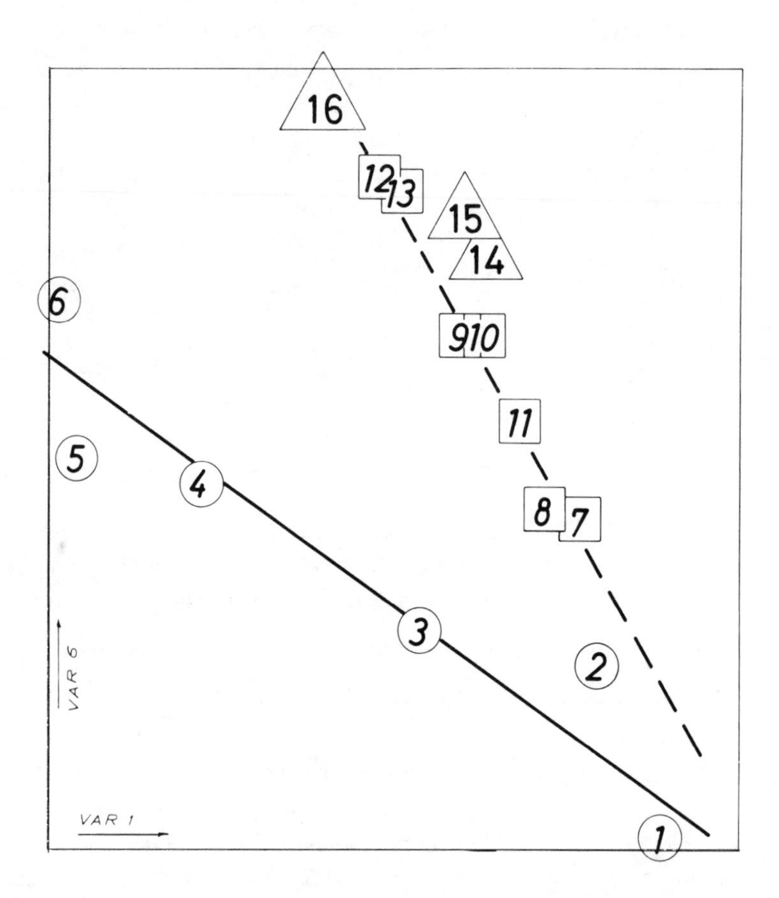

Figure 26. The same variables plotted against each other reveal a resolution of the two classes trans (circles) and sterically hindered (squares). The three "locked" cis compounds (triangles) are close to the sterically hindered ones.

Figure 27. In the p-dimensional space -- "p-space" -- formed by the orthogonal coordinate axes of the p variables, classes are situated in separate regions if data contains any information. The methods of PARC in one way or another describe mathematically either the location of the classes or their separation.

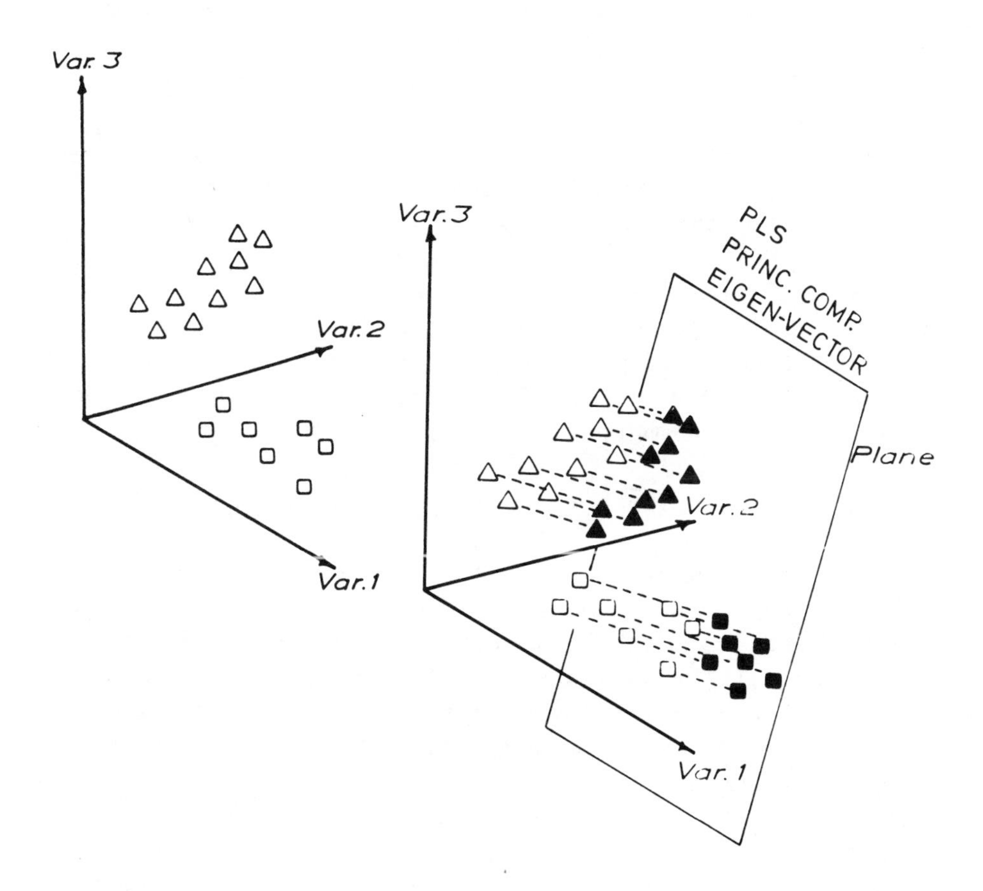

Figure 28. A linear projection of the data set down on a two-dimensional plane provides a picture of the object point configuration in p-space. Principal components planes -- often called eigenvector planes because the loading vectors p_f are eigenvectors to the covariance matrix of X -- give the best overall mapping, while PLS planes (see below) give a slightly improved class separation in the projection.

We note that the PC projection no longer gives the optimal window displaying the class separation. Rather one should use a projection which both separates the classes if such a separation exists and which well describes the X-data. Such a projection is provided by the PLS plot shown above in fig.21 for the brain tissue data. This plot has similar properties to the PC plot but is, as discussed in a separate section below, much faster to compute and gives a somewhat better view of the class separation.

PARC methods as geometrical constructs in p-space

We shall now briefly discuss the PARC methods commonly used in chemistry. They can all be seen as geometrical constructions in p-space, either separating the classes by surfaces or by enclosing the classes by closed boundaries.

We shall not divide the methods into parametric or non-parametric. This because all methods are based on certain assumptions and hence to some extent are parametric. Rather, we shall discuss each method with respect to its modelling aspects of data in p-space, thus bringing assumptions out into the open.

Linear discriminant analysis (LDA) and related methods.

The historically first method is linear discriminant analysis, LDA The linear learning machine, LLM, is very similar, but slightly less dependent on assumptions about the distribution of the objects in p-space and also faster for large data sets. Assuming that the classes have the same multinormal distribution, LDA has certain optimality properties. The fact that it in real cases works less well than other methods shows that these assumptions are not fulfilled in practise.

Both these methods, moreover, have severe operational drawbacks making them little useful in chemical data analysis. They assume that the classes are linearly separated which they often are not (see asymmetric data structures below). They operate on level one and cannot cope with outliers in a good and general way. And they work only when the number of variables, p, is small compared to n. The methods get into great difficulties when the number of classes gets larger than about five. Finally, they are sensitive to collinearities in the X-matrix like all methods related to multiple regression.

It can be shown that for two class data sets, the LDA model can be computed by multiple linear regression (MR). See Mardia et al. (1979). A "y-variable" is constructed being +1 and -1 for training objects in class one and two, respectively. MR with this y-variable linearly modelled by the p x-variables gives the LDA solution. Hence various PARC methods called "least squares" and "regression" methods are nothing but inefficient variants of LDA

In p-space these methods operate by calculating a planar p-1 dimensional surface which separates the classes as well as possible (figure 29). The number of estimated parameters is p and the number of degrees of freedom in the data is n. Hence the condition n>>p. The relation to MR makes LDA independent of the scaling of the X-data.

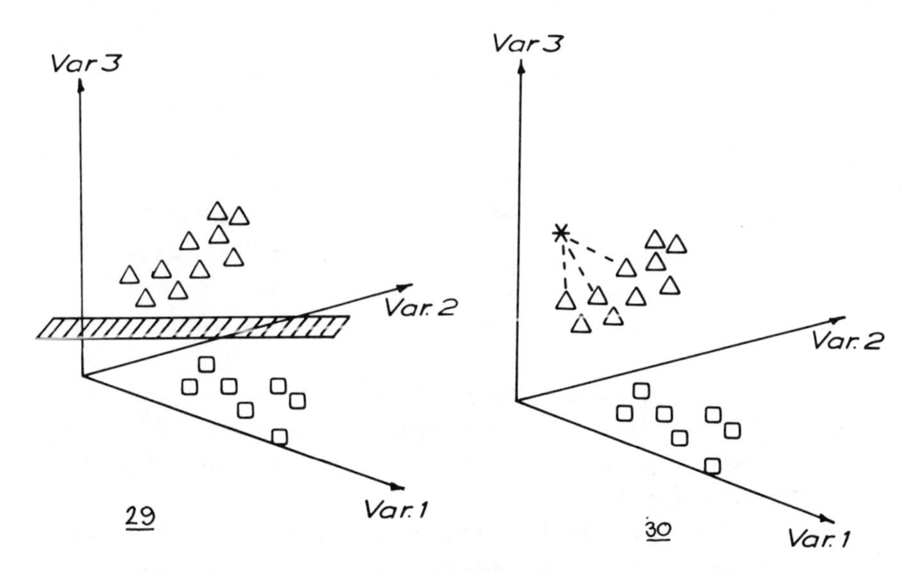

Figure 29. Linear discriminant analysis (LDA), the linear learning machine (LLM) and similar methods separate the classes by p-1 dimensional planes. These methods work only when the number of training objects (n) is considerably larger that the number of variables (p).

Figure 30. With the K nearest neighbor (KNN) method, a test object is classified according to the class of its K nearest neighbors in the training set. K is usually one or three.

K nearest neighbor methods (KNN)

This is a method with better properties than LDA and LLM, because the only assumption is that of the existance of a local metric being related to the similarity between objects. KNN classifies a new object according to its distance to the training set objects in p-space. The K nearest neighbors in the training set are found -- K is usually one or three -- and the object is assigned to the class of the majority of these K nearest neighbors (figure 30).

The KNN method as originally formulated operates on level one. From the distribution of the distances of the K nearest neighbors in the training set, it may be possible to construct tolerance levels so that an object with its nearest neighbors further away than a critical distance is labeled as an outlier.

When a dependent y-variable is given for some classes in the training set, the value for this variable may be predicted for test set objects as weighted averages of the values of the K nearest neighbors. Hence, the KNN method can be made to work as a level three method. See e.g. Martens and Russwurm (1983).

The KNN method works with any number of variables (p) for a given number of objects (n). When p exceeds n/4, however, the data must not be scaled to enhance the class separation (Fisher scaling, variance scaling) because the class separation is then grossly overestimated.

The KNN method is the only method that works also when classes in the training set are strongly subgrouped. This makes the method a good complement to other methods such as SIMCA which assume class homogeneity.

When used alone, the KNN method has the drawback to give little information about the internal class structure and about the relevance of the individual variables. It is not graphically oriented in the presentation of the classification results. With large training sets (more than 100 objects) the KNN methods becomes computationally slow.

KNN generalizations

The KNN method is based on the distances to the nearest training set objects. By calculating the distances to all training set objects and weighting them together according to weights quadratically or exponentially decreasing with these distances one gets a potential method. The ALLOC method of Coomans, Massart, et.al. (1981) is one example. Thus in a point (i) in p-space the potential of class g is, for instance (a is an arbitrary constant):

$$\sum_{j=\text{objects in class}} \left\{ 1/(1 + a \cdot d_{ij}^{2}) \right\}$$

With well represented classes in the training set, this class potential approximates the class probability density function, pdf. In this way and in the practical performance, the potential methods resemble the KNN method and work well when p is small or moderate. They are even more computationally demanding, however.

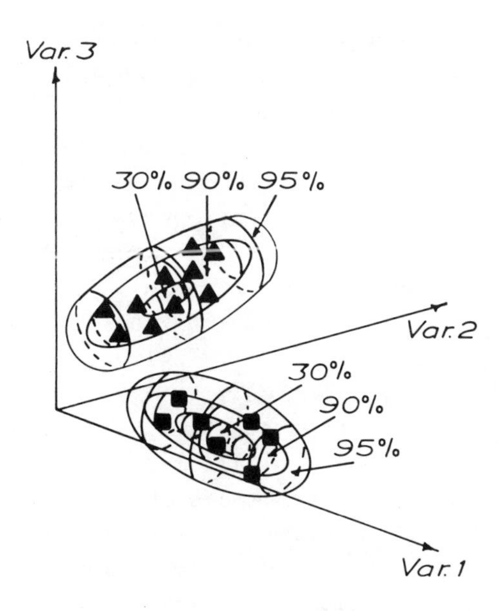

Figure 31. With probability density functions (PDF) calculated for each class, objects are classified to the class with the highest probability density in the object point. These methods are often called Bayesean methods.

PDF (Bayesean) methods

From the distribution of the training set objects in p-space, one can calculate empirical probability density functions (pdf) of each class (figure 31). Test set objects are then assigned to the class with the highest probability density in the object point -- PARC level one -- or to the class with sufficiently high density -- PARC level two. Since even the simplest pdf needs the specification of p averages and $p \cdot (p-1)/2$ covariances for each class with $p \cdot n_g$ data elements, one must have more than 2p training objects in each class. This makes pdf methods less useful in chemistry, but in applications with large numbers of objects, such as remote sensing, they are much used because of their optimal properties.

SIMCA (soft independent modelling of class analogy)

In part one we saw that the data observed on a class of similar objects can always be well approximated by a few-component PC model. This provided that most of the variables indeed express this similarity This is the basis of the SIMCA method (figure 32). The number of significant components in each class model is determined by crossvalidation (see part one).

From the scatter around the models one can construct tolerance intervals in the form of hyper cylinders around the PC hyper planes (figure 33). New test set objects are assigned to a class if they are inside its class "cylinder" or as outliers if they are outside all cylinders (figure 34).

As described in part one, the residuals of the class models can be used to calculate the relevance of the variables (modelling power) and the distances of the training objects to their class models. The latter can be used to find moderate outliers in the classes. If the objects in one class are fitted to a second class model and vice versa, the resulting residuals can be used to calculate the distance between the classes. The statistical significance of the class separation can be measured by an approximate F-test of the inter and intra class residual variances. This class distance can be split into the contribution of each variable and is then called the discrimination power of the variables.

Like other methods based on distances, SIMCA is scaling

Figure 32. In the SIMCA method (soft independent modelling of class analogy) each class is separately modelled by a principal components (PC) model. Here the simplest type of PC models is shown in the form of straight lines (the number of components, F, is one). In reality planes or hyperplanes are often used as determined by cross validation.

Thus each class, g, is modelled as $X_g = 1 \cdot \bar{x}_g + T_g P_g + E_g$.

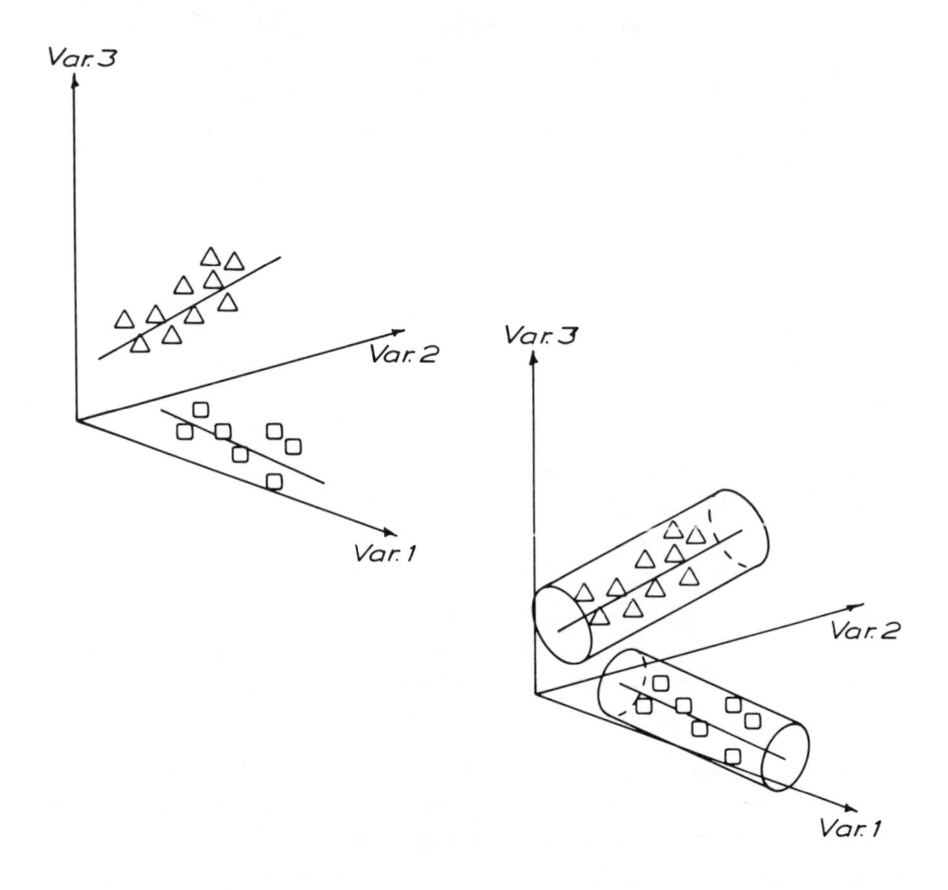

Figure 33. From the spread of the object points around the class model one can calculate a tolerance interval for any probability level, usually 95 percent. Future objects similar to the class fall inside this hypercylinder with the given probability, say, 95 %.

dependent. To make the class models really independent, the data are scaled differently for each class. When no apriori information is available, the variables are scaled to unit variance in each class (Derde et.al., 1982).

SIMCA is graphically oriented. An initial PCA or PLS discriminant plot (see below) of the training set gives an overview and an indication of the class separation (figures 21 and 35). Then the PCA of each class gives indications of outliers, subgroupings, etc., which may be used to "polish" the data. This polishing must, of course, not be done to inappropriately enhance the class separation. The final class PC models can then be graphically presented as score and loading plots revealing the inter class data structures.

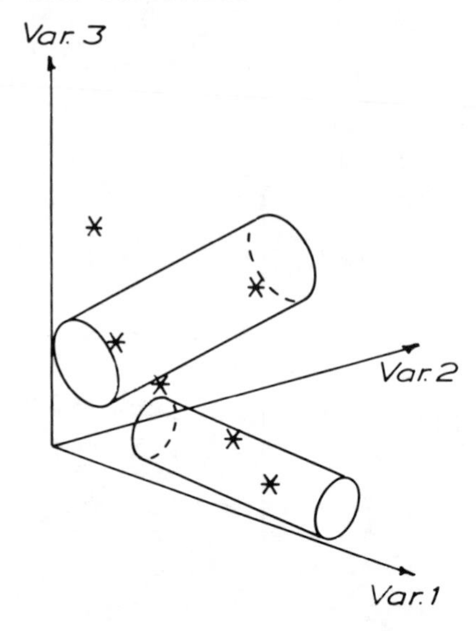

Figure 34. An object is classified in SIMCA according to which tolerance interval the object point falls in. Objects outside all "SIMCA cans" are labelled as outliers. In regions where the cans overlap, objects are not uniquely classified.

The calculation of the distance from an object point to a class model is made by relating the object data vector (x_j) to the class model: $x_j - \bar{x}_g = t_j P_g + e_j$. The coefficients in the vector t_j (with F elements) are determined by a simple linear multiple regression since the class parameters P_g are fixed. The standard deviation (SD) of the residuals e_{jk} directly corresponds to the distance between the object and the class.

With the ketone data, the SIMCA model parameters are so few that they are as well presented in the table below.

Results of a SIMCA analysis of the ketone data. The data were scaled separately to unit variance for each class training set using the scaling weights w_k. The cross validation indicates one significant PC in class one and two in class two. The modelling power for each variable in each class, i.e. the modelled amount of the data standard deviation is given as mp_k and the discrimination power of each variable as d_k.

All parameters refer to the scaled data.

k=	1	2	3	4	5	6	7
d_k	3.4	14.9	3.2	2.8	9.4	9.1	9.1
class 1							
w	.077	1.33	.14	.11	1.29	.011	.00044
x	129	11.4	1.71	171	1.40	24.3	5.63
p	.39	-.37	-.33	-.33	-.41	-.40	-.42
mp_k	.40	.34	.21	.23	.48	.46	.53
class 2							
w	.27	1.95	.26	.26	.59	.015	.00034
x	464	8.32	2.73	428	2.0	34.5	3.0
p	.41	-.47	-.44	-.33	.12	-.43	.34
p	.31	.07	.06	-.20	-.72	-.29	-.50
mp_k	.62	.75	.52	.17	.92	.71	.74

The results of the classification are nicely presented in a Coomans plot (figure 36) where the distance of each object to the two classes is plotted. Tolerance levels are also shown.

Conditions for PARC

We must remember that PARC does not solve all chemical problems. It only helps us to more efficiently extract the information inherent in the data.

The primary condition for success is a well designed data set. The classes must be well chosen with respect to the given problem. They must be homogeneous, i.e. not contain objects strongly clustered into less similar subgroups. Claims that PARC works with classes of diverse objects, e.g. chemical compounds of diverse structures, are wrong and based on incorrect data analysis (Wold and Dunn, 1983).

Each class must have at least five, and preferably ten to twenty, representative training objects. The large majority of the training objects must be assigned to the correct class

Several relevant variables must have been measured on each object. Relevant here means related to the given classification problem and monotonously related to within class similarity. The larger the number of relevant variables, the better the results. This as long as a method such as KNN or SIMCA is used where many variables can be utilized in an efficient way.

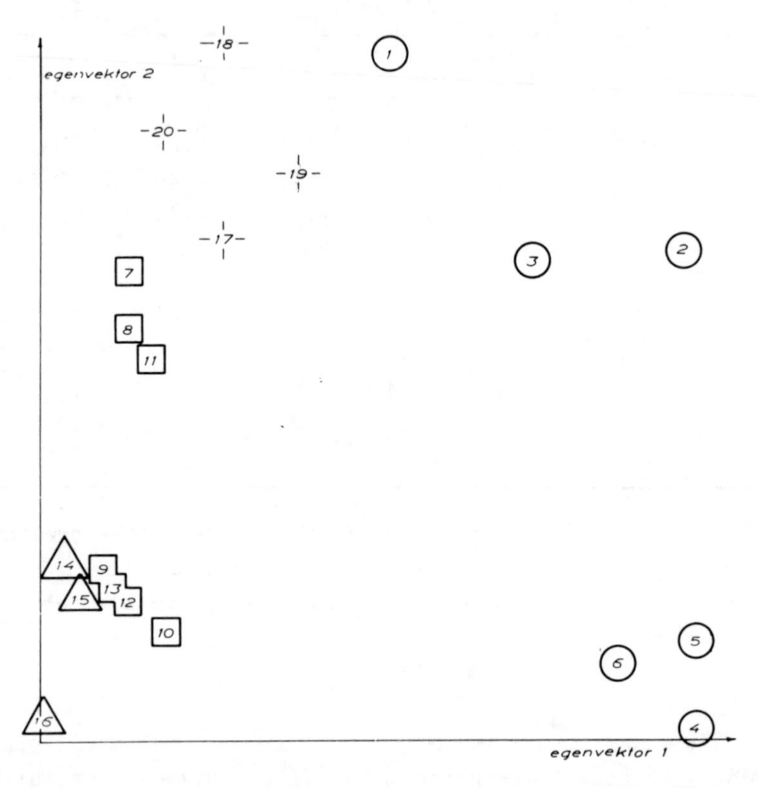

Figure 35. An eigenvector plot (PC plot) of the ketone data set. For notation, see fig. 26. The crosses are four test objects of uncertain class.

The variables should be transformed as to reveal within-class similarity. Mass spectra should be converted to autocorrelation or Fourier transforms.

The data should be collected in the same way in the whole data set and not in a class wise order. If there is a drift in

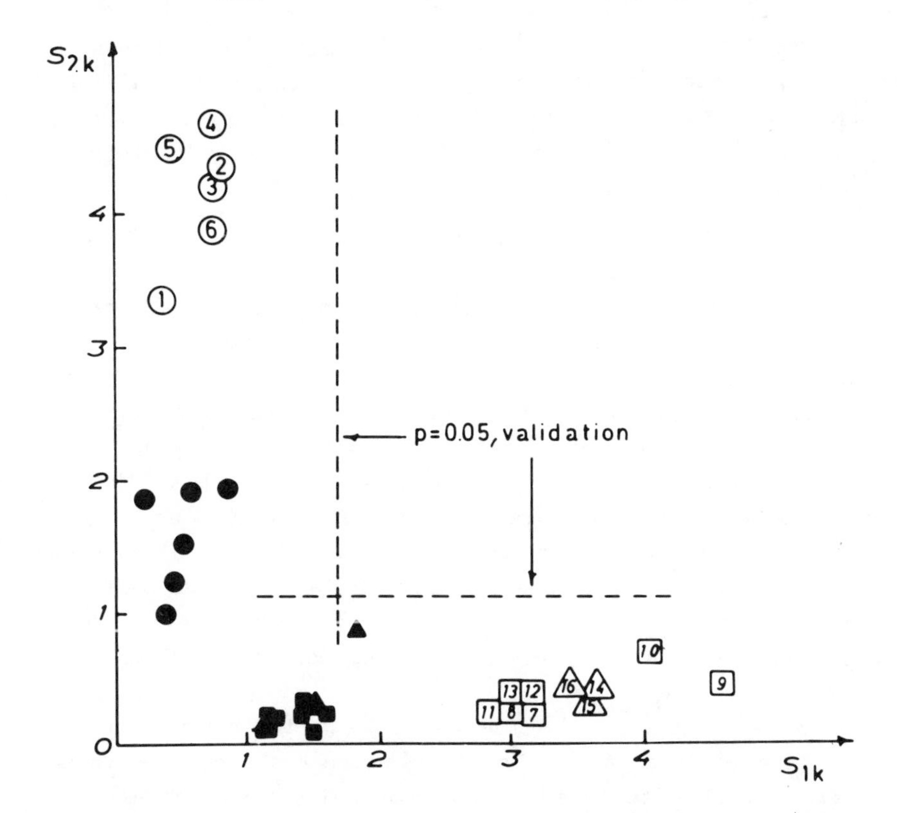

Figure 36. A Coomans plot of the SIMCA classification results of the ketone data set. The coordinates are the calculated SIMCA distances (residual standard deviations) of the objects with respect to the two class models. We see that the classes are well separated and that the three cis compounds are are classified uniquely to the same class as the sterically hindered compounds.

The tolerance interval limits of 5 % are calculated from a validation run with each object kept out from the training set once and reclassified. The open symbols refer to data scaled separately for each class, the filled symbols to data autoscaled over the training set.

measuring instruments or any other sampling conditions, the measuring of data on class one training objects before class two training objects will produce a class difference which is just the result of this drift. It will then be difficult to say how much of the found class separation that is "real and how much is just an artifact due to the drift.

The data should be analysed with a single method selected to be appropriate for the given problem and the given data. The custom to run many different PARC methods and then select the results that "look best" is deplorable and just a way to increase the risk for spurious results.

The asymmetric case

The PARC problem frequently is formulated so the resulting data are asymmetric (Dunn and Wold, 1980). This happens when one well specified class -- say, compounds with some specified type biological activity, e.g. carcinogenic in a given test system -- is contrasted to objects without this class property -- say, non-active compounds. The well specified class then is homogeneous and occupies a small domain in p-space. Hence, it can be modelled by a few term PC model. In contrast, the other class is not homogeneous with inherent similarity. The absence of a property is not class defining. Hence, these objects are more or less randomly spread out over p-space and cannot be PC modelled (figure 37).

By modelling the homogeneous class and constructing the class tolerance interval, we still get a classification scheme, however. New objects inside this interval are assigned to the class and objects far outside the interval are assigned to the "non-class" cathegory. Hence, only level two methods such as SIMCA work at all in this situation.

Any attempt to separate the two classes by a discriminant plane will obviously fail (figure 38). This is probably one reason why LDA and LLM are so unsuccesful in QSAR and medical diagnosis where the problem formulation often is asymmetric.

We realize that objects far outside the homogeneous class do not necessarily lack the class specifying property, e.g. carcinogenicity Objects having this property according to another mechanism of action, and thus being dissimilar to the first class, also fall outside the homogeneous class domain. This is a

Figure 37. The asymmetric PARC data structure. One class is well defined and occupies a small and regular part of p-space, while the second class is inhomogeneous and fairly "randomly" spread over p-space. With SIMCA the proper class can still be modelled and contained in a tolerance interval.

New objects are classified as belonging to this class if they are inside this tolerance interval (SIMCA-can). Objects far outside this "class can" are assigned to the second improper class.

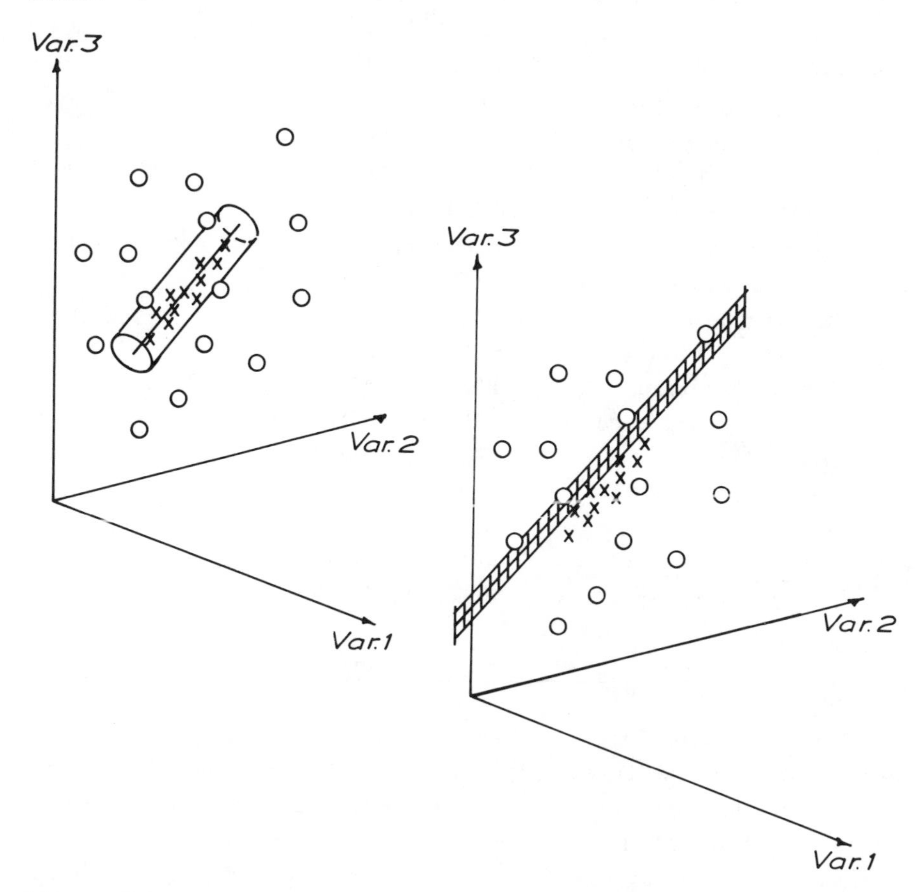

Figure 38. An attempt to use linear discriminant functions to separate the proper class from the rest is bound to fail in the asymmetric PARC situation.

Quadratic discriminant analysis will work in case the number of objects (n) is large compared to the number of variables in square (p^2).

philosophical problem, however, which cannot be solved simply by proper data analysis. Only a recognition and representation of the second type of "mechanism" will give data that can be used to "train" a method to give a correct classification.

Summary of PARC methods

Below we give a summary of the properties a chemically useful method of PARC should have. Positive properties are indicated with "+" and negative with "-" after the name of the method. LDA and LLM are bunched under the label LDA.

Simple to apply:	KNN +
Easy to understand:	KNN +, SIMCA +
Many variables	KNN +, SIMCA ++, LDA -, PDF -
Small training set	SIMCA +, PDF -
Feature selection	SIMCA +, LDA and PDF + when n>>p
Outlier detection	SIMCA +, PDF +
Missing data	SIMCA +
Probabil. classif.	SIMCA +, PDF +
Class structure	PDF +, SIMCA +, KNN and LDA -
Uncertain train.set	KNN (+), SIMCA (+)
Level 3,4	KNN (+), SIMCA ++
Graphical orient.	SIMCA +
Flexible	SIMCA +
Inhom. train.set	KNN +, LDA --
Computationally fast	SIMCA +, PDF +, KNN -, LDA -

Typical applications

Pattern recognition is increasingly applied in chemistry, in particular in the investigation of more complicated systems. Forina in this volume provides a nice illustration with the classification of food samples. Varmuza (1980) gives a good review of the use of PARC in chemistry before 1980. The references to this article include SIMCA applications to the classification of chemical compounds according to their type of biological activity, chemical reactivity, chemical structure (variables from NMR, IR or mass spectra) and mineral samples. Nonchemical SIMCA applications include the classification of ecological systems, economical systems and accounting tables.

Part 3, Dependencies between variables: Multiple regression (MR), partial least squares modelling with latent variables (PLS) and PARC levels 3 and 4

We shall now look at data analytical problems where dependencies between two blocks of variables are modelled. For convenience, we divide the data table into two parts, the blocks X and Y. In the general case we assume also that the data still are divided into a class-wise partitioned training set and a test set (figure 39). In many practical cases the number of classes is just one, however.

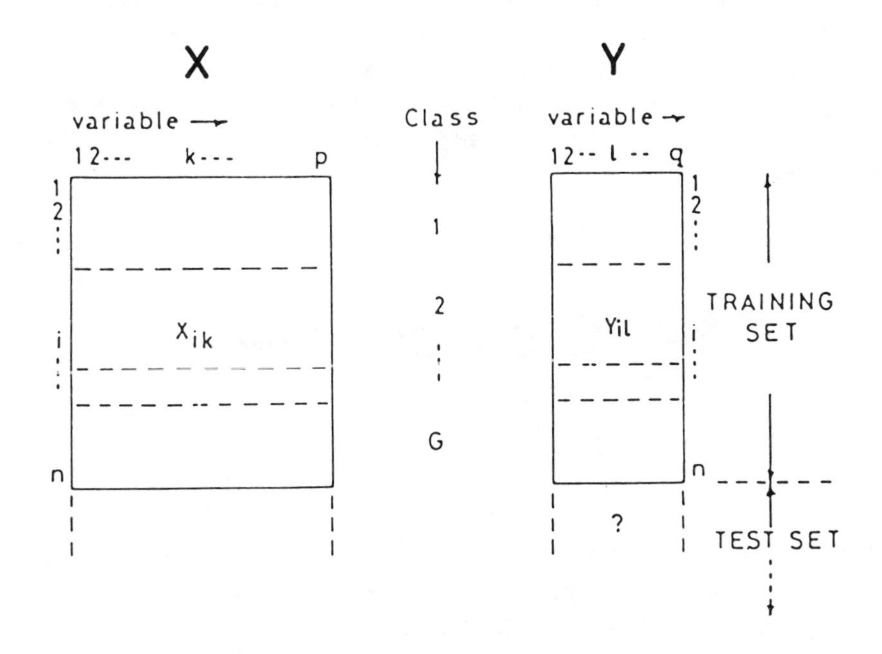

Figure 39. The data set of pattern recognition (PARC) levels three and four contains data for "dependent variables", Y for some classes in the training set. When the number of y-variables is one we have PARC level three. This case is similar to that of multiple regression (MR) if the number of x-variables (p) is small in comparison to the number of objects (n).

With many y-variables (two or more) we have PARC level 4.

This type of problems typically arise in all types of calibrations. X then contains signals observed on samples with known concentrations of interesting chemicals, Y. The analytical chemical case is discussed by Martens in this volume.

We here see the problem as an extension of the pattern recognition problem. We wish both to classify objects into one of the given G classes and to predict their values of "dependent variables", Y. To achieve this, values of Y are given for the training objects in, at least, some of the classes.

We call the problem one of PARC level 3 when there is just a single y-variable. With multiple (q>1) y-variables we have PARC level 4 (Albano et.al. 1978).

This desire to make quantitative predictions in addition to a classification is common. In medicine, for instance, we wish to classify "patients" as having one disease or another (or several) or as being "normal". In addition it is helpful to know how sick a person is, if he is severely, moderately or just a little affected.

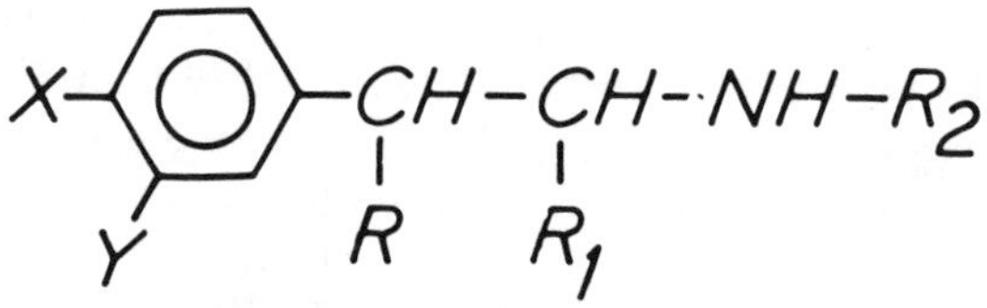

Figure 40. The structure of the n=33 substituted phenethylamines SIMCA-analysed by Dunn, Wold and Martin (1978). For different substituent combinations of R, R_1, X and Y the compounds are either beta receptor antagonists (class 1), agonists (class 2) or neither (inactive). For the class 1 compounds data for receptor binding (y_1) and antagonist activity (y_2) were given. For the class 2 compounds a data for a third y-variable, intrinsic activity, was also given.

The variation in structure, i.e. the variation of the substituents R, R_1, X and Y was described by 13 variables related to the size, lipophilicity and electronic properties of these substituents (see fig.41).

In studies of chemical reactivity, one often attempts to classify compounds as reactive or not in a certain reaction (asymmetric). If a compound is classified as reactive it is also valuable to know if it is very reactive or just a little.

The biological activity of chemical compounds may be seen as a special type of reactivity. The classification and quantitative prediction of the biological activity from the chemical structure -- QSAR = quantitative structure activity relationships -- is an area where multivariate data analysis has found use as indicated with the example below.

The beta-QSAR example

These data were first analysed by Dunn, Wold and Martin (1978). They concern the beta receptor activity of substituted phenethyl amines (figure 40). The problem is partly one of classification since some of the compounds are receptor agonists (stimulators) while others are antagonists (blockers). In addition, two measures of the degree of biological activity were given for the antagonists and three for the agonists.

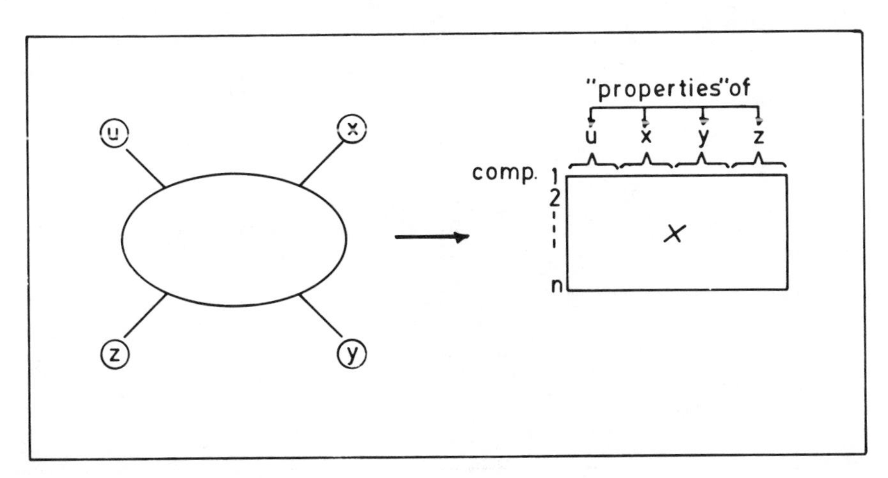

Figure 41. When the molecules in a structure-activity study can be described as a common skeleton and substituents (u,x,y,z), the structural variation can be translated to data by the use of "properties" of the substituents such as steric size, electron demand and lipohilicity. These properties have previously been determined by measuring chemical properties of standard series of compounds modified by the same substituents.

By translating the variation in chemical structure among the 33 compounds to the values of 13 variables, a table X is obtained (fig.41). When this table is related class wise to the table Y containing the measured biological activities, two PLS models with three significant dimensions (cross validation) were obtained. These models classify 30 of the 33 compounds correctl" as agonists or antagonists (fig.42). The confidence of the classification is better for the compounds with high activity.

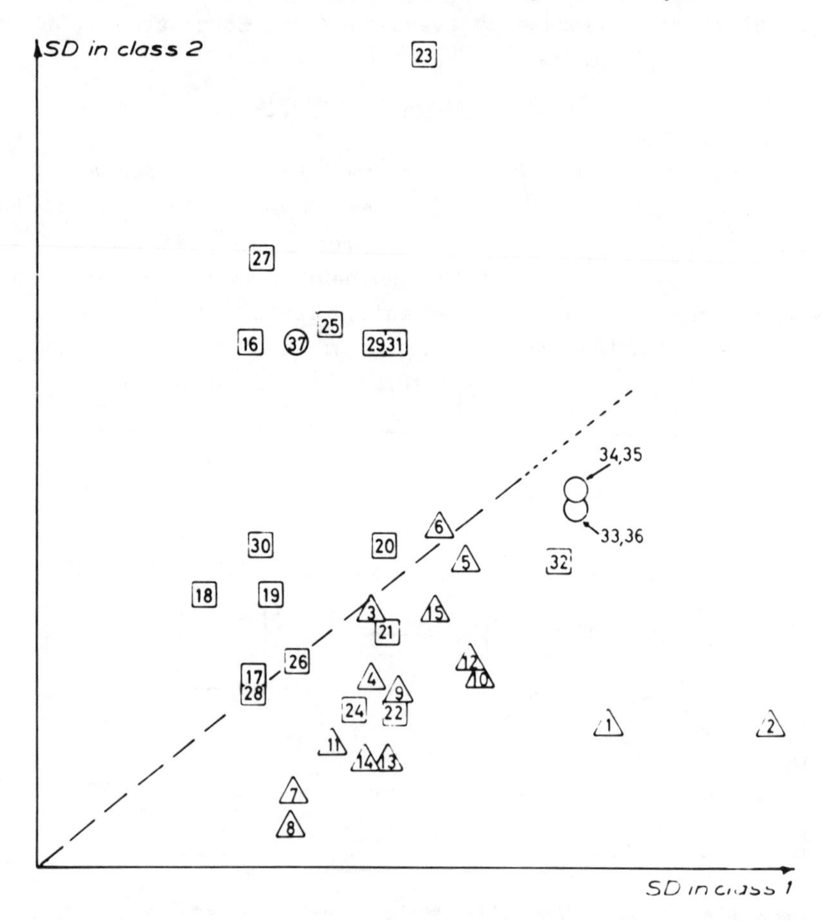

Figure 42. A Coomans plot of the SIMCA classification of the phenethyl amines. The classes are fairly well separated with the highly active compounds well classified while the compounds with low acitivity are close to both class models. Triangles are agonists and squares antagonists. Rings indicate predictions for the five test compounds.

Moreover, a strong relation between the measured biological activity and the latent variable of the structure matrix X is obtained (figure 43). The interpretation of the direction t_1 allows the medicinal chemist to construct compounds which are predicted by the model to have increased biological activity.

Hence new compounds of similar structure can be (a) classified as agonist, antagonist or neither and (b) their level of activity can be predicted if they are inside one of the classes.

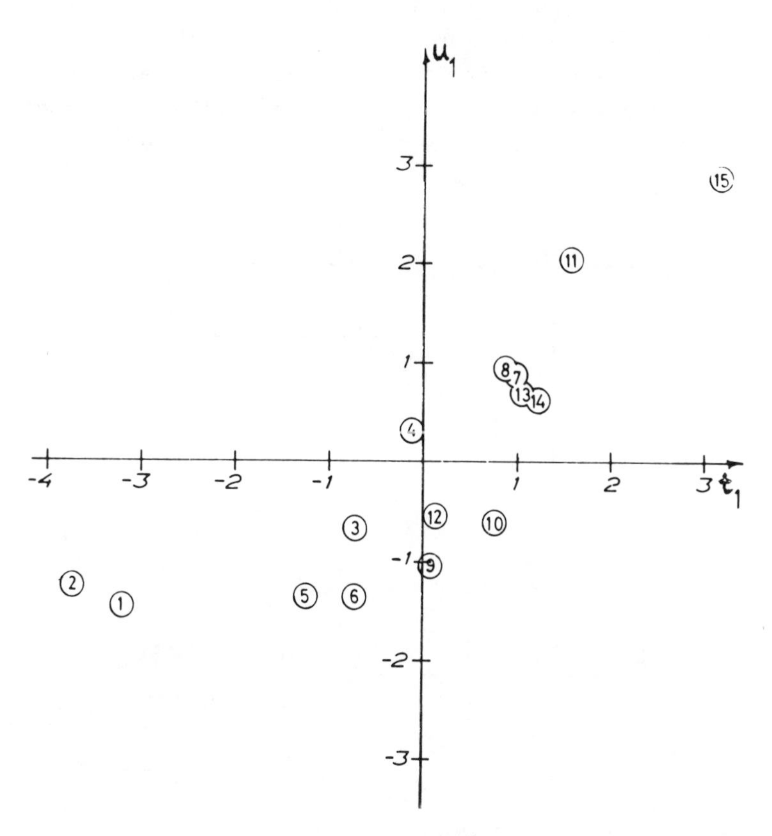

Figure 43. Relation for class two compounds (agonists) between the measured activities (summarized by the latent variable u_1) and the first PLS component for the X-block, t_1).

The traditional way: Multiple Regression (MR)

Laplace and Gauss developed methods relating one y-variable to a number of X-variables. The case with a linear relation is called multiple linear regression (MR). This method may be seen as the natural choice also in the present case, i.e. one would develop a separate PARC model to classify objects and then for each class develop one multiple regression model for each y-variable.

We shall here just note that the assumptions underlying MR (figure 44) make the method inappropriate in chemical data analysis except in one specific case, namely that of designed experimentation where all x-variables are under control and the columns in X (in each class) are orthogonal to each other or almost so.

MR cannot and must not be used when the number of x-variab-

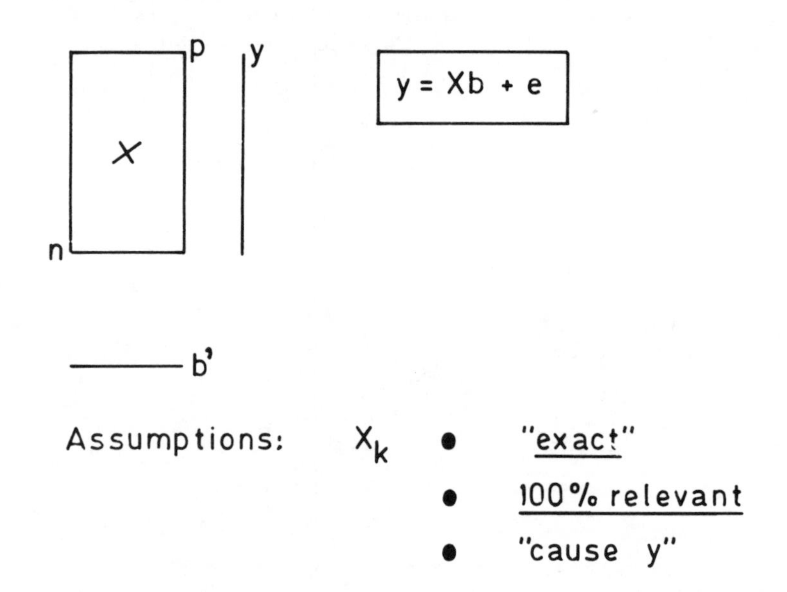

Figure 44. The multiple regression (MR) model and the underlying assumptions. Since these assumptions rarely are fulfilled except in data from an experimental design, MR can only be recommended for analysis of statistically designed data. For all other data, only rank reducing methods such as PCR and PLS are appropriate.

les is large. Stepwise procedures for variable selection are used to try to bring the situation back to one where MR is appropriate. However, as shown by numerous authors (e.g. Topliss and Edwards, 1979) this is usually just selfdeceptive.

MR cannot and must not be used when the data are just passively monitored, because the data then invariably are collinear and a unique and interpretable MR solution cannot be computed.

Remembering the typical case of chemical data sets today with many variables, some possibly irrelevant, we see that MR indeed is inappropriate. But then, what to do ?

The projection of X down on a small matrix T

In part one we learnt that whenever we have a data matrix X observed on a set of similar objects, this matrix can be well approximated by a matrix T with few and orthogonal columns. Thus we might convert the data set with many diffuse x-variables to one with few orthogonal t-variables and then use T instead of X in a MR model. This is indeed a good approach, and used much in applied statistics under the name of principal components regression (PCR).

However, as pointed out by Jolliffe (1982) and others, there is a risk that numerically small structures in the x-data which "explain" Y disappear in the PC modelling of X. This will then give a bad prediction of Y from T.

The partial least squares (PLS) models recently developed by H.Wold et.al (1982) have properties that partly circumvent this difficulty. Thus the matrix T -- still a projection of X -- is calculated both to approximate X and to predict Y; the latter matrix is used to "tilt" the PC plane to improve the relation betwen T and Y. Cross validation is used to control the dimensionality of the projection to be predictively relevant. See S.Wold et al. (1983)

The PLS method has the further advantage before PCR to handle also the case with several Y-variables as in the QSAR example above. This is accomplished by having a separate space for the Y data and computing a projection U which is well predicted by T at the same time as it well approximates Y.

In the case with several classes, one PLS model is constructed for each class and the tolerance intervals in the X-space are then used in the ordinary SIMCA way for the classifiation of new

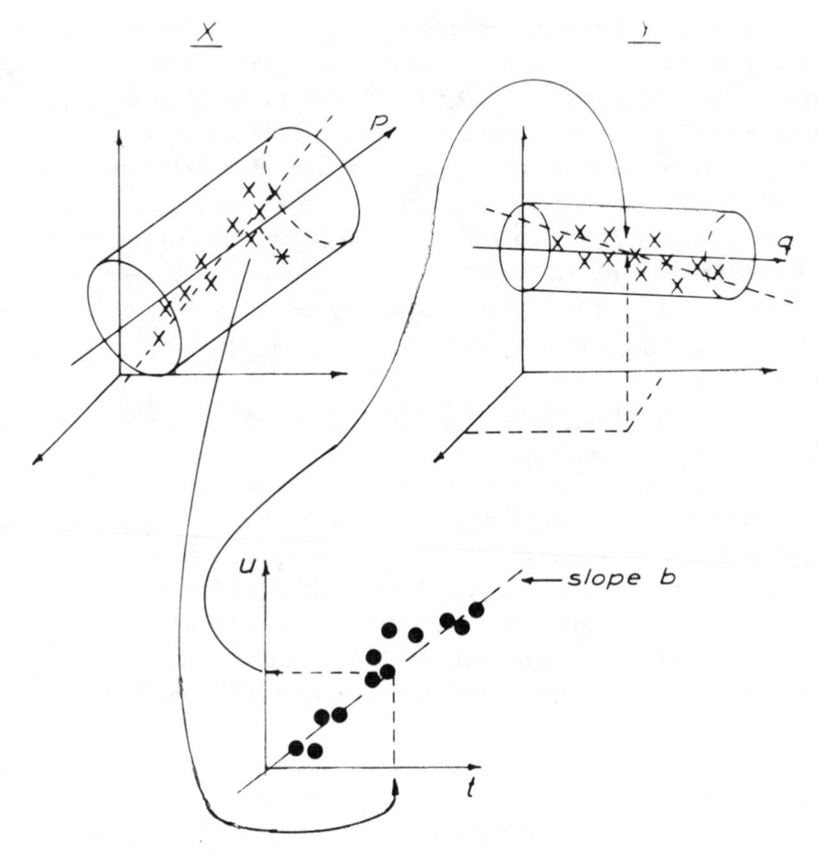

Figure 45. The PLS model in its geometrical interpretation. The X-block and Y-blocks are represented as points in separate spaces, X-space and Y-space. In each space the data are modelled as lines or planes or hyperplanes (with F dimensions), here shown as lines. The projections of the object points down on these models have the coordinates T and U for the X and Y parts of the data, respectively. Each column in U is related to the corresponding column in T by a linear model (bottom).

A new object is classified as belonging to the class or not on the basis of its X-data and thereby its position inside or outside the class tolerance interval in X-space (upper left). If the object indeed belongs to the class, it projects down on the class model to get the class coordinates t. Each of these t-values connects to a u-value by means of the inner PLS U-T relation (bottom). These u-values define a point on the model in Y-space, which, in turn, corresponds to a value of each y-variable.

objects, to find outliers, etc. (figure 45). Computationally, the PLS method is extremely fast, making it applicable even with 8 bit micro computers. As PCA, PLS can operate with missing data and with many X and Y-variables compared to the number of objects. The projection in matrix form is indicated in figure 46. See also the Martens article in this volume.

Multivariate data analysis and projections

The reader´s conclusion might now be that all multivariate data analytic problems are readily solved by projections. We

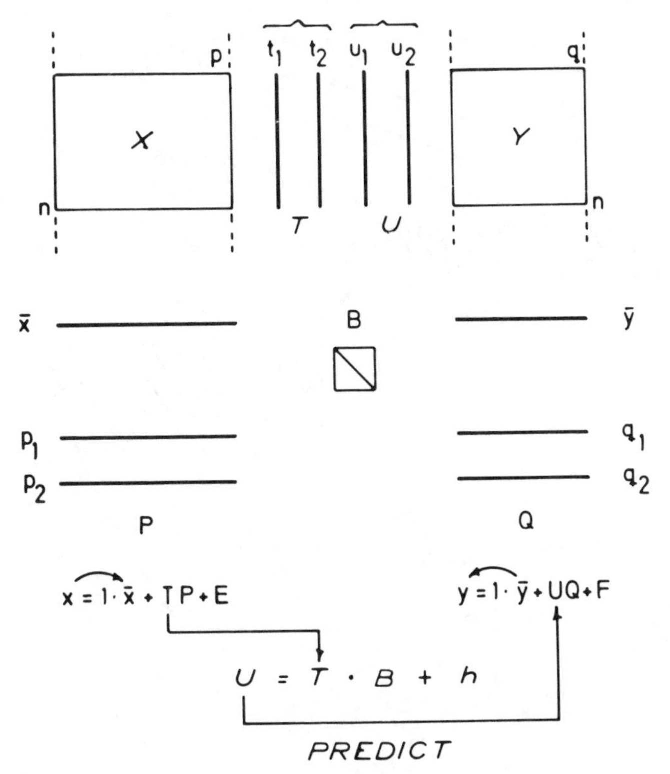

Figure 46. The PLS model in matrix form. The matrices T and P model the X block similarly to an ordinary PC model, while the matrices U and Q model the Y-block. The connection between the blocks is modelled as a relation between the U and T matrices by means of a diagonal matrix B. The block residual matrices are E and F and the vector of residuals for the inner relations is h.

agree. Projections are simple to use and interpret and lend themselves to graphical presentation. Bi-linear projections such as PC and PLS projections are rapidly computed and 8-bit micros with simple BASIC programs are prefectly adequate for data matrices up to about 40 objects per class with up to about 75 variables (X and Y together).

We shall end by showing a few additional examples of specific projections and how to reformulate some common problems to various forms of projections.

I. A picture of a data matrix: PC (part one, above).

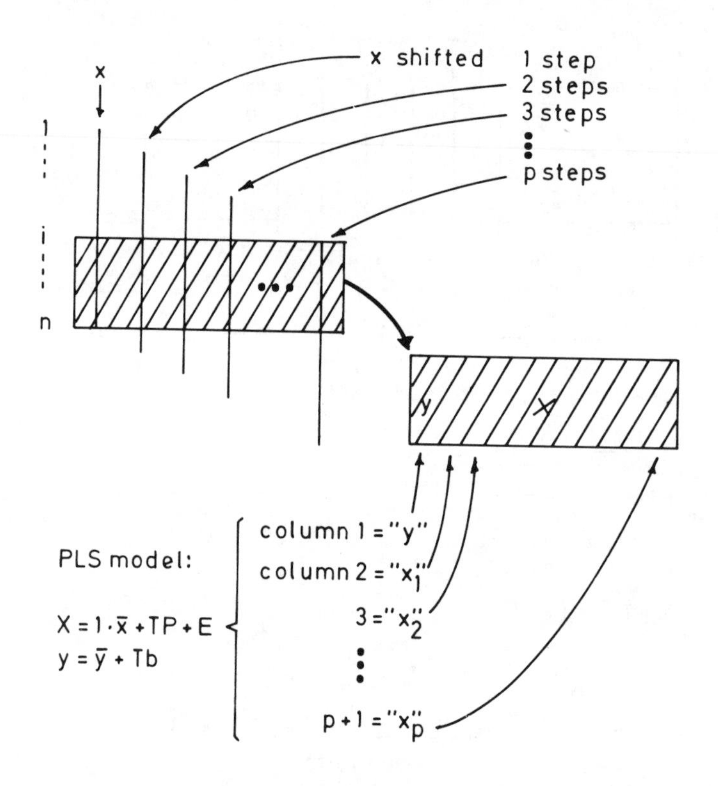

Figure 47. Time series models $X_t = b_1 \cdot X_{t-1} + b_2 \cdot X_{t-2} + \ldots$ can be handled by PLS modelling. An expanded data matrix is created by shifting the original series X_t (can be a vector or a matrix) one step, two steps, ... , up to , say, n/4 steps. This expanded matrix is used as the X-block and the non-shifted series as the y-block in a standard PLS analysis.

II. Classification (PARC level 2): PC modelling of each class (SIMCA).

III. Plus quantitative prediction (PARC levels 3 and 4): PLS models of each class (SIMCA-MACUP).

IV. Regression-like models when the number of x-variables is large or when X has collinearities: PLS.

V: Time series analysis: PLS (we don´t claim that PLS is better than Box and Jenkins analysis, but rather that the projection formalism might be useful also here), see figure 47.

VI. Discriminant plot: PLS (figure 48). This gives an overview of the X-matrix in the same way as PCA, but with classes

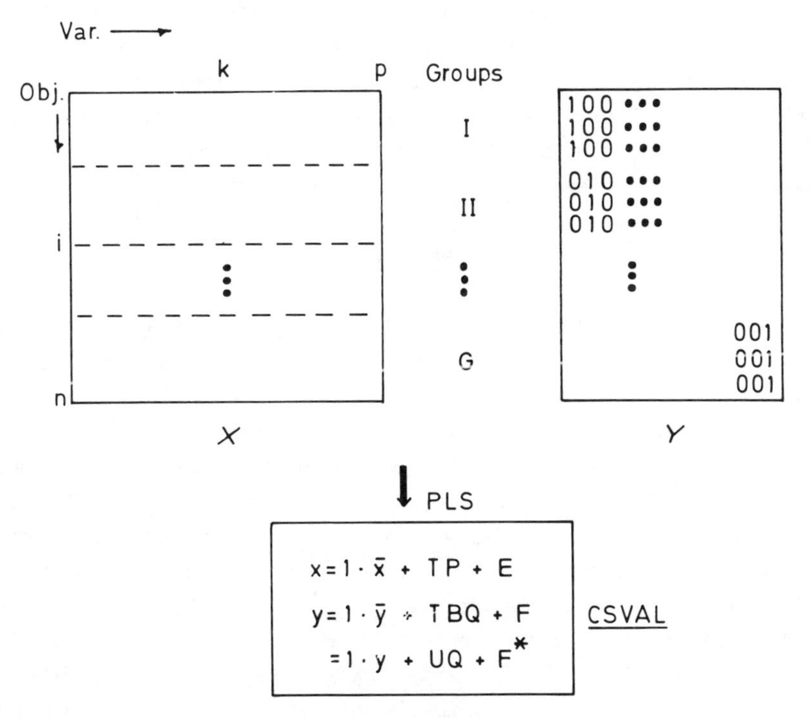

Figure 48. PLS discriminant plots and PLS canonical variate plots are obtained by constructing a G column dummy matrix Y with ones and zeroes (column j has ones only for objects of group or class j), scaling the X and Y matrices, and then analysing the X+Y blocks by an ordinary PLS two-block model with crossvalidation and finally plotting the resulting t-vectors against each other.

better separated. The Brain Tissue plot (fig.21) is an example. The loading plot directly gives the discrimination power of the x-variables, e.g. figure 22. This projection is useful also when one has measured samples in replicate and wishes to maximize the between sample variation compared with the within replicate variation.

VII. K-way tables: PCA or PLS with unfolding. Nowadays data arrays with more than two "directions" are often produced. As an example, think of food samples -- say fruits -- analysed by two-dimensional chromatography once a day to monitor their ripening or decay. This gives a four directional data array, object x time x two chromatographic dimensions. As discussed by Esbensen and Wold (1983), such a K-way table can be "unfolded" to an ordinary two-directional array and modelled by PC or PLS with additional constraints derived from the folding structure. The solution -- the matrices T and P -- can then be "folded back" to give the desired projection of the original K-way table. This can, of course, be done class-wise in a case of PARC.

Conclusions

We have, in three parts, discussed the analysis of data tables, matrices X (and Y). In chemistry, these tables often contain collinearities because they are not collected according to a strict experimental design and the variables are often not controlled. Moreover, in the way the tables are produced in chemical practice, they often have many variables p in comparison with the number of objects, n. The tables often contain errors or odd cases (outliers) and often are incomplete (missing data).

Simple bilinear projections, PC and PLS, are useful to extract information from these tables. We see this as generalized pattern recognition (PARC), ways to find regularities in the data and separate these from "noise". We note that the division of X and Y into these two parts depends on the given problem, a very regular data matrix may contain very little information about the problem.

The methods have found increasing use in chemistry as shown by the survey in Malinowski and Howery (1980) and references in this article and those of Martens and Forina in this volume. In particular, the classification of complicated samples characterized by a multitude of variables allows a "new type of scientific approach" to the study of complicated systems. This is examplified by the classification of oil-spills, food samples, mineral samples, micro-organisms, clinical chemistry samples, mosquitoes, irises, tobaccoes, PCB samples, brain tissue, NMR, IR and mass spectra, and compounds according to their chemical reactivity or biological activity.

The concept of multivariate calibration (see Martens in this volume) provides the quantitative extension of this idea to "continuous classification", i.e. the prediction where an object is on one or several continuous scales. When the samples are diverse, one must divide them into separate homogeneous classes to make the models work well. One then has a combination of classification and quantitative modelling, PARC levels 3 and 4.

Final words

One may now think that PCA and PLS and PARC solve all chemical problems. This is not so. On the contrary, it is as difficult as before to reformulate chemical problems so that they can be approached by a combination of experiments, data collection and data analysis. Only the last piece is made more efficient by the use of multivariate methods, but the essential parts still remain the design of the experiments and the precise measurement of pertinent data.

This brings us to what we think is the second greatest problem in chemometrics, namely to make the few who use "statistical" methods in their chemical practise to accept that:

IN A DATA SET THERE IS OFTEN NO INFORMATION WHATSOEVER ABOUT THE GIVEN PROBLEM

Some data analytical methods are particularly dangerous in that they are based on the assumption that there is information. Among these methods we have the commonly used multiple regression and linear discriminant analysis, particularly in their step wise forms, the linear learning machine and Fisher variance scaling.

Bilinear projection methods don´t make this assumption and therefore are less likely to mislead the user. This is by some considered a drawback since the results usually are apparently less beautiful than when the former methods are used. In the real world, however, methods which tend to confirm ones prejudice are more dangerous in the long run than methods that are conservative with respect to the amount of extracted information.

Acknowledgements

We are grateful for financial support from the Swedish Natural Science Research Council (NFR), the Swedish Council for Planning and Coordination of Research (FRN) and the National Swedish Board for Technical Development (STU). Herman Wold's never failing enthusiasm and statistical knowledge has been of greatest value.

References

C.Albano, W.J.Dunn, U.Edlund, E.Johansson, B.Nordeń, M.Sjöström and S.Wold (1978). Four levels of pattern recognition. Anal.Chim.Acta Comput.Tech. Optim. 103, 429-443.

C.Albano and S.Wold (1980). Multivariate analysis of solvolysis kinetic data. An empirical classification parallelling charge delocalization in transition state. J. Chem. Soc. Perkin 2, 1980, 1447-51.

G.E.P. Box, W.G. Hunter, J.S. Hunter (1978). Statistics for experimenters. Wiley, New York.

R.N.Carey, S.Wold and J.O.Westgard (1975). Principal Components Analysis: An Alternative to "Referee" Methods in Method Comparison Studies. Anal.Chem. 47, 1824.

D. Coomans, D.L. Massart, I. Brockaert, A. Tassin (1981). Potential methods in pattern recognition. Part 1. Classification aspects of the supervised method ALLOC. Analyt. Chim. Acta Comp. Tech. Optim., 133, 215 - 224

M.P.Derde, D.Coomans and D.L.Massart (1982). Effect of scaling on class modelling with the SIMCA method. Anal.Chim.Acta 141, 187.

P.Diaconis and B.Efron (1983). Computer intensive methods in statistics. Scientific American, May 1983, 96-108.

N.R. Draper and H. Smith (1981). Applied regression analysis, 2.nd edition. Wiley, New York.

W.J.Dunn III and S.Wold (1980). Structure - activity analyzed by pattern recognition: The asymmetric case. J. Med. Chem. 23, 595.

W.J.Dunn III and S.Wold (1980). Relationships between chemical structure and biological activity modelled by SIMCA pattern recognition. Bioorg. Chem. 9, 505-23.

H.T.Eastment and W.J.Krzanowski (1982). Cross-validatory choice of the number of components from a principal component analysis. Technometrics 24, 73-77.

K.H.Esbensen and S.Wold (1983). SIMCA, MACUP, SELPLS, GDAM, SPACE & UNFOLD: The ways towards regionalized principal components analysis and subconstrained N-way decomposition -- with geological illustrations. Proc. Conf. Applied Statistics, Stavanger (O.Christie, Ed.).

R.A.Fisher and W.A.MacKenzie (1923). Studies in Crop Variation. II. The manurial response of different potato varieties. J.Agr.Sci. 13, 311-320.

I.E.Frank and B.R.Kowalski (1982). Chemometrics. Anal. Chem. 54, 232R - 243R.

R.Gnanadesikan (1977). Methods for Statistical Data Analysis of Multivariate Observations. Wiley, New York.

J.D.F.Habbema (1983). Some useful extensions of the standard model for probabilistic supervised pattern recognition. Anal.Chim.Acta 150, 1-10.

E.Jellum, I.Björnson, R.Nesbakken, E.Johansson and S.Wold (1981). Classification of human cancer cells by means of capillary gas chromatography and pattern recognition analysis. J.Chromatogr. 217, 231 - 237.

E.Johansson, S.Wold and K.Sjödin (1983). Closure or the constant sum problem in analytical chemistry, with examples from gas chromatography. Submitted to Anal.Chem. 1983.

D.Johnels, U.Edlund, E.Johansson and S.Wold (1983). A Multivariate Method for Carbon-13 NMR Chemical Shift Predictions Using Partial Least-Squares Data Analysis. J.Magn.Reson. 55, 316-21.

I.T.Jolliffe (1982). A note on the Use of Principal Components in Regression. Appl. Statist. 31, 300-303.

K.G. Jöreskog, J.E. Klovan and R.A. Reyment (1976). Geological factor analysis. Elsevier, Amsterdam.

K.G.Jöreskog and H.Wold, Ed.s (1982). Systems under indirect observation, Vol. I and II. North Holland, Amsterdam.

B.R. Kowalski and C.F. Bender (1972). Pattern recognition. A powerful approach to interpreting chemical data. J. Amer. Chem. Soc., 94, 5632 - 5639.

B.R. Kowalski and C.F. Bender (1973). Pattern recognition. II. Linear and nonlinear methods for displaying chemical data. J. Amer. Chem. Soc., 95, 686 - 693.

B.R.Kowalski (1974). Pattern Recognition in Chemical Research. In Computers in Chemical and Biochemical Research, Vol.2 (C.E.Klopfenstein and C.L.Wilkins, Ed.s). Academic Press, N.Y.

B.R. Kowalski (1980). Chemometrics. Anal. Chem., 52, 112R.

B R.Kowalski and M.A Sharaf (1981). Extraction of individual mass spectra from gas chromatography-mass spectrometry data of unseparated mixtures. Anal.Chem. 53, 518-22.

B.R.Kowalski and S.Wold (1982). Pattern Recognition in Chemistry. In: Classification, Pattern Recognition and Reduction of Dimensionality, (P.R.Krishnaiah and L.N.Kanal, Ed.s), North-Holland, Amsterdam.

W.Lindberg, J-Å.Persson and S.Wold (1983). Partial Least-Squares Method for Spectrofluorimetric Analysis of Mixtures of Humic Acid and Ligninsulfonate. Anal. Chem. 55, 643-648.

E.R. Malinowski and D.G. Howery (1980). Factor analysis in chemistry. Wiley, New York.

K.V.Mardia, J.T.Kent and J.M.Bibby (1979). Multivariate Analysis. Academic Press, New York.

H.Martens and H.Russworm, Jr., Ed.s (1983). Food Research and Data Analysis. Applied Science Publ., London 1983.

D.L.Massart, A.Dijkstra and L.Kaufman (1978). Evaluation and Optimization of Laboratory Methods and Analytical Procedures. Elsevier, Amsterdam.

R.Mecke und K.Noack (1960). Strukturbestimmungen von ungesättigen Ketonen mit Hilfe von Infrarot- und Ultraviolett-Spektren. Chem.Ber. 93, 210.

G.Musumarra, S.Wold and S.Gronowitz (1981). Application of principal component analysis to C-13 NMR shifts of chalcones and their thiophene and furan analogues: a useful tool for the shift assignment and for the study of substituent effects. Org.Magn.Resonance 17, 118-123.

G.Musumarra, G.Scarlata, G.Romano and S.Clementi (1984). Identification of drugs by principal components analysis of Rf data obtained by TLC in different eluent systems. J.Anal.Toxicol. 1984, in press.

K.Pearson (1901). On lines and planes of closest fit to systems of points in space. Phil.Mag. (6) 2, 559-72.

Schiffman, S.S., Reynolds, M.L. and Young, F.W (1981). Introduction to multidimensional scaling: Theory, methods and applications. Academic Press, New York 1981.

H.F Shurvell and A.Dunham (1978). The application of factor analysis and Raman band contour resolution techniques to the study of aqueous Zn(II) chloride solutions. Can.J.Spectrosc. 23, 160-5.

M.Sjöström and U.Edlund (1977). Analysis of C-13 NMR data by means of pattern recognition methodology. J.Magn.Reson. 25, 285.

M.Sjöström and B.R.Kowalski (1979). A comparison of five pattern recognition methods based on the classification results from six real data bases. Anal.Chim.Acta 112, 11-30.

B.Söderström, S.Wold and G.Blomqvist (1982). Pyrolysis-Gas Chromatography Combined with SIMCA Pattern Recognition for Classification of Fruit-bodies of Some Ectomycorrhizal Suillus Species. J.Gen.Microbiol. 128, 1783-1784.

J.G. Topliss, R.P. Edwards (1979). Chance factors in studies of quantitative structure-activity relationships. J. Med. Chem., 22, 1238 - 1244.

K. Varmuza (1980). Pattern recognition in chemistry, Springer-Verlag, Berlin.

H.Wold (1982). Soft Modeling. The Basic Design and Some Extensions. In Jöreskog and Wold, Ed.s (1982).

S.Wold (1976). Pattern recognition by means of disjoint principal components models. Pattern Recognition 8, 127-139.

S.Wold and M.Sjöström (1977). SIMCA, a method for analyzing chemical data in terms of similarity and analogy. In Chemometrics, Theory and Application (B.Kowalski, Ed.). Amer.Chem.Soc.-Symp.Ser. 52.

S.Wold (1978). Cross validatory estimation of the number of components in factor and principal components models. Technometrics 20, 397-406.

S.Wold and W.J.Dunn III (1983). Multivariate quantitative structure activity relationships (QSAR): Conditions for their applicability. J.Chem.Inf Comput.Sci. 23, 6-13.

S.Wold, W.J.Dunn and S.Hellberg (1982). Survey of applications of pattern recognition to structure-activity problems. Research Group for Chemometrics, Tech.Rep. 1-1982, Inst. of Chemistry, Umeå University, S-901 87 Umeå, Sweden

S. Wold et.al.(1983). Pattern Recognition: Finding and Using Regularities in Multivariate Data. In Martens and Russwurm, Ed.s (1983).

S.M.Wolfrum and G.Kateman (1983). A survey of chemometric publications of 1982. Chemometric Newsletter No.9 (Jan.1983).

STATISTICS AND CHEMISTRY, AND THE LINEAR CALIBRATION PROBLEM

William G. Hunter

University of Wisconsin

Statistics is the study of how we learn from data: how data can be collected efficiently and analyzed effectively. In this paper, the role of statistics in chemistry is discussed. After model-building and Bayesian analysis are briefly described, a Bayesian analysis of the linear calibration problem is presented.

1. STATISTICS AND SCIENCE

Statistics is the study of (1) the efficient collection and (2) the effective analysis of data. Data are collected from experiments, sample surveys, and censuses. Analysis is the attempt to extract all the relevant information from a set of data. Of these two parts of statistics, (1) is more important. The damage of poor experimental design, for example, is irreparable. No amount of fancy analysis can extract much information from a set of data that contains little or none.

Four stages of typical experimental investigations are the following: conjecture, design, experiment, and analysis. Conjecture is a new idea (hypothesis, model, theory). The design of the experiment is the plan to be followed in collecting the data. In the experiment, data are collected. They are then analyzed. The analysis usually shows some ways in which the initial conjecture might be modified. This modified conjecture may lead to a new design, a new experiment, new analysis, and so forth. This particular way of thinking about the iterative nature of experimental work was first presented by Box and Youle (1955).

B. R. Kowalski (ed.), Chemometrics. Mathematics and Statistics in Chemistry, 97–114.

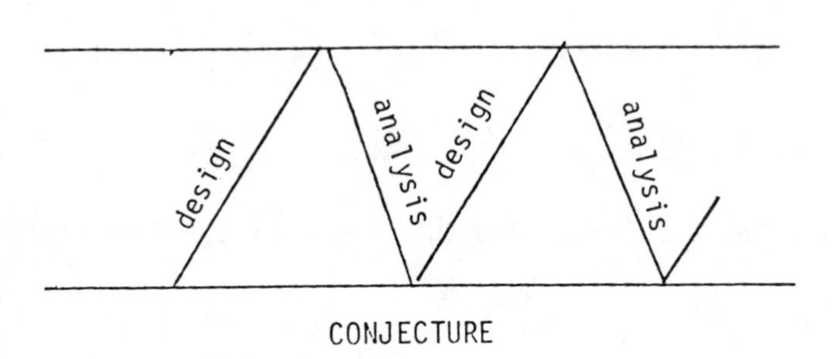

Figure 1. Four stages of experimental investigations: conjecture, design, experiment, and analysis.

Note that the two parts of statistics - design and analysis - appear in the center of Figure 1. Statistics is important in scientific work. Statistics is not something that experimenters decide to use or not to use – it is what they do. They collect data and analyze data. Statistics, in a sense, is the science of science. More generally, it is the study of how all of us - scientists and non-scientists - learn from data.

2. MODEL-BUILDING

Figure 2 represents the major steps in model-building. The purpose of model-building is to develop an equation or set of equations that adequately describes the data. There are two key questions: Is the model adequate to describe the data? What are the best estimates of the parameters in the model? If the model is found to be inadequate, it has to be modified or perhaps scrapped all together. If the model is adequate, the parameter estimates, unfortunately, may be found to be too imprecise to permit use of the model in practice. In that situation, the investigator may elect to collect additional data. If the model is discovered to be adequate and the parameters are estimated with satisfactory precision, the investigator may want to collect additional data to validate the model. That is, predictions are made on the basis of the fitted model and these predictions are checked against new data.

Experimenters collect the data. One or more chemists, physicists, biologists, and other subject matter experts create the model. The model may be empirical or mechanistic. It may combine both empirical and mechanistic elements. The two questions

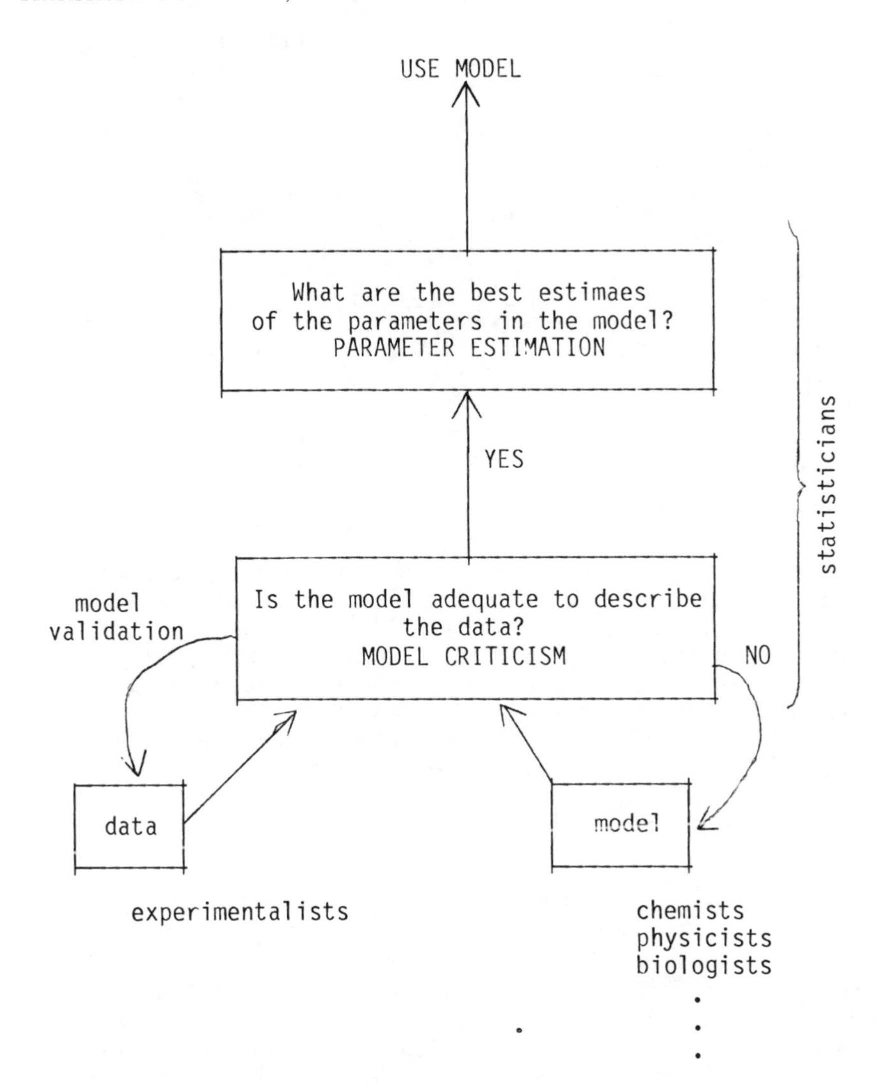

Figure 2. Model-building.

concerning parameter estimation and model criticism are in the domain of statistics.

Recent work by Box (1980) indicates that Bayesian analysis should be used for answering questions related to parameter estimation and classical sampling theory (significance tests, etc.) should be used for answering questions related to model criticism. To check model adequacy, the investigator must carefully examine

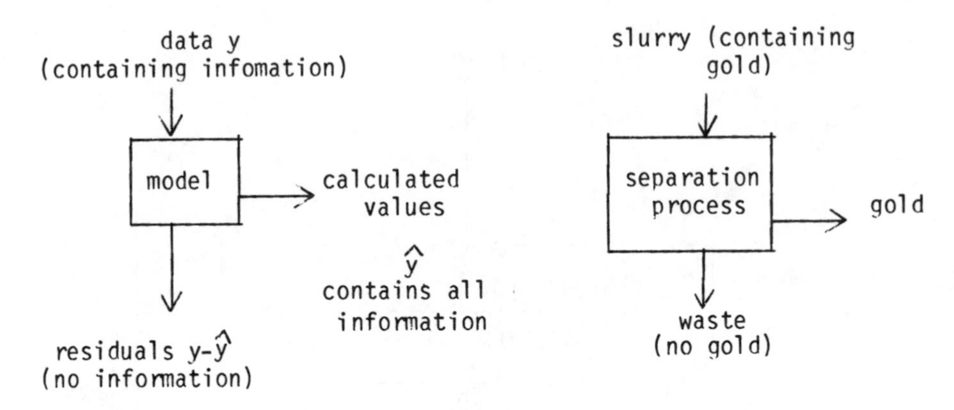

Figure 3. Analogy between a perfect model and perfect separation process.

the residuals, which are the differences between the observed and fitted values. The logic is that the residuals should contain no information. The object of model-building is that all the useful information in the data should be "trapped" in the model. See Figure 3, which shows an analogy between model-building and a separation process the purpose of which is to extract gold from a slurry. In the ideal situation, no gold would be contained in the waste stream. Likewise, in the ideal situation, no information would be contained in the residuals. The most useful way to examine residuals is to plot them.

3. BAYESIAN ANALYSIS AND CLASSICAL SAMPLING THEORY

Bayesian analysis and classical sampling theory are built on contradictory bases. Figure 4 illustrates the essential differences.. In Bayesian analysis, the data are regarded as fixed and the parameters are regarded as variable. In classical sampling theory, the data are regarded as variable and the parameters are regarded as fixed.

For more information on Bayesian methods, see Box and Tiao (1973). Classical sampling theory is prevalent, and it is described in virtually all introductory statistics texts; it concerns testing hypotheses, confidence intervals, and related concepts. Scientists find Bayesian methods more natural than sampling theory methods because they allow scientists to think and talk in ways that they are used to thinking and talking. The language and thought processes associated with sampling theory, however, are rather unnatural.

	Bayesian analysis	Classical sampling theory
data	fixed	variable
parameters	variable	fixed

Figure 4. Comparison between Bayesian analysis and classical sampling theory.

4. A BAYESIAN ANALYSIS OF THE LINEAR CALIBRATION PROBLEM

The remainder of this paper is reprinted from Technometrics with kind permission of the copyright holder, the American Statistical Association. The original article, by W.G. Hunter and W. F. Lamboy, appeared in 1981 in Volume 23, No. 4, pages 323-328. A lengthy discussion was published with the original article, but it is not reproduced here.

1. THE CALIBRATION PROBLEM

A chemist wants to establish a calibration line to use in measuring the amount of molybdenum in samples sent to an analytical laboratory. There are two stages in the calibration process.

In the first stage, for each of ten samples with "known" amounts of molybdenum, two replicate measurements are made with the relatively quick, inexpensive *test method* being calibrated. The "known" amounts of molybdenum have been determined by an extremely accurate *standard method* that is slow and expensive. A straight line is fitted to these 20 observations. This calibration line is now ready for use in the second stage of the calibration process, in which samples with unknown amounts of molybdenum are analyzed with the test method. For a given sample, one or more measurements may be made.

Mathematically, this common calibration procedure can be formulated as follows. Letting the true values associated with the standard and test methods be designated by ξ and η, respectively, we suppose there is a straight line relationship between them:

$$\eta = \beta_0 + \beta\,\xi, \tag{1.1}$$

where β_0 and β are the intercept and the slope, respectively. In the first stage, n pairs of observations (x_u, y_u) are obtained, where x_u and y_u are the observed values of ξ_u and η_u, respectively,

$$\begin{aligned} y_u &= \eta_u + \varepsilon_u \qquad u = 1, 2, \ldots, n, \\ x_u &= \xi_u + \delta_u \qquad u = 1, 2, \ldots, n, \end{aligned} \tag{1.2}$$

where ε_u and δ_u are experimental errors. It is a useful approximation in many situations to assume $\delta_u = 0$ for all u, for example, when the variance of δ is much smaller than that of ε, as is usually the case in practice. Making this assumption and substituting (1.2) into (1.1), we get

$$y_u = \beta_0 + \beta\, x_u + \varepsilon_u \qquad u = 1, 2, \ldots, n. \tag{1.3}$$

We assume the ε_u's are independently and identically distributed according to a normal distribution with mean zero and variance σ^2. The calibration line is determined by the unknown parameters β_0 and β. Bayes' theorem can be used to make inferences about these parameters, in particular, to derive their posterior distribution $p(\beta_0, \beta \mid \text{data})$. (Because the data after the second stage include the data from the first stage, the posterior distribution $p(\beta_0, \beta \mid \text{data})$ may differ at the end of the two stages when σ^2 is unknown. We discuss this point further in Sec. 4.)

Having established the calibration line, we are now ready to proceed to the second stage. A sample is presented with a specific unknown value η, and one or more measurements are made using the test method, from which are obtained the observations

$$y_u = \eta + \varepsilon_u \qquad u = N + 1, N + 2, \ldots, N + m, \tag{1.4}$$

where N is the number of observations y_u previously made after the calibration line was established. For the first test sample, $N = n$. We assume that the errors ε_u in the second-stage measurements have the same distribution as in the first-stage measurements. (In some applications, however, this assumption will be violated. The theory described in this article can be extended to cover that case.) Using Bayes' theorem, one can compute the posterior distribution $p(\eta \mid \text{data})$.

Given the data from both the first and the second stages, and hence the posterior distributions $p(\beta_0, \beta \mid \text{data})$ and $p(\eta \mid \text{data})$, we can now make inferences about the unknown value ξ that corresponds to η for the sample being measured,

$$\xi = (\eta - \beta_0)/\beta. \tag{1.5}$$

This equation follows directly from (1.1).

For non-Bayesian treatments of the calibration problem see Berkson (1969), Halperin (1970), Kalotay (1971), Krutchkoff (1969), Martinelle (1970), Scheffé (1973), Shukla (1972), Williams (1969), and references listed therein.

In Section 2 of this article we present a Bayesian solution to the problem by deriving the posterior distribution $p(\xi \mid \text{data})$. In Section 3 we consider the chemical example previously introduced. In Section 4 we compare our results to others, both Bayesian and non-Bayesian, that have appeared in the literature, and we comment on the controversy concerning inverse regression.

2. A BAYESIAN SOLUTION

In this section we first present a Bayesian solution for the calibration problem on the assumption that σ^2 is known. We then consider the situation in which σ^2 is unknown. In each case we derive the posterior distribution $p(\xi \mid \text{data})$.

2.1 σ^2 Known

Suppose n pairs of data (x_u, y_u) are available from the first stage and σ^2 is known. Assume that the prior distribution $p(\beta_0, \beta)$ is locally uniform. The posterior distribution $p(\beta_0, \beta \mid \text{data}_1)$ is the bivariate normal distribution with mean $\mathbf{b} = (b_0, b)$ and variance-covariance matrix $[\mathbf{X}'\mathbf{X}]^{-1}\sigma^2$, where b_0 and b are the usual least squares estimates of β_0 and β, and $\mathbf{X}$ is the $n \times 2$ matrix, the uth row of which is $(1, x_u)$. Note that "data_1" represents all the information available in the first stage: the n pairs of observations (x_u, y_u); the model (1.3); the prior $p(\beta_0, \beta)$; and the known value σ^2. The calibration line is given by $\hat{y} = b_0 + bx$. Suppose a sample is submitted for analysis and that m replicate readings y are then made with the test method. Assume that the prior distribution $p(\eta)$ is locally uniform. The posterior distribution $p(\eta \mid \text{data})$ is $N(\bar{y}, \sigma^2/m)$, that is, a normal distribution whose mean is $\bar{y}$, the average of the m values of y collected in the second stage, and whose variance is σ^2/m. Note that "data" represents all the information available after the observations have been obtained in the second stage: the model (1.4), the $N + m$ observations y, and the prior distribution of η, in addition to data_1.

The posterior distribution of (β_0, β, η), where η is the specific value associated with the sample being measured, is

$$p(\beta_0, \beta, \eta \mid \text{data}) = |\mathbf{S}|^{-1/2}(2\pi)^{-3/2} \times \exp\left[-\frac{1}{2}\begin{pmatrix}\beta_0 - b_0 \\ \beta - b \\ \eta - \bar{y}\end{pmatrix}' \mathbf{S}^{-1}\begin{pmatrix}\beta_0 - b_0 \\ \beta - b \\ \eta - \bar{y}\end{pmatrix}\right] \tag{2.1}$$

where

$$\mathbf{S} = \{s_{ij}\} = \begin{bmatrix} s_{11} & s_{12} & 0 \\ s_{21} & s_{22} & 0 \\ 0 & 0 & s_{33} \end{bmatrix} = \begin{bmatrix} [\mathbf{X}'\mathbf{X}]^{-1}\sigma^2 & 0 \\ 0 & \sigma^2/m \end{bmatrix},$$

since the observations taken at the second stage are independent of those taken at the first stage. From the properties of the normal distribution we can obtain the joint density of $(\eta - \beta_0, \beta)$ as

$$p(\eta - \beta_0, \beta \mid \text{data}) = |\mathbf{V}|^{-1/2} 2\pi^{-1} \times \exp \left[-\frac{1}{2}\begin{pmatrix}\eta - \beta_0 - \bar{y} + b_0 \\ \beta - b\end{pmatrix}' \mathbf{V}^{-1}\begin{pmatrix}\eta - \beta_0 - \bar{y} + b_0 \\ \beta - b\end{pmatrix}\right] \tag{2.2}$$

where

$$\mathbf{V} = \begin{bmatrix} s_{11} + s_{33} & -s_{12} \\ -s_{12} & s_{22} \end{bmatrix}.$$

From this density we can obtain the density of $\xi = (\eta - \beta_0)/\beta$ (see Hunter and Lamboy 1979a),

$$p(\xi \mid \text{data}) = k_1 \exp(-k_4)\left\{\frac{1}{k_2} - \frac{\pi^{1/2}}{2k_2^{3/2}} \exp\left(\frac{-k_3^2}{4k_2}\right) \times k_3[1 - 2\Phi(k_3/\sqrt{2k_2})]\right\}, \tag{2.3}$$

where $\Phi(\cdot)$ is the normal cumulative distribution function, and, with $\{v^{ij}\} = \mathbf{V}^{-1}$,

$$k_1 = 2\pi^{-1}|\mathbf{V}|^{-1/2},$$
$$k_2 = (v^{11}\xi^2 + 2v^{12}\xi + v^{22})/2,$$
$$k_3 = [-2v^{11}(\bar{y} - b_0)\xi - 2v^{12}(\bar{y} - b_0) - 2v^{12}b\xi - 2v^{22}b]/2,$$
$$k_4 = [v^{11}(\bar{y} - b_0)^2 + 2v^{12}(\bar{y} - b_0)b + v^{22}b^2]/2.$$

As new sets of m replicate observations $\{y_u\}$ become available, (2.3) can be used in conjunction with the calibration line established from the first stage to obtain appropriate posterior distributions for ξ.

2.2 σ^2 Unknown

We now derive the Bayesian posterior distribution for ξ on the assumption that σ^2 is unknown. Assume again that data are available from both stages of the experiment. Assume that locally the prior distribution $p(\eta, \beta_0, \beta, \sigma^2)$ is proportional to σ^{-2}. Then the posterior distribution $p(\eta, \beta_0, \beta)$ is a multivariate t density with v degrees of freedom:

$$p(\eta, \beta_0, \beta \mid \text{data}) = \frac{\Gamma\left(\frac{v+3}{2}\right)}{v\pi^{3/2}\Gamma\left(\frac{v}{2}\right)|\mathbf{S}|^{1/2}} \times \left[1 + \begin{pmatrix} \beta_0 - b_0 \\ \beta - b \\ \eta - \bar{y} \end{pmatrix}' \frac{\mathbf{S}^{-1}}{v} \begin{pmatrix} \beta_0 - b_0 \\ \beta - b \\ \eta - \bar{y} \end{pmatrix}\right]^{-[(v+3)/2]}, \tag{2.4}$$

where

$$\mathbf{S} = \{s_{ij}\} = \begin{bmatrix} s_{11} & s_{12} & 0 \\ s_{21} & s_{22} & 0 \\ 0 & 0 & s_{33} \end{bmatrix} = \begin{bmatrix} [\mathbf{X}'\mathbf{X}]^{-1}s^2 & \mathbf{0} \\ \mathbf{0} & s^2/m \end{bmatrix}$$

and s^2 is the estimate of σ^2 with $n - 2 + \sum_{i=1}^{l}(m_i - 1)$ degrees of freedom, where m_i is the number of measurements made on the ith unknown sample using the test method ($i = 1, 2, \ldots, l$). The pooled estimate s^2 is the weighted average of the estimates from the first and second stages. Contribution from the second-stage measurements comes from those samples on which replicate readings are made, that is, those for which $m_i > 1$. If only single measurements ($m_i = 1$) are made on all unknown samples, s^2 has $n - 2$ degrees of freedom.

The joint density of $\eta - \beta_0$ and β (see Box and Tiao 1973, p. 118) is

$$p(\eta - \beta_0, \beta \mid \text{data}) = \frac{\Gamma\left(\frac{\nu + 2}{2}\right)}{\pi\Gamma\left(\frac{\nu}{2}\right) |\mathbf{V}|^{1/2}} \times \left[1 + \begin{pmatrix} \eta - \beta_0 - \bar{y} + b_0 \\ \beta - b \end{pmatrix}' \mathbf{V}^{-1} \times \begin{pmatrix} \eta - \beta_0 - \bar{y} + b_0 \\ \beta - b \end{pmatrix}\right]^{-[(\nu+2)/2]}, \tag{2.5}$$

where

$$\mathbf{V} = \nu \begin{bmatrix} s_{11} + s_{33} & -s_{12} \\ -s_{12} & s_{22} \end{bmatrix}.$$

The derivation from this point on is tedious and follows that given in Hunter and Lamboy (1979a). The cases ν even and ν odd need to be treated separately. The resulting density of $\xi = (\eta - \beta_0)/\beta$ for ν even is

$$p(\xi \mid \text{data}) = \frac{k_1}{(n-1)k_2 k_4^{n-1}} + \frac{k_1 k_3^2}{(2n-1)k_2 k_4^{n-1}} \times \sum_{k=0}^{n-2} \frac{(2k_2 k_4)^k (2n-1) \cdots (2n-2k-1)}{(n-1) \cdots (n-k-1)\, \Delta^{k+1}} + \frac{2^{n+1}(2n-3)!!k_1 k_3 k_2^{n-2}}{(n-1)!\, \Delta^{n-1/2}} [\arctan(k_3/\Delta^{1/2})], \tag{2.6a}$$

and for ν odd is

$$p(\xi \mid \text{data}) = \frac{2k_1}{(2n-1)k_2 k_4^{(n-1)/2}} + \frac{2k_1 k_3^2}{(2n-1)\, \Delta k_2 k_4^{(2n-1)/2}} \times \left\{1 + \sum_{k=1}^{n-1} \frac{(8k_2 k_4)^k (n-1) \cdots (n-k)}{(2n-3) \cdots (2n-2k-1)\, \Delta^k}\right\}, \tag{2.6b}$$

where

$$k_1 = \Gamma\left(\frac{v+2}{2}\right)|\mathbf{V}|^{-1/2} \Big/ \pi\Gamma\left(\frac{v}{2}\right),$$

$$k_2 = v^{11}\xi^2 + 2v^{12}\xi + v^{22},$$

$$k_3 = -2v^{11}b_0\xi - 2v^{12}b\xi - 2v^{12}b_0 - 2v^{22}b,$$

$$k_4 = 1 + v^{11}b_0^2 + 2v^{12}b_0 b + v^{22}b^2,$$

$$\Delta = 4k_2 k_4 - k_3^2,$$

$$(2n-3)!! = (2n-3)(2n-5)\ldots 1.$$

3. AN EXAMPLE

We now consider an example, the data for which are given in Table 1. For the regression line fit to these

Table 1. Data for Establishing Calibration Line (data courtesy of G. E. P. Box)

"known" amount of molybdenum	1	2	3	4	5	6	7	8	9	10
measured amount of molybdenum	1.8 1.6	3.1 2.6	3.6 3.4	4.9 4.2	6.0 5.9	6.8 6.9	8.2 7.3	8.8 8.5	9.5 9.5	10.6 10.6

data, $b_0 = .7600$ and $b = .9873$. For the sake of illustration, . 072 is assumed to be the value of σ^2 for the σ_2-known case. Actually, it is the usual estimate $s^2$of σ^2 having 18 degrees of freedom, so this same value is used in the σ^2-unknown case ($s^2 = .072$). For an observed value of $y = 6.0$ at an unknown value of ξ, a graph of the posterior density of $\xi = (\eta - \beta_0)/\beta$ is given in Figure 1 for the σ^2-known case. A useful approximation to it is provided by the normal distribution $N((\bar{y} - b_0)/b, (v_{11}v_{22} - v_{12}^2)/(v_{22}b^2))$, that is, $N(5.3074, .0776)$. The analogous graph when σ^2 is unknown is similar, but the peak is shorter. The density in that situation is approximated by a suitably scaled t distribution with 18 degrees of freedom and scale factor $(v_{11}v_{22} - v_{12}^2)^{1/2}/\sqrt{v_{22}}\,b = \sqrt{.0776}$; it is centered at $(\bar{y} - b_0)/b = 5.3074$.

Table 2 shows highest posterior density (HPD) in-

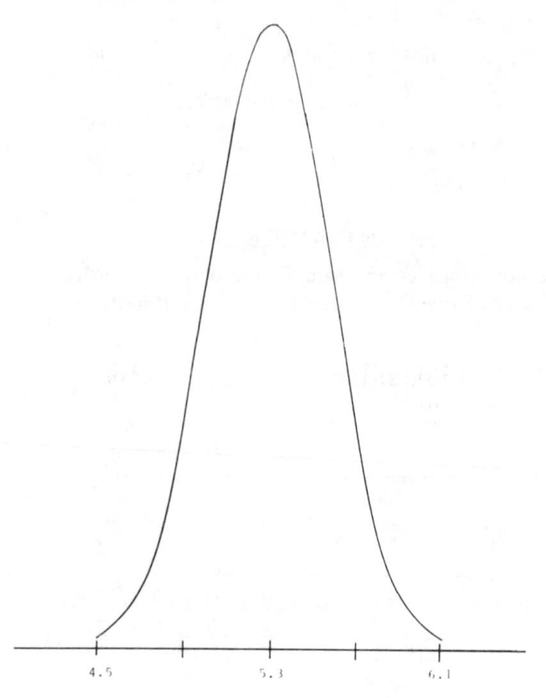

Figure 1. *Posterior Density for ξ, σ^2 - Known (the graph of the density for the normal approximation N(5.3074, .0766) is virtually identical)*

tervals for the σ^2-known and unknown cases, along with intervals obtained both from Fieller's theorem (1954) and from the normal and t approximations.

To calculate intervals using Fieller's theorem is relatively simple. Consider the 95 percent interval for the σ^2-known case. The limits are obtained from this expression:

$$L/(b^2 - t^2 v_{22}) \tag{3.1}$$

where

$$L = (\bar{y} - b_0)b - t^2 v_{12} \pm \{[(\bar{y} - b_0)b - t^2 v_{12}]^2 - [(\bar{y} - b_0)^2 - t^2 v_{11}][b^2 - t^2 v_{22}]\}^{1/2}$$

with $t = 1.96$. Substituting our values, we get

$$\frac{5.1642 \pm (26.6693 - 26.3866)^{1/2}}{.9731} \tag{3.2}$$

or (4.761, 5.853), the result given in Table 2. We conjecture that Fieller's theorem will provide approximate $1 - \alpha$ intervals differing by not more than

$$2 \text{ Prob } (t(v) \geq |b|/\sqrt{v_{22}}) \tag{3.3}$$

from the probability actually contained in the corresponding $1 - \alpha$ HPD intervals, and that whenever

$$|b|/\sqrt{v_{22}} \geq 10 \quad \text{and}$$
$$|v_{22}b_0 - v_{12}b| \leq |\mathbf{V}|^{1/2}|b|/10, \tag{3.4}$$

Table 2. Exact and Approximate HPD Intervals for ξ, for σ^2-Known and σ^2-Unknown Based on Data in Table 1 and the Observed Value $y = 6.0$.

σ^2-known ($\sigma^2 = 0.0720$)

	exact HPD interval	interval supplied by Fieller's theorem	approximate HPD interval based on N(5.3074,0.0776)
.90	(4.849, 5.766)	(4.849, 5.766)	(4.849, 5.766)
.95	(4.761, 5.854)	(4.761, 5.853)	(4.762, 5.853)
.99	(4.588, 6.025)	(4.589, 6.025)	(4.590, 6.025)

σ^2-unknown ($s^2 = 0.0720$)

	exact HPD interval	interval supplied by Fieller's theorem	approximate HPD interval based on t_{18}(5.3074,0.0776)
.90	(4.824, 5.790)	(4.824, 5.790)	(4.824, 5.790)
.95	(4.721, 5.893)	(4.721, 5.893)	(4.722, 5.893)
.99	(4.504, 6.111)	(4.506, 6.110)	(4.506, 6.109)

the normal or t approximation already noted will provide accurate approximations to the corresponding HPD intervals. Note that for this example, for 90, 95, and 99 percent intervals, Fieller's theorem and the normal or t approximations are extremely accurate. As far as conditions (3.4) are concerned,

$$|b|/\sqrt{v_{22}} = (.9873)/\sqrt{.0004364}$$
$$= 47.26 \geq 10;$$

$$|v_{22}b_0 - v_{12}b|$$

$$= |(.0004364)(5.24) - (.0024)(.9873)|$$
$$= .000083;$$

and

$$|v|^{1/2}|b|/10$$
$$= \sqrt{v_{11}v_{22} - v_{12}^2}\,|b|/10$$
$$= \sqrt{(.0888)(.004364) - (.0024)^2}(.9873)/10$$
$$= .00057,$$

values that indicate that the normal and t approximations are good.

4. DISCUSSION

4.1 Inverse Regression

The history of the calibration problem has been marked by a controversy that started with the publication of a paper by Krutchkoff (1967). In it the author concluded that the inverse estimator had uniformly smaller mean squared error than the classical estimator and so was preferable. (The classical estimator is obtained when y is regressed on x, and the inverse estimator is obtained when x is regressed on y.) Unfortunately, he left unspecified the conditions under which this conclusion is true. He did, however, in a succeeding paper (Krutchkoff 1969) state conditions under which the classical estimator is superior.

Berkson (1969), Martinelle (1970), Halperin (1970), and Shukla (1972) pointed out that Krutchkoff's conclusion held only in very restrictive circumstances. Perhaps the most incisive reply to Krutchkoff was produced by Williams (1969), who showed that any unbiased estimator of ξ has *infinite* variance, and, although Krutchkoff's estimator has finite variance, it is biased. From his analysis, Williams concluded that minimum variance or minimum mean squared deviation are not suitable criteria to use in a problem of this kind. We agree with Williams's conclusion.

Our Bayesian analysis yields a posterior distribution that has infinite variance. This fact is not disturbing in the least because one can always plot the posterior density of ξ and readily interpret it. One may choose, for example, to find its mode and selected

HPD regions. From Bayesian point of view, the existence of a finite variance is unnecessary. Theoretically, if one's model is adequate, all the relevant information is contained in the appropriate posterior distribution, whether its variance happens to be finite or not. Furthermore, on practical grounds, we believe it is easy and natural for practitioners to use posterior distributions in making inferences. Therefore, our (Bayesian) view is that for standard calibration problems, arguments about which estimator is best that are based on mean squared error or related criteria are simply irrelevant.

4.2 Previous Results

The Bayesian posterior densities for ξ, (2.3) and (2.6), are the posterior densities for a ratio of bivariate normal random variables (in the σ^2-known case) and of bivariate t random variables (in the σ^2-unknown case). A more extensive analysis of their properties may be found in Hunter and Lamboy (1979a, b). Equation (2.6) is a generalization of the distribution for ξ found by Kalotay (1971). In the special case that he considers $[\sum x_u = 0 \text{ and } \boldsymbol{\varepsilon} \sim N(\mathbf{O},\mathbf{I})]$, (2.6) reduces to his result. It differs from the Bayesian posterior density found by Hoadley (1970) because while we find the density of $\xi = (\eta - \beta_0)/\beta$, Hoadley does not treat ξ as an explicit function of η, β_0, and β, but rather directly uses a prior distribution for ξ. In our analysis, the prior for ξ is implicitly given by the priors for η, β_0, and β.

4.3 Practical Considerations

In actual calibration problems σ^2 is unknown. Data from both stages can be used in estimating σ^2. Suppose that, after a calibration line has been established, k unknown samples are presented for analysis and m_i measurements are obtained for sample i using the test method ($i = 1, 2, \ldots, k$). Note that all such measurements for which $m_i > 1$ (that is replicate measurements) contain information about σ^2. The posterior distribution of ξ will now depend not only on the most recent set of m_k measurements, *but also on the replicate measurements from previous unknown samples.*

Accordingly, if results are reported after each sample is analyzed, the later results will tend to be more precise than earlier ones because v, the degrees of freedom associated with the values s^2, will increase as data are collected. Hence, the variances of the marginal posterior density $p(\sigma\,|\,\text{data})$ and the posterior density $p\,(\sigma\,|\,\text{data})$ will tend to decrease, as will the lengths of the HPD intervals for ξ.

Theoretically, these facts make sense. Nevertheless, practitioners might find them puzzling. One, for instance, might say, "Look, if I have six samples on the bench, from what you have just said the one I choose to measure last will tend to give the most precise result. But, if my skill stays the same for all six samples, I see no reason to believe that ξ will tend to be measured best in the last sample." A possible response to this comment is that if one makes measurements on all six samples first and *only then* calculates the posterior distributions and the HPD intervals, the general precision for ξ will be the same for all six samples in the sense that v will be the same for all the samples.

The practitioner might then reply, "That's more bother than I should have to put up with, but even if I did agree in principle to follow such a procedure, there will be many times when it will be impossible to do so in practice. Take one instrument, for instance. We may recalibrate it every month, but we use the calibration line for many samples a day, and we need to send the results out as soon as we get them. We can't wait till the end of the month when we have all the data so that we can report all results with equal precision. So, according to your theory, our reported results will tend to get better as the month goes along, and I just don't buy that. Based on my experience, I believe that once the calibration line is fixed, the results I report for a given sample should depend only on measurements made on that particular sample, and not on other measurements." The practitioner is saying that the specific implication of the theory does not seem sensible because the results reported for a given sample should depend only on the measurements made on that sample. So, the theory and practice appear to disagree.

In this case, the simplest way to achieve agreement is to change the assumption about σ^2: if it is assumed that σ^2 is known, the results reported for a given sample will then depend only on the measurements for that sample. But, of course, it is unacceptable to change the assumption on σ^2 only because it is the simpler way to resolve the difficulty. The key question is whether it makes sense to use a procedure based on the assumption that σ^2 is known. In most common situations, we think it does. If the measurement process is in a state of statistical control, the precision of the measurement process will remain constant over time. Consequently, all past readings that carry information about σ^2 can be used to estimate this parameter. These readings include all first-stage readings and all replicated second-stage readings from the past (recall we are assuming σ^2 is the same for the first- and second-stage readings). For most common situations, therefore, the point will have been reached that σ^2 is essentially known and the σ^2-known solution will provide an adequate approximation to the correct solution. Such a solution will satisfy all concerns expressed by the practitioner quoted above. (Incidentally, for standard analytical methods, the approximate known values for σ^2 will usually be available before any calibration runs have been made in a particular laboratory, for example, in published form from the American Society for Testing Materials.)

The use of the σ^2-known solution in practice can be justified using Bayes' theorem as follows. For a test method that is not entirely new, the appropriate prior density to use for σ^2 is not the "noninformative" prior but rather the posterior density $p(\sigma^2 \mid \text{data})$, where "data" here refer to all past data. (As indicated above, these data may even include those obtained in other laboratories.) As the degrees of freedom increase, the σ^2-known case is rapidly approached.

The conclusion we reach is that although, strictly speaking, in all actual calibration situations σ^2 is unknown, it will usually be acceptable from a theoretical point of view and preferable from a practical point of view to compute the posterior density $p(\xi \mid \text{data})$ by using (2.3) for the σ^2-known case rather than (2.6) for the σ^2-unknown case. Moreover, when conditions (3.4) hold, Fieller's theorem via (3.1) provides useful approximations to the relative Bayesian HPD intervals for ξ. These approximate intervals are relatively easy to compute.

5. ACKNOWLEDGMENTS

We are grateful to Conrad A. Fung for his help in getting us started on this work, and for many useful discussions.

[*Received March 1980. Revised June 1981.*]

REFERENCES

BERKSON, J. (1969), "Estimation of a Linear Function for a Calibration Line: Consideration of a Recent Proposal," *Technometrics*, 11, 649–660.

BOX, G. E. P., and TIAO, G. (1973), *Bayesian Inference in Statistical Analysis*, Reading, Mass.: Addison Wesley.

FIELLER, E. C. (1954), "Some Problems in Internal Estimation," *J. Roy. Statist. Soc.*, Ser. B, 16, 175–185.

HALPERIN, M. (1970), "On Inverse Estimation in Linear Regression," *Technometrics*, 12, 727–736.

HOADLEY, B. (1970), "A Bayesian Look at Inverse Linear Regression," *J. Amer. Statist. Assoc.*, 65, 357–369.
HUNTER, W. G., and LAMBOY, W. F. (1979a), "Bayesian Analysis of Ratios," Technical Report No. 587, Statistics Dept., University of Wisconsin.
HUNTER, W. G., and LAMBOY, W. F. (1979b), "Making Inferences About Ratios: Some Examples," Technical Report No. 588, Statistics Dept., University of Wisconsin.
KALOTAY, A. J. (1971), "Structural Solution to the Linear Calibration Problem," *Technometrics*, 13, 761–769.
KRUTCHKOFF, R. G. (1967), "Classical and Inverse Regression Methods of Calibration," *Technometrics*, 9, 425–439.
———, (1969), "Classical and Inverse Regression Methods of Calibration in Extrapolation," *Technometrics*, 11, 605–608.
MARTINELLE, S. (1970), "On the Choice of Regression in Linear Calibration," *Technometrics*, 12, 157–161.
SCHEFFÉ, H. (1973), "A Statistical Theory of Calibration" *Ann. Statist.*, 1, 1–37.
SHUKLA, G. K. (1972), "On the Problem of Calibration," *Technometrics*, 14, 547–553.
WILLIAMS, E. J. (1969), "A Note on Regression Methods in Calibration," *Technometrics*, 11, 189–192.

ADDITIONAL REFERENCES

Box, G.E.P. (1980), "Sampling and Bayes' Inference in Scientific Modelling and Robustness" (with discussion), J. Roy. Statist. Soc., Ser. A, 143, pp. 383-430.
Box, G.E.P. and Youle, P.V. (1955), "The Exploration and Exploitation of Response Surfaces: An Example of the Link Between the Fitted Surface and the Basic Mechanism of the System", Biometrics, Vol. 11, No. 3, pp. 287-323.

CHEMOMETRICS AND ANALYTICAL CHEMISTRY*
Societal Issues, Structure of the Measurement Process, Detection, and Validation

Lloyd A. Currie

Center for Analytical Chemistry
National Bureau of Standards
Washington, D.C. 20234
USA

INTRODUCTION

Modern Analytical Chemistry has become intrinsically tied to the exposure, understanding, and resolution of many of today's sociotechnical problems, in areas ranging from medical diagnostics to monitoring our climate. Chemometrics is central in deriving adequate responses to these needs in terms of the design, control, evaluation and validation of the Analytical Measurement Process. The substance of this lecture is based on two key references dealing with this theme (1, 2). Reference (1) is a rather extensive review of sociochemical and mathematical-statistical issues which influence the Quality of Analytical Results, while reference (2) comprises a case study from the relatively new discipline of "Chemometric Intercomparison", or interlaboratory (numerical) validation via simulated analytical data. The paragraphs which follow are offered to introduce these topics and to summarize some recent observations on the subject of detection.

SOCIETY AND CHEMISTRY

A significant fraction of the problems facing society today have chemical origins and, in many cases, chemical solutions. Sound approaches to these problems require accurate communication of the societal issues (sociopolitical, legal, institutional,

* Contribution from the National Bureau of Standards;

B. R. Kowalsi (ed.), Chemometrics. Mathematics and Statistics in Chemistry, 115–146.

economic) to the scientific community, and the converse. Though the scientific issues are generally less complicated, their full complexity, nevertheless, is often unappreciated because of inadequate attention (even among scientists) to the nature and magnitude of uncertainty associated with both measurement error and our models of chemical, biological, and environmental systems (8). In fact, intelligent members of the lay public sometimes believe that scientific measurements and conclusions can be generated with little or no residual uncertainty (given a sufficient budget). Such is the case, for example, for the issue of "Acid Rain", where some believe that a comprehensive research program should first be completed in order to "get all the facts" before environmental (regulatory) controls are instituted (12).

Adequate estimation and communication of scientific uncertainty is thus one of the most important tasks facing chemists (and especially chemometricians). This issue is at the core of public misunderstanding of: residual scientific uncertainty, the fact that zero contamination can neither exist nor be measured (4, 6, 27, 28), and that "detection" in the regulatory sense is *at best* a probabilistic concept. Inadequate communication of uncertainties leads to distrust, inappropriate legislative or regulatory action (or inaction), and erosion of credibility. It was with this perspective that the Joint Board/ Council Committee on Science of the American Chemical Society recently published recommendations on analytical data used for public purposes (4). Topical highlights from the recommendations ("Rogers Committee") are indicated in Figure 1.

A second area of recent (and current) involvement for this author relates to the matter of detection limits as applied to the regulation of nuclear utilities. Information gathered for the preparation of a Nuclear Regulatory Commission (NRC) document on the Lower Limit of Detection (60) has revealed a number of problems common to this basic concept in Analytical Chemistry (Figure 2). Though these issues were raised in the context of detection capabilities for effluent and environmental radioactivity measurement processes, most are generic and they will be briefly discussed in the following section on detection. Further specific discussion of societal issues as related to modern trace inorganic and organic analysis may be found in Reference 1.

DETECTION

History and Principles

In this section we offer a few comments and illustrations to supplement the presentation in (1). (See also the introductory discussion of Hypothesis Testing in the next subsection--quoted from Ref. 1.) First, as indicated in Figure 3, which refers only to detection capability (not detection decision levels), the development of detection terminology and formulations for Analytical Chemistry covers an extended period of time and it has been characterized by diverse and not always consistent approaches. (Besides alternative terms for the same concept, one occasionally finds the same term applied to different concepts -- viz., Kaiser's "Nachweisgrenze", which refers to the test or detection decision level, is commonly translated "detection limit"; yet, in english, "detection limit" generally relates to the inherent detection capability of the Chemical Measurement Process (CMP).) For information concerning the detailed assumptions and formulations associated with the terms presented in Figure 3, the reader is referred to the original literature. The principal approaches, however, are represented by: (a) Feigl -- selecting a more or less arbitrary concentration (or amount), based on expert judgment of the current state of the art; (b) Kaiser -- grounding detection theory on the principles of hypothesis testing; (c) St. John -- using signal/noise (assumed "white") and considering only the error of the first kind; (d) Nicholson -- considering detection from the perspective of a specific assumed probability distribution (Poisson); (e) Liteanu -- treating detection in terms of the directly observed (empirical) frequency distribution, and (f) Grinzaid -- applying the weaker, but more robust approaches of non-parametric statistics to the problem. The widespread practice of ignoring the error of the second kind is epitomized by Ingle in his inference that it is too complex for ordinary chemists to use and comprehend! Treatment of detection in the presence of possible systematic and/or model error is considered briefly in (1) and Ref. (26).

A condensed summary of the principal approaches to *signal detection* in Analytical Chemistry is presented in Figure 4. The hypothesis testing approach, which this author favors, serves also as the basis for the more familiar construction of confidence intervals for signals which are detected (17). For more information on the relationship between the power of an hypothesis test and the significance levels and number of replicates (for normally distributed data) the reader may refer to OC (Operation Characteristic) curves as compiled by Natrella (18). In Figure 5 we see, for example, that five replicates are necessary if one wishes to establish a detection limit which is no greater than 2σ, taking α and β risks at 5% each. (Note the inequality statement; this arises because of the discrete nature of replication.) Once we leave the domain of simple detection of signals, and face the questions of *analyte detection*, we encounter numerous added problems and difficulties with assumption validity. That is assumptions concerning the calibration function or functions -- i.e., the full analytical model -- and the "propa-crucial. A summary of some of these issues is given in Figure 6; further discussion will be found in the following two subsections. Finally, for more detailed coverage of a number of the issues identified above, one may refer to an excellent collection of texts and reviews (13-28).

Detection Theory and Hypothesis Testing
[Introductory material quoted from Ref. 1]

"The basic issue is whether one primary hypothesis (the "null hypothesis", H_0) describes the state of the system or whether one or more "alternative hypotheses" (H_A) describe it. The actual test is one of consistency - i.e., given the experimental sample, are the data consistent with H_0, at the specified level of significance, α? That is the first question, and if we draw (unknowningly) the wrong conclusion, it is called an error of the first kind. This is equivalent to a false positive in the case of trace analysis - i.e., although the (unknown) true analyte signal S equals zero (state H_0), the analyst reports, "detected".

"The second question relates to discrimination. That is, given a decision- (or critical-) level S_C used for deciding upon consistency of the experimental sample with H_0, what true signal level S_D can be distinguished from S_C at a level of significance β? If the state of the system corresponds to H_A ($S=S_D$) and we falsely conclude that it is in state H_0, that is called an error of the second kind, and it corresponds in trace analysis to a false negative. The probabilities of making correct decisions are therefore $1-\alpha$ (given H_0) and $1-\beta$ (given H_A); $1-\beta$ is also known as the "power" of the test, and it is fixed by $1-\alpha$ (or S_C) and S_D. One major objective in selecting a CMP for trace analysis is thus to achieve adequate detection power ($1-\beta$) at the signal level of interest (S_D), while minimizing the risk (α) of false positives. Given α and β (commonly taken to be 5% each), there are clearly two derived quantities of interest: S_C for making the detection decision, and S_D the detection limit.

" An assumption underlying the above test procedure is that the estimated net signal $\hat{S}$ is an independent random variable having a known distribution. (This is identical to the prerequisite for specifying confidence intervals.) Thus knowing (or having a statistical estimate for) the standard deviation of the estimated net signal $\hat{S}$, one can calculate S_C and S_D, given the form of the distribution and α and β . If the distribution is normal, and $\alpha=\beta=0.05$, $S_D \simeq 3.29\ \sigma_{\hat{S}}$ and $S_C \simeq S_D/2$. Thus, the relative standard deviation of the estimated net signal equals 30% at the detection limit (33). Indidentally, the theory of *differential detection* follows exactly that of detection, except that ΔS_{JND} (the "just noticeable difference") takes the place of S_D, and for H_0 reference is made to the base level S_0 of the analyte rather than the zero level (blank). A small fractional change $(\Delta S/S)_D$ thus requires even smaller relative imprecision.

"Obviously, the smallest detection limits obtain for interference-free measurements and in the absence of systematic error. Allowance for these factors not only increases S_D, but (at least in the case of systematic error) distorts the probabilistic setting, just as it does with confidence intervals. Special treatments for these questions and for non-normal distributions are needed, but are beyond the scope of this paper. Not so obvious perhaps is the fact that S_D depends on the specific algorithm selected for data reduction. As with interference effects on S_D, this dependence comes about because of the effect on $\sigma_{\hat{S}}$, the standard deviation of the estimated net signal. Some more explicit coverage of these matters appears in reference 26, and some will be given below.

"Hypothesis testing is extremely important for other phases of chemical analysis, in addition to the question of analyte detection limits. Through the use of appropriate test statistics, one may test data sets for bias, for unexpected random error components, for outliers, and even for erroneous evaluation (data reduction) models (26). Because of statistical limitations of such tests, especially when there are relatively few degrees of freedom, they are somewhat insensitive (lack power) except for quite large effects. *For this reason it is worth considerable effort on the part of the analyst to construct his CMP so that it is as free from or resistant to bias, blunders, and imperfect models as possible."*

Two Case Studies (Assumptions and Practices)

Explicit practices and difficulties involving detection will be illustrated by brief reference to two current studies: detection (by GC) of trace levels of the pesticide fenvalerate, and detection-regulation in the nuclear (utility) industry.

The question posed in the first example is represented in Figure 7. Here, we find the calibration data for fenvalerate presented as a two-way table, where the columns represent replicates, and the rows, selected concentrations. These data were presented to participants at the 1983 (Amer. Chem. Soc.)

Chemometrics in Pesticide Residue Analysis Symposium for the purpose of examining alternative approaches and difficulties in estimating "real-life" detection limits (19). A glance at Figure 8 shows that the variance was not only unknown (estimated with 4 degrees of freedom at each concentration), but it was not homogeneous; so estimated statistical weights were in order for the regression. The figure shows clearly that the calibration function was non-linear; but it does not reveal another problem: that there were non-random errors in the calibration standards.

Most serious of all, from the point of view of detection, was the fact that the observations did not extend to zero concentration -- i.e., they did not include the blank. Exclusion of the blank was due in part to the fact that the instrument was used as a threshold detector; it was set to give no reading if the signal (area, cm^2) fell short of 1.00. Such a threshold is legitimate, of course; it is equivalent to preselecting the critical level instead of the α-risk. A detection limit corresponding to said threshold and a given β-risk may still be estimated. The drawback is that the resultant detection limit far exceeds that which the CMP is inherently capable of -- in this case ~46 pg in contrast to ~4 pg (19). This example deserves one additional comment: although it does not conform to the best experimental design for detection limit estimation and optimization, it is representative of a large class of imperfect experiments whose detection capabilities we must estimate despite departures from ideal practices. Also, the needlessly high threshold, whether due to software or hardware, is a matter of increasing concern as instrumental "black boxes" take over the (detection) decision-making function.

The Nuclear Regulatory Commission (NRC) study (Figure 2) revealed some of the same difficulties as the foregoing example -- e.g., with respect to "black boxes". That is, self-contained hardware and software in modern gamma ray spectrometers often have algorithms and nuclear parameters which are hidden from the users, with the result that incorrect or at the very least non-consistent data and models may be employed without the knowledge of the operator. Cross-check (intercomparison) samples help, but these are not always representative (of the real samples), they are seldom "blind", and most importantly they confound "lab" (experimental) and data evaluation error. Simulated intercomparison data, however, such as discussed in the last section of these remarks, may serve as a key method for algorithmic interlaboratory quality control.

Other facets of the NRC study cannot be covered in detail in this brief summary (60), but attention should be given to the widespread misuse of the critical level (used for detection decisions) as the detection limit -- this forces the β-risk to take on a value of 50%! Also, diverse schemes for reporting results, especially when they are "not detected" yield non-comparable data and frequently biased averaging. As with the preceding example, inadequate attention to the magnitude and distribution of the blank is a common failing, occasionally resulting in much too optimistic estimates for the detection limit.

The Blank

The distribution of the blank is the most important determining factor in detection (apart from the calibration constant). Yet, at extreme trace levels, one seldom finds an adequate study of sampling, procedural, or instrumental blanks (and baselines) which is needed to gain reliable information on this topic. (See Figure 9). We shall give two brief illustrations here; for further discussion the reader is urged to consult reference 1.

The first illustration is taken from a just-completed investigation of trace element pollution in one of our largest estuaries, the Chesapeake Bay (71). Unlike the common practice where detection (decisions and limits) are based on one or just a few observations of blanks (or surrogate blanks) and assumed one (Poisson) or two (Normal) parameter probability distribution functions (pdf's), this study sought to experimentally assess the nature of the blank distribution functions. For each of 15 trace metals in both solution and particle phases, 24 samples of high purity water were treated on shipboard and subsequently in the laboratory in exactly the same manner as the samples taken from the Bay. Very interesting and important conclusions followed from the examination of the empirical pdf's: it was seen, for example, that while many of the species had normally-distributed blanks, some exhibited pdf's which were not Normal and which implied several sources of contamination, leading to a convolution of distributions (81). Referring to Figures 10 and 11, for example, it was found that blanks for dissolved cobalt (and iron) could be modeled with three-parameter distributions, equivalent to a log-Normal pdf having a zero offset (τ). This suggested two sources for the blank (e.g., reagents plus the sampling process), one at contamination level - τ underlying a second, log-Normally distributed source. Chromium, on the other hand, reflected only a normal component. Obviously, such skew distributions, which can be discerned only when many blanks are assayed, have a profound influence on detection limits; assumed normality would generally lead to overoptimistic estimates.

The second illustration relates to counting statistics or the time series of pulse arrival times which characterizes many experiments in the physical sciences, in particular the measurement of radioactivity. The assumption, upon which the vast majority of counting experiments is based, is that the counts (including background counts) are Poisson-distributed. As with the trace element blanks in the Chesapeake Bay, if this (Poisson) assumption is invalid for the background pdf, then overly optimistic detection limits will be estimated. Because of its relevance to the NRC study as well as to our own program involving low-level counting of environmental radiocarbon (83), we chose to apply some tests to the observed background pulse interval pdf's in our laboratory, and at the same time consider possible (physically-based) alternative hypotheses (78). Seven such hypotheses are listed in Figure 12, where we have indicated that the null hypothesis would yield counts which are Poisson-, arrival times which are Uniform-, and interarrival times which are Exponential- in distribution. (See also Figure 13.) The simplest procedure is to test replicate sets of counts for Poisson behavior (73), but this has limited power against several of the alternatives.

Because of limited space we shall show results for just two of the more interesting tests of our data -- one in which the Kolmogorov-Smirnov (K-S) test was applied to the empirical cumulative distribution function (cdf) for the pulse arrival times (Figure 14), and the other where the chi square test was applied to the grouped empirical probability density function for interarrival times (Figure 15). These tests of two different data sets gave some interesting insight into the background characteristics (for those particular runs); the first suggest an initial high rate transient, while the second is consistent with a queuing process which delays (in a buffer) pulses having quite short interarrival times. Further work (78) involves the application of the K-S test to transformed inter-arrival times, using a special version which avoids the estimation of nuisance parameters (79, 80). The large majority of our counting data are consistent with the null hypothesis. The aim of these examples has been to illustrate some of the more powerful techniques for testing assumptions about background distributions and, hence, estimated detection limits.

CHEMOMETRIC INTERCOMPARISON

Intercomparison of reference samples is an accepted means for assessing and establishing control among laboratories. With the increased capacity for acquiring and processing analytical data and with the increased sophistication of models and algorithms, however, a new type of control is called for: the Chemometric Intercomparison. By "chemometric intercomparison", I am referring to the use of reference data sets which simulate analytical measurements (possibly tied to simulated biological or geophysical processes), for the purpose of assessing the quality of data analytic methods as performed in participants' laboratories. The merit of such a scheme is that the data evaluation step of the CMP can be evaluated for interlaboratory (and interalgorithmic) precision and accuracy, free from confounding experimental measurement effects. Of special importance is the fact that error distributions and parameters can be rigorously controlled, and the participants can carry out their numerical analyses with full knowledge of this matter. From the perspective of detection, such chemometric intercomparisons offer the intriguing opportunity to include components which are subliminal -- and direct evaluation of false positives and false negatives can take place. Finally, as with reference sample intercomparisons, the reference data intercomparisons will generate surprises (see below) and they will give a clear indication of the algorithm dependence of precision, accuracy and detection limits.

An illustration relating to the analysis of gamma-ray spectra, as devised by the International Atomic Energy Agency (IAEA) (87) will now be described. The gamma spectrum exercise is reviewed in part in reference (1), but will be briefly summarized here. This exercise, undertaken by IAEA in 1977, was one of the most revealing tests of γ-ray peak evaluation algorithms to date. Some 200 participants including this author were invited to apply their methods for peak estimation, detection and resolution to a simulated data set constructed by the IAEA. The basis for the data were actual Ge(Li) γ-ray observations made at high precision. Following this, the intercomparison organizers systematically altered peak positions and intensities, added known replicate Poisson random errors, created a set of marginally detectable peaks, and prepared one spectrum comprising nine doublets. The advantage was that the "truth was known" (to the IAEA), *so the exercise provided an authentic test of precision and accuracy of the crucial data evaluation step of the CMP.*

While most participants were able to produce results for the six replicates of 22 easily detectable single peaks, less than half of them provided reliable uncertainty estimates. Two-thirds of the participants attacked the problem of doublet resolution, but only 23% were able to provide a result for the most difficult case. (Accuracy assessment for the doublet results was not even attempted by the IAEA because of the unreliability of participants' uncertainty estimates.) Of special import from the point of view of trace analysis, however, was the outcome for the peak detection exercise. The results were astonishing. Of the 22 subliminal peaks, the number correctly detected ranged from 2 to 19. Most participants reported at most one spurious peak, but large numbers of false positives did occur, ranging up to 23! Considering the modeling and computational power available today, it was most interesting that the best peak detection performance was given by the "trained eye" (visual method).

Figure 16 is offered to give an overview of the planning and structure of the gamma spectrum intercomparison. Noting the planning stages shown there, it is appropriate to emphasize the considerable value of the pilot study stage. Allowing for such a step in the source apportionment data intercomparison (described below) would have greatly enhanced that experiment.

A second illustration, which relates to the apportionment of air particulate pollutant sources from multivariate chemical signatures (95), is described in reference 2 and is recommended to the reader. This example is more complicated because of its multivariate character, and because we sought to simulate the production and dispersion of atmospheric particles in a realistic manner -- using fixed city plans and emissions inventories, plus reasonable operating schedules and real meteorological data. A summary of the structure of the three data sets prepared for this exercise is given in Figure 17. Both Factor Analysis (FA) and Chemical Mass Balance (CMB), a form of least squares taking into account errors in both observations and in source profiles, were applied to these data by the participants. For data set I, where there was a relatively small number of sources, one of whose profiles was unknown, FA gave better performance; for set II, with its many sources, all of whose profiles were known to the participants, CMB was the better choice; set III proved too difficult for either approach.

The detailed outcomes of the two exercises can be read in (1) and (2). One point, however, is clear: very significant differences among laboratories and numerical methods were exposed. Perhaps the most severe deficiency noted in both studies was the inability to report reliable estimates of either random or systematic components of uncertainty. Positive outcomes were that direct bounds (sometimes quite acceptable ones) resulted for interlaboratory (and inter-method) error, and that the simulation data sets are currently serving as substantial tools for upgrading laboratory performance and refining computational techniques.

To conclude this brief discussion of models and computational methods and sociochemical problems and truth -- and the plight of the chemometrician -- it seems fitting to share a statement I came upon while perusing the scientific literature of the '20's:

> "As far as the laws of mathematics refer to reality, they are not certain; and as far as they are certain, they do not refer to reality." (A. Einstein, Sidelights on Relativity, 1922).

ACKNOWLEDGMENT

Appreciation is expressed to H.M. Kingston and W.S. Liggett

for valuable discussions of the blank in the Chesapeake Bay study, and for permission to quote some of their data.

Portions of Reference 1 are quoted by IUPAC permission from Pure Appl. Chem., Vol. 54, No. 4 (1982), pp. 714-754. This official IUPAC journal is published monthly by Pergamon Press, Oxford.

Exhibit 5 has been reproduced from Ref. 18 with permission of the author (M. Natrella) and the original publisher. It was adapted by permission of Prentice-Hall, Inc., from A. H. Bowker, G.J. Lieberman, ENGINEERING STATISTICS, c1959, p. 132.

Exhibit 10 is adapted from data in Ref. 71 (Table 31), with permission of the authors.

Exhibit 17 has been adapted from Ref. 2 with permission of the publisher, Air Pollution Control Association.

SELECTED REFERENCES

1) Currie, L.A., Pure & Appl. Chem. 54(4): 715-754, 1982.

2) Gerlach, R.W.; Currie, L.A.; Lewis, C.W., Air Pollution Control Association Specialty Conference on Receptor Models Applied to Contemporary Air Pollution, 96 (A.P.C.A. SP-48, 1983).

Societal, Philosophical and Policy Issues

3) Einstein, A. Sidelights on Relativity. London: Methuen & Co., 1922, p. 280.

4) Rogers, L.B. Subcommittee Dealing with the Scientific Aspects of Regulatory Measurements, American Chemical Society, 1982.

5) Guidelines for Data Acquisition and Data Quality Evaluation in Environmental Chemistry, 1980. Anal. Chem., 52, p. 2242.

6) Horwitz, W.; Kamps, L.R.; Boyer, K.W. J. Assoc. Off. Anal. Chem. 1980, 63, 1344.

7) Nalimov, V.V. Faces of Science, ISI Press, Philadelphia, 1981.

8) Currie, L.A. Anal. Letts. 13, 1 (1980).

9) Handler, P. Pangs of Science, 1978 Washington, D.C. Meeting of the Amer. Assoc. Adv. Science (Reprinted in Chem. Eng. News, 29 (April 17, 1978).)

10) Brooks, H. Potentials and Limitations of Societal Response to Long-Term Environmental Threats, Global Chemical Cycles and Their Alterations by Man, W. Stumm, ed., Dahlem Konferenzen, Berlin (1977).

11) Williams, J., ed., Workshop on Carbon Dioxide, Climate and Society, IIASA, Pergamon Press (1978)

12) Marshall, E., "Acid Rain, A Year Later," Science (1983)221:241.

Detection Limits

a) Important reviews and texts

13) IUPAC Comm. on Spectrochem. and other Optical Procedures for Analysis. Nomenclature, symbols, units, and their usage in spectrochemical analysis. II. Terms and symbols related to

analytical functions and their figures of merit. Int. Bull. I.U.P.A.C., Append. Tentative Nomencl., Symb., Units, Stand.; 1972, 26; see also: II. Data Interpretation, Pure Appl. Chem. 45: 99; 1976; Spectrochim. Acta 33B: 241; 1978.

14) Nalimov, V.V. The application of mathematical statistics to chemical analysis. Oxford: Pergamon Press; 1963.

15) Svoboda, V.; Gerbatsch, R. Fresenius' Z. Anal. Chem. 242 (1): 1-13; 1968.

16) Boumans, P.W.J.M. Spectrochim. Acta 33B: 625; 1978.

17) Ku, H.H., Edit., Precision Measurement and Calibration, NBS Spec. Public. 300 (1969). [See Chapt. 6.6 by M.G. Natrella, "The Relation between Confidence Intervals and Tests of Significance."]

18) Natrella, M.G., Experimental Statistics, NBS Handbook 91, Chapt. 3 (1963).

19) Kurtz, D.A., Role of Chemometrics in Pesticide/Environmental Residue Analytical Determinations, Amer. Chem. Soc. Sympos. Series, to be published (1984), [See Chapter by L.A. Currie, "The Many Dimensions of Detection in Analytical Chemistry"].

20) Liteanu, C.; Rica, I. Pure Appl. Chem. 44(3): 535-553; 1975.

21) Massart, D.L.; Dijkstra, A.; Kaufman, L. Evaluation and optimization of laboratory methods and analytical procedures. New York: Elsevier Scientific Publishing Co.; 1978.

22) Kateman, G.; Pijpers, F.W. Quality control in analytical chemistry. New York: John Wiley & Sons; 1981.

23) Frank, I.E. and Kowalski, B.R., Chemometrics (Review), Analytical Chemistry 54 (1982) 232R.

24) Winefordner, J.D., ed. Trace analysis. New York: Wiley; 1976. (See especially Chap. 2, Analytical Considerations by T.C. O'Haver.)

25) Liteanu, C.; Rica, I. Statistical theory and methodology of trace analysis. New York: John Wiley & Sons; 1980.

26) Currie, L.A. Sources of error and the approach to accuracy in analytical chemistry, chapter 4 in treatise on analytical chemistry, Vol. 1 P. Elving and I.M. Kolthoff, eds. New York: J. Wiley & Sons; 1978.

27) Wilson, A.L. The performance-characteristics of analytical methods. Talanta 17: 21; 1970; 17; 31: 1970; 20: 725; 1973; and 21: 1109; 1974.

28) Hirschfeld, T. Anal. Chem. 48: 17A; 1976.

b) Basic analytical articles

29) Feigl, F. Mikrochemie 1: 4-11; 1923.

30) Kaiser, H. Two papers on the Limit of Detection of a Complete Analytical Procedure, London: Hilger; 1968.

31) St. John, P.A.; Winefordner, J.D. Anal. Chem. 39: 1495; 1967.

32) Altshuler, B.; Pasternack, B. Health Physics 9: 293; 1963.

33) Currie, L.A. Anal. Chem. 40(3): 586-593; 1968.

34) Frank, I.E., Pungor, E., and Veress, G.E., Anal. Chim. Acta 133 (1981) 433.

35) Gabriels, R. Anal. Chem. 42: 1439; 1970.

36) Crummett, W.B. Ann. N.Y. Acad. Sci. 320: 43-7; 1979.

37) Grinzaid, E.L.; Zil'bershtein, Kh. I.; Nadezhina, L.S.; Yufa, B. Ya J. Anal. Chem. - USSR 32: 1678; 1977.

38) Winefordner, J.D.; Ward, J.L. Analytical Letters 13(A14): 1293-1297; 1980.

39) Liteanu, C.; Rica, I.; Hopirtean, E. Rev. Roumaine de Chimie 25: 735-743; 1980.

40) Blank, A.B. J. Anal. Chem. - USSR 34: 1; 1979.

41) Liteanu, C.; Rica, I. Mikrochim. Acta 2(3): 311; 1975.

42) Tanaka, N. Kyoto-fu Eisei Kogai Kenkyusho Nempo 22: 121; 1978.

43) Ingle, J.D., Jr. J. Chem. Educ. 51(2): 100; 1974.

44) Pantony, D.A.; Hurley, P.W. Analyst 97: 497; 1972.

45) Hubaux, A.; Vos, G. Anal. Chem. 1970, 42, 849.

46) Ingle, J.D., Jr.; Wilson, R.L. Anal. Chem. 1976, 48, 1641.

47) Long, G.L.; Winefordner, J.D. Anal. Chem. Vol. 55, No. 7, June 1983, 712A.

48) Lub, T.T. and Smit, H.C., Anal. Chim. Acta 112 (1979) 341.

49) Eckschlager, K. and Stepanek, V., Mikrochim. Acta II (1981) 143.

c) Basic nuclear articles

50) Currie, L.A. IEEE Trans. Nucl. Sci. NS-19: 119; 1972.

51) Upgrading environmental radiation data. Health Physics Society Committee Report HPSR-1: 1980.

52) Donn, J.J.; Wolke, R.L. Health Physics, Vol. 32: 1-14; 1977.

53) Lochamy, J.C. Nat. Bur. Stand. Spec. Publ. 456; 1976, 169.

54) Nakaoka, A.; Fukushima, M.; Takagi, S. Health Physics, Vol. 38: 743; 1980.

55) Pasternack, B.S.; Harley, N.H. Nucl. Instr. and Meth. 91: 533; 1971.

56) Guinn, V.P. J. Radioanal. Chem. 15: 473; 1973.

57) Tschurlovits, V.M. Atomkernenergie. 29: 266; 1977.

58) Robertson, R.; Spyrou, N.M.; Kennett, T. J. Anal. Chem. 47: 65; 1975.

59) Nicholson, W.L., "What Can Be Detected," Developments in Applied Spectroscopy, v.6, Plenum Press, p. 101-113, 1968.

60) Currie, L.A., Lower Limit of Detection: Definition and Elaboration of the Proposed NRC Position, to be published (1984). [See also Nuclear Regulatory Commission, Branch technical position on regulatory guide 4.8; 1979; Regulatory Guide 4.8 "Environmental Technical Specifications for Nuclear Power Plants". 1975.]

61) Head, J.H. Nucl. Instrum. & Meth. 98: 419; 1972.

62) Obrusnik, I.; Kucera, J. Radiochem. Radioanal. Lett. 32(3-4): 149; 1978.

63) Heydorn, K.; Wanscher, B. Fresenius' Z. Anal. Chem. 292(1): 34; 1978.

64) Mundschenk, H. Nucl. Instruments & Methods 177: 563; 1980.

65) Currie, L.A. in Modern trends in activation analysis, Vol. II, J.R. DeVoe; P.D. LaFleur, eds. Nat. Bur. Stand. Spec. Publ. 312; p. 1215, 1968.

66) Currie, L.A. Anal. Lett. 4: 873; 1971.

67) Pazdur, M.F. Int. J. Appl. Rad. & Isotopes 27: 179; 1976.

68) Rogers, V.C. Anal. Chem. 42: 807; 1970.

69) Sterlinski, S. Nuclear Instruments and Methods 68: 341; 1969.

70) Gilbert, R.O.; Kinnison, R.R. Health Phys. 40(3): 377; 1981.

d) The blank and distributional issues

71) Kingston, H.M.; Greenberg, R.R.; Beary, E.S.; Hardas, B.R.; Moody, J.R.; Rains, T.C.; and Liggett, W.S., "The Characterization of the Chesapeake Bay: A Systematic Analysis of Toxic Trace Elements," National Bureau of Standards, Washington, D.C., 1983, NBSIR 83-2698.

72) Cox, D.R.; Lewis, P.A.W., 1966, The statistical analysis of series of events: New York, Wiley & Sons.

73) Currie, L.A., 1972, Nuclear Instruments and Methods, v. 100, p. 387-395.

74) Berkson, J. Int. J. Appl. Rad. Isot., 1975, Vol. 26, pp. 543.

75) Cannizzaro, F.; Greco, G.; Rizzo, S.; and Sinagra, E. Int. J. Appl. Rad. Isot. Vol. 29, pp. 649.

76) Scales, B. Anal. Biochem. 5 (1963) 489.

77) Patterson, C.C.; Settle, D.M. 7th Materials Res. Symposium, NBS Spec. Publ. 422, Washington, D.C., 321 (1976).

78) Currie, L.A., Tompkins, G.B., and Spiegelman, C.H., Basic Tests of Poisson and Alternative Distributions of Low-Level Counting Data (1984), to be published.

79) Pyke, R., J. Roy. Statist. Soc. B. 27: 395, 1965.

80) Durbin, J., Biometrika 62: 5, 1975.

81) Liggett, W. ASTM Conf. on Quality Assurance for Environmental Measurements, 1984, Boulder, CO, in press.

82) Murphy, T.J., NSB Special Pub. 422, Vol. II, Washington, D.C., 1976, 509.

83) Currie, L.A.; Gerlach, R.W.; Klouda, G.A.; Ruegg, F.C.; and Tompkins, G.B. Radiocarbon, Vol. 25, No. 2, 1983, p. 553.

Chemometric Intercomparisons and Related Literature

a) Gamma-ray spectroscopy

84) Filliben, J.J. (1975), Technometrics, 17, 111.

85) Currie, L.A. J. of Radioanalytical Chemistry 39, 223 (1977).

86) Rice, J. An Approach to Peak Area Estimation, J. of Research of NBS vol. 87, no. 1, p. 53-65 (1982).

87) R. Parr, H. Houtermans and K. Schaerf, The IAEA Intercomparison of Methods for Processing GE(Li) Gamma-ray Spectra, in Computers in Activation Analysis and Gamma-ray Spectroscopy, CONF-780421, p. 544 (1979).

88) Ritter, G.L.; Currie, L.A. in Computers in Activation Analysis and Gamma-ray Spectroscopy, CONF780421, p. 39 (1979).

89) Quittner, P. Gamma Ray Spectroscopy, Halsted Press, New York, 1973.

90) J. Op de Beeck, Gamma Ray Spectrometry Data Collection and Reduction by Simple Computing Systems, At. Energy Rev., 13(4): 743 (1975).

91) Gunnik, R.; Niday, J. Computerized Quantitative Analysis by Gamma-Ray Spectrometry. USAEC Report UCRL-51061 (Vol. 1), University of California, Livermore, NTIS, 1972.

92) Routti, J.T.; Prussin, S.G. Nucl. Instrum. Methods, 72: 125 (1969).

b) Multivariate environmental data

93) Stevens, R.K., and Pace, T.G. (1984) Overview of the Mathematical and Empirical Receptor Models Workshop (Quail Roost II). Atmospheric Environment (in press).

94) Henry, R.C., Lewis, C.W., Hopke, P.K., and Williamson, H.J. (1984) Review of receptor model fundamentals, Atmospheric Environment (in press).

95) Currie, L.A., Gerlach, R.W., Lewis, C.W., Balfour, W.D., Cooper, J.A., Dattner, S.L., De Cesar, R.T., Gordon, G.E., Heisler, S.L., Hopke, P.K., Shah, J.J., Thurston, G.D., and Williamson, H.J. Interlaboratory comparison of source apportionment procedures (1984) Atmospheric Environment (in press).

96) Kowalski, B.R., Edit., Chemometrics, Theory and Application, ACS Sympos. Series 52 (Amer. Chem. Soc., 1977).

97) Malinowski, E.R.; Howery, D.G. Factor Analysis in Chemistry; J. Wiley & Sons: New York, 1980.

98) Lawton, W.H.; Sylvestre, E.A. Technometrics 1971, 13, 617; 1972, 14, 3.

99) Full, W.E., Ehrlich, R., and Klovan, J.E. (1981) Math. Geol. 13, 331.

100) Gordon, G.E. (1980) Envir. Sci. Technol. 14, 792.

ROGERS COMMITTEE[a]

IMPROVING THE RELIABILITY AND ACCEPTABILITY OF ANALYTICAL CHEMICAL DATA USED FOR PUBLIC PURPOSES (1982)

* Perceived Needs of the Public
* Analytical Chemical Characteristics of Regulations
* Analytical Chemical Considerations
    - Methods
    - Sampling
    - Collaborative Testing
    - Quality Assurance
* Problems Encountered in Some Specific Regulatory and Clinical Samples
* Analytical Uncertainty and the Courts

(a) Reference 4.

Figure 1.

NUCLEAR REGULATORY COMMISSION STUDY [a]

OBJECTIVE: To Develop a Document and Position on the Lower Limit of Detection (LLD) (Nomenclature, Formulation, Evaluation)

MEETINGS: NRC Headquarters: Regional Offices (Inspector, Mobile Laboratory)
Power Reactors
Contractors: Commercial, University
Environmental Laboratory
EPA Cross-check Laboratory
Trade Association (Atomic Industrial Forum)

SOME PRELIMINARY FINDINGS

*LLD Manual Needed--Practical Guidance for Diverse Educational Backgrounds

*Wide Ranging Nomenclature, Formulations (LLD,MDA....)

*Policy Issues Regarding Minor Components (when high interference); Biased Reporting

*Detection Decisions--ranged from 5% to 50% false negatives

*Blank--Ambiguity in Present Draft Documentation; Blank variations excluded from LLD by most workers.

*Non-counting Errors Important, especially Sampling, and Calibration when non-specific ("gross") radionuclides

*QA and Cross-check Samples--"Blind" and Varying Compositions Needed

*De Minimis Reporting (LLD unspecified for some materials,but "report if detected")

*Subtle Problems Involving Uncertainties, reporting levels, litigation

*Multiple Detection Decisions, effect on α-, β-risks (γ-spectrometry)

*Monitoring Modes and Averaging Practices

*Hidden Algorithms, Bad Nuclear Parameters (automated, commercial instrumentation)

*Treatment of Data from Continuous (effluent) Monitors (time constants, transients,...)

(a) Reference 60.

Figure 2.

HISTORICAL PERSPECTIVE

Feigl ('23)	- Limit of Identification [Ref.29]
Kaiser ('65-'68)	- Limit of Guarantee for Purity [Ref.30]
St. John ('67)	- Limiting Detectable Concentration (S/N_{rms}) [Ref. 31]
Currie ('68)	- Detection Limit [Ref. 33]
Nicholson ('68)	- Detectability [Ref. 59]
IUPAC ('72)	- Sensitivity; Limit of Detection...[Ref. 13]
Ingle ('74)	- ("[too] complex...not common") [Ref. 43]
Lochamy ('76)	- Minimum Detectable Activity [Ref.53]
Grinzaid ('77)	- Nonparameteric...Detection Limit [Ref. 37]
Liteanu ('80)	- Frequentometric Detection [Ref. 25]

Figure 3.

DETECTION LIMITS: Approaches, Difficulties

Signal/Noise (S/N) [Ref's: 22, 24, 31, 48]

Detection Limit $\equiv 2N_{p-p}$, $2N_{rms}$, 3s (n = 16-20)

$[N_{rms} \simeq N_{p-p}/5.]$

DC: white noise assumed, β-error ignored

AC: must consider noise power spectrum, non-stationarity, digitization noise

Simple Hypothesis Testing [Ref's 16, 17, 30, 33, 63]

$\hat{S} = y - \hat{B}$

H_o: significance test (α-error) ~ cf 1-sided confidence interval

H_A: power of test (β-error) ~ cf O.C. Curve

Determination of S_D requires accurate knowledge of the distribution function for $\hat{S}$. If $\hat{S} \sim N(S, \sigma^2_{\hat{\sigma}})$, and $\alpha, \beta = 0.05$, then $S_D = 2S_c = 3.29\sigma_{\hat{\sigma}}$, where $\sigma_o = \sigma_{\hat{S}}$ when $S = 0$.

Other Approaches [Ref's 7, 21, 34, 49]

Decision Analysis (uniformly best, Bayes, minimax), Information and Fuzzy Set Theories.

Figure 4.

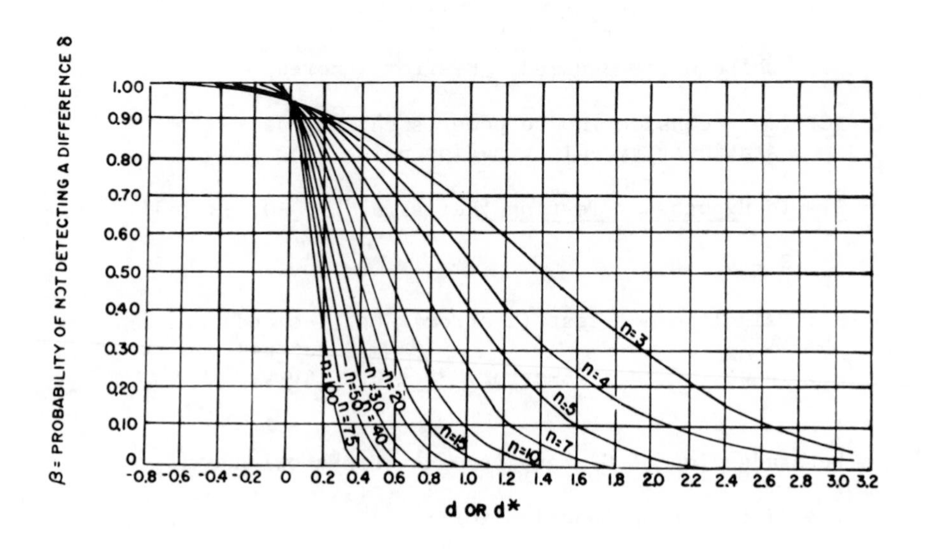

a) Reference 18

Figure 5.

Operating Characteristic Curves for the One-Sided t-Test. The error of the 2nd kind (β) is shown as a function of the standardized difference (d) and the number of replicates (n), for the error of the 1st kind (α) equal to 0.05. (Normality assumed.) (Reprinted from Natrella (18) who adapted the original from Engineering Statistics by A.H. Bowker and G.J. Lieberman (1959), Prentice-Hall, Inc.)

Concentration Detection Limits - Some Problems

- σ^2 only estimated; H_o-test ok ($t\ s/\sqrt{n}$), but x_D is uncertain
- Calibration function estimated, so Normality not exactly preserved:

  $$\hat{x} = (y - \hat{B}) / \hat{A} \neq \text{linear function (observations)}$$

- B-distribution (or even magnitude) may not be directly observed
- Effects of non-linear regression; effects of "errors in x- and y" (calibration)
- Systematic error, blunders -- e.g., in the shape, parameters of A

  [random errors become systematic, without continual re-calibration]
- Non-additivity (matrix effects)
- Uncertain number of components (and identity)

  [Lack of fit tests lose power under multicollinearity]

- Multiple detection decisions: $(1 - \alpha) \rightarrow (1 - \alpha)^n$

Figure 6.

Fenvalerate GC Data (D. Kurtz)[(a)]

Calibration: $y(cm^2)$ vs $x(ng)$

Replicate x	A	B	C	D	E
0.05	1.13	1.23	1.22	1.20	1.12
0.25	6.55	7.98	6.54	6.37	7.96
1.00	29.70	30.00	30.10	29.50	29.10
5.00	211.	204.	212.	213.	205.
20.00	929.	905.	922.	928.	919.

Question: What is the detection limit?

(a) Data from Ref. 19

Figure 7.

Fenvalerate (GC) Data - Set B
(average of 5 replicates)

Response ($\bar{y}$, cm^2)[a]	Amount (x, ng)	$\Delta\bar{y}/\Delta x$
$1.18 \pm 0.02_4$	0.05	23.6
$7.08 \pm 0.06_8$	0.25	29.5
29.68 ± 0.23	1.00	30.1
$209.0 \pm 1.1_1$	5.00	44.8
$920.6 \pm 4.4_0$	20.0	47.4

a) Uncertainties are estimated standard errors.
Figure 8.

THE BLANK

- Direct observation - crucial for detection limit
- Adequate No. of measuerments needed; with but two, σ_B may be $16|\Delta_B|$
- Efficiency correction may differ between blank and analyte (Scales, 1963) [Ref. 76]
- Yield corrections must recognize point(s) of introduction of blank (Patterson, 1976) [Ref. 77]
- Multisource blanks generate strange probability distributions shape and parameters important
- Poisson Hypothesis must be tested for counting background

Figure 9.

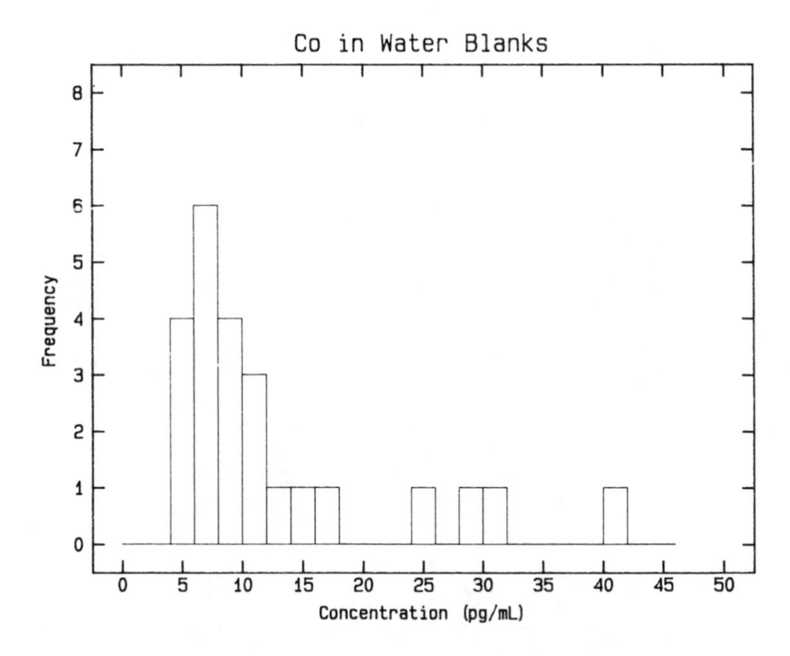

Frequency distribution observed for dissolved Co in Water blanks -- Chesapeake Bay trace element study (71).

Figure 10.

CHESAPEAKE BAY - TRACE ELEMENT BLANKS (ng/mL)*

24 samples of high-purity sub-boiling distilled H_2O, treated on shipboard in parallel to samples

15 trace elements analyzed by NAA or AAS

	Log (x - τ)	N(μ, σ²)	Point Prediction	.96 PI
Co (Diss.)	$\hat{t} = 0.0045$	-	0.0091	0.040
Fe (Diss.)	$\hat{t} = 0.4$	-	1.12	3.70
Cr (Diss.)	-	$\hat{\sigma} = 0.10$	1.55	

* Ref's. 71 and 81.

Figure 11.

HYPOTHESES -- COUNTER PULSES

H_0 - N: Poisson, t_i: Uniform, Δt: Exponential
H_1 - Periodic Disturbances
H_2 - Bursts; Delayed Multiples
H_3 - Counter, Circuit Paralysis
H_4 - Dead-time, Gate (μs - clock)
H_5 - Software Blunder
H_6 - Buffer, t-distortion without loss
H_7 - start-up transients

(a) Ref. 78

Figure 12.

COUNTER BACKGROUND TESTS
(Poisson Hypothesis)

- Radioactivity Tests (Am-241): previously χ^2 used to distribution of time intervals [Ref. 74, 75] (Berkson; Cannizzaro et al.)

- NBS low-level background data tested similarly, but also to test a) with improved exponential class resolution, and b) for complete distribution of individual pulse occurrence times.

Hypothesis (null)

a) Prob $(N_{T,\ T+dT} = 1) = dT/\tau$ (Occurrence Times--Uniform)

b) Prob $(N_T = x) = \frac{\mu^x e^{-\mu}}{x!}$ [$\mu = T/\tau$ (Counts -- Poisson)

c) Prob $(\Delta t > t) = e^{-t/\tau}$ (Intervals -- Exponential)

where $\tau = \frac{1}{\text{rate}}$ = mean interval between counts

Kolmogorov-Smirnov Test [cdf]

$\chi^2 = \sum \frac{(f_o - f_e)^2}{f_e}$ Test [pdf - classes]

Figure 13.

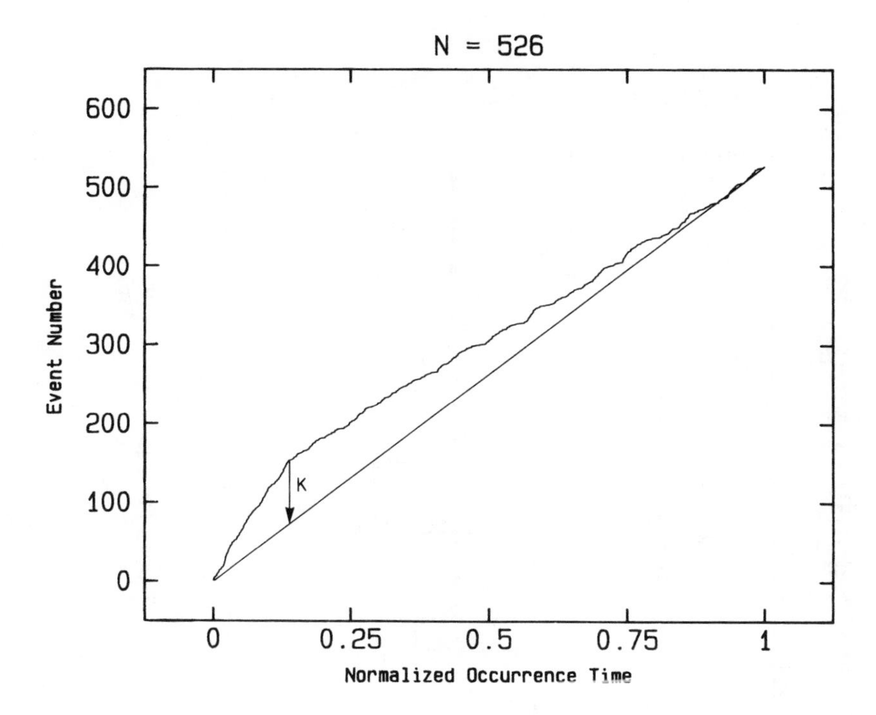

Figure 14.

Kolmogorov-Smirnov test of the empirical cumulative distribution function for low-level (radiocarbon) counting pulse occurrence times (78). The test was significant $D_n = K/526 = 0.152$

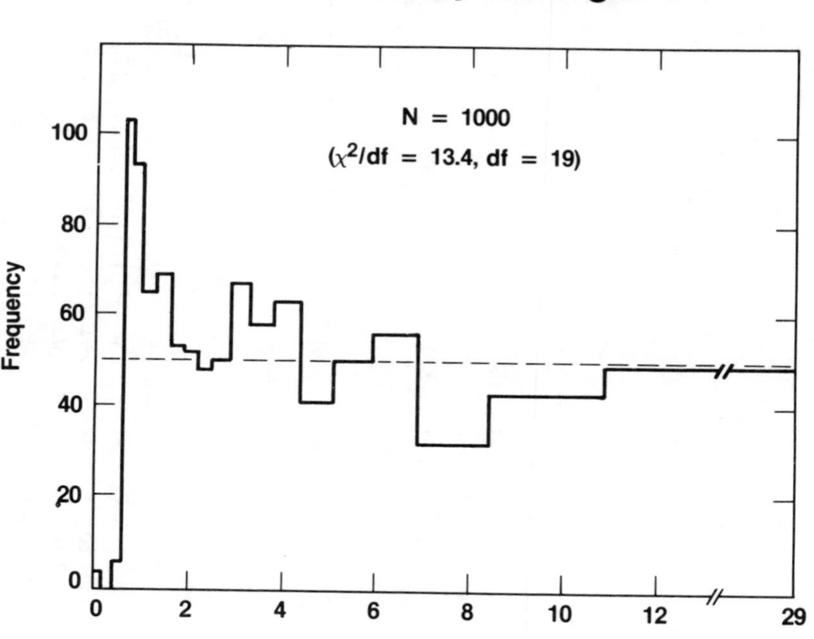

Figure 15.

Chi-Square test of the empirical equal probability histogram for low-level counting pulse inter-arrival times (78). The test was significant ($\chi^2 = 255$).

IAEA Gamma-Ray Intercomparison
[Ref. 87]

. Objective ~ to evaluate methods for processing gamma-ray spectra

. Organization ~ proposed (consultants meeting) 1973; Pilot study, 1975-76; full intercomparison, 1976-77(+): 163 labs/34 member states.

. Data ~ derived from high-precision, high-quality Ge(Li) γ-ray spectra of pure radionuclides
Adjusted → known locations, peak areas, Poisson errors.

Spectrum 100 ~ [Reference], 20 independent peaks

Spectrum 200 ~ [Detection], N - subliminal peaks, including "Compton edge" (Judgment regarding α-, β-errors left up to the participants)

Spectra 300-5 ~ [Precision], 6 spectra with 20 known + 2 unknown peaks, isolated and with high intensity

Spectrum 400 ~ [Resolution], 9 doublets

. Outcome ~ wide range of performance; common absence of uncertainty estimates; best method for unknown peaks appeared to be "visual"!

Figure 16.

STRUCTURE OF THE SIMULATED DATA SETS (Ref. 2)

Generating Equation

$$\tilde{C}_i^{(k)} = \sum^{P} [\tilde{A} - e_m - e_H^{(k)}]_{ij} S_j^{(k)} + e_i^{(k)}$$

where: k = sampling period $[1 \le k \le 40]$

$\tilde{C}_i^{(k)}$ = "observed" concentration of species-i, period-k $[1 \le i \le N, N \le 20]$

$S_j^{(k)}$ = true intensity (at receptor) of source-j $[1 \le j \le P, P \le 13]$

$\tilde{A}_{ij}$ = "observed source profile matrix (element-i,j)

e_i = random measurement errors, independent and normally distributed

e_m = systematic source profile errors, independent and normally distributed (systematic, because fixed over the 40 sampling periods)

e_H = random source profile variation errors, independent and log-normally distributed

Data Set Characteristics

P = No. of active sources; P_o = No. in the "world list" = 13]

Set I : $P = 9$ (including one unknown source)*; errors = e_i, e_m; City Plan -1

Set II : $P_o = P$ (all known); errors = e_i, e_m; City Plan -2

Set III : $P = P_o$ (all known); errors = e_i, e_m, e_H; City Plan -2

standard deviations (σ's) were made known for all errors (i, m, H).

* For data Set I, participants were told only that $P \le P_o$

Figure 17.

MULTIVARIATE CALIBRATION

Harald Martens and Tormod Næs

Norwegian Food Research Institute, Ås, Norway

Determination of chemical concentrations can be made more rapid and reliable by combining information from several measurement variables, e.g. light absorbances at several wavelengths. Systematic errors can thereby be eliminated so that very unspecific measurement data can be used for quantitative determinations. Automatic warning of unexpected errors is also possible. In addition, multivariate calibration can give new information on how the chemical components behave in situ in the sample. Near InfraRed (NIR) reflectance analysis of wheat flour is used as an example.

Quantitative instrumental analysis (e.g. of a given chemical constituent by spectroscopic measurement of light absorbance) can be affected by several types of systematic noise:

1) Interferences from other chemical constituents, (e.g. other constituents with overlapping absorbance spectra)
2) Interferences from physical phenomena in the samples (e.g. light scattering)
3) Interferences from changes in the measurement process itself (e.g. temperature changes in a monochromator or detector)

In traditional analytical techniques one relies on measuring one single variable, e.g. the light absorbance at a single wavelength. To avoid errors due to interferences one therefore

B. R. Kowalski (ed.), Chemometrics. Mathematics and Statistics in Chemistry, 147–156.

has to purify and standardize the samples first, in order to get sufficiently specific measurements. This is often time-consuming and may create artifacts. In some cases the required purification is too expensive or even physically impossible.

By multivariate data analysis one can use measured data more efficiently (1). Multivariate calibration is a chemometric method designed for quantitative analysis. Multivariate calibration can reduce the need for sample preparation in chemical analysis, because various systematic noise types can be eliminated mathematically. Unexpected problems in unknown samples can be detected automatically. Reliable quantitative measurement therefore becomes possible even in "dirty" systems, such as diffuse reflectance determination of protein in wheat flour.

Multivariate calibration (2,3,4) means that you use a set of calibration samples with known composition to "teach" the mathematical model in the microcomputer how to remove systematic errors due to interferences in the measured data. Then the mathematical results are used to get a rapid determination of the chemical composition of similar unknown samples, regardless of the level of these interferences.

The best example of multivariate calibration today is in diffuse light spectroscopy of foods. Rapid determination of chemical composition in foods by Near-Infra-Red (NIR) multi-wavelength reflectance measurement has attained a wide acceptance over the past decade (5). The method can be used in many areas. We shall illustrate this by determination of protein in wheat flour.

In order to calibrate the NIR-instrument you measure protein in the conventional manner by Kjeldahl-nitrogen analysis, in e.g. 50 representative flour samples. The NIR reflectance spectra are also measured for these calibration samples, and a computer program for multivariate statistical modelling, e.g. using the Partial Least Squares regression, PLS (6,7) is used to determine the coefficients that relate NIR-measurements to protein content. Then the protein content in new, unknown flour samples can be determined by measuring their NIR-spectrum and multiplying this with the estimated calibration coefficient spectrum.

Advantage No. 1: Quantification from unspecific data

The NIR method would have been impossible to use with traditional univariate calibration. The reason is that <u>all</u> the wheat's main components water, protein, starch, etc. absorb at <u>all</u> NIR wavelengths, causing gross chemical interferences as shown in Figure 1. This means that even if the main peaks of

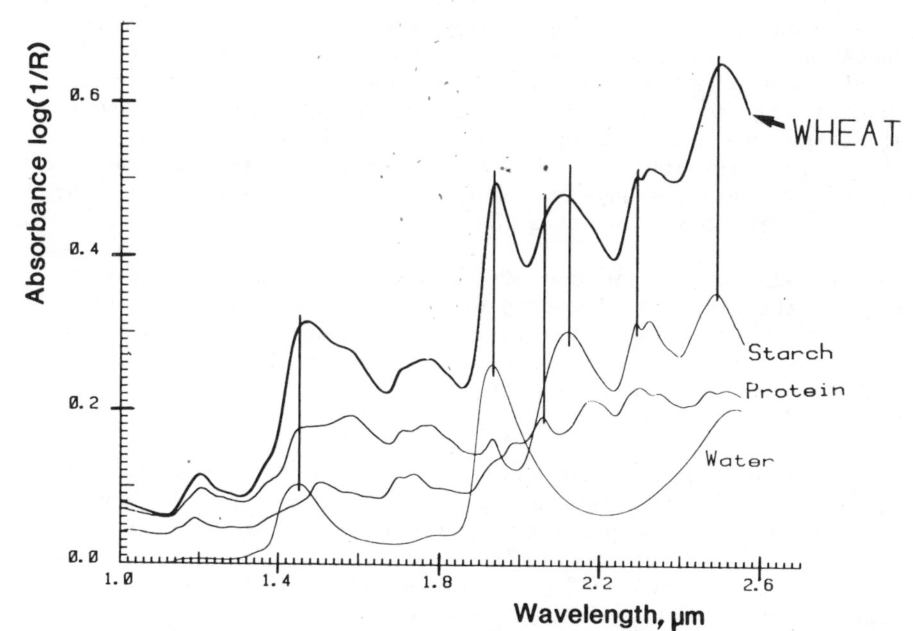

Figure 1. NIR absorbance spectra of wheat and its primary pure constituents starch, protein and water, as functions of wavelengths (K.H. Norris, USDA, pers. comm. 1982).

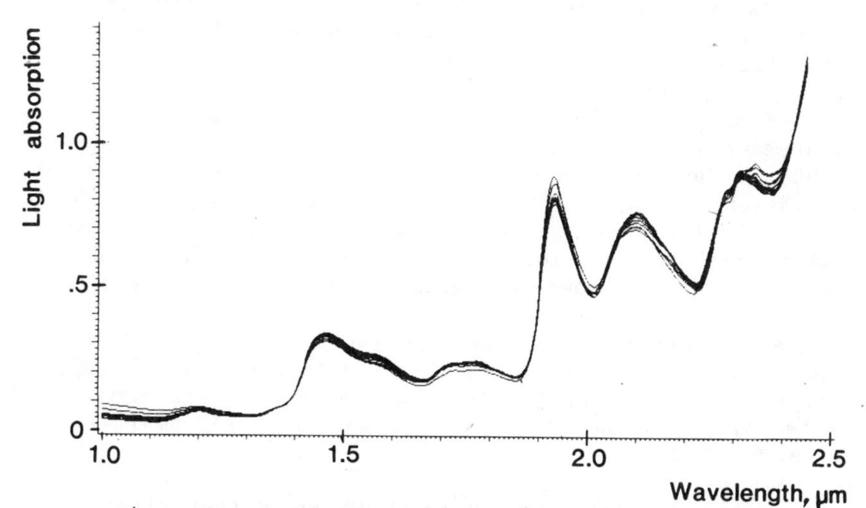

Figure 2. Light absorption for 10 wheat flour samples in the NIR-area, measured by a reflectance instrument.

water and starch can be recognized in a flour sample's spectrum, it is impossible to measure e.g. protein with one single wavelength. This is illustrated in Figure 2 by the spectra for 10 wheat flour samples that have varying contents of starch, protein, fiber, water, etc., and widely varying particle size distributions. The light absorption is expressed as light scattering-corrected Kubelka-Munk transform of the measured reflectance spectra (9).

Most of the variations seen in Figure 2 are caused by physical light scattering interferences due to particle size variations. The chemically related variations can hardly be seen at all! Since the interferences from e.g. light scatter is unknown, conventional multicomponent analysis cannot be used for this type of data.

Still, these NIR data can give very good protein results (8). In Figure 3 the correlation between NIR determined and conventionally determined protein (Kjeldahl-N x 6.25) in a set of "unknown" flour samples of the same type is shown to be very high ($r > 0.99$). Standard deviation between the two methods for protein determination is approximately 0.15 % of the sample weight in the range 10 - 18 % protein.

The spectroscopic rapid measurements were made directly on flour samples. How is this precision possible with such fuzzy data? The key lies in measuring <u>many</u> variables simultaneously in each sample. In this case the reflectances at some 500 different wavelengths were used simultaneously in the protein determination. Different unknown "background" interference effects, e.g. from other chemical components (water, starch, etc.), from physical phenomena (light scattering) and from the measurement itself (temperature) can then be removed by multivariate mathematical modelling. This is called multivariate calibration, because several measurement variables are being combined in the same calibration model. In the present case a 7-dimensional PLS calibration solution was used (modelling protein and 6 unknown background interference phenomena).

A number of different statistical techniques can be used for estimating the multivariate calibration coefficients (2,3). The multivariate calibration itself can be rather laborious, and the technique is therefore not recommendable for small sample series.

The advantage of multivariate calibration is that after it is done one is able to replace cumbersome, slow chemical analysis methods by simple, fast instrumental measurements, provided that the unknown samples belong to the same type of samples used for calibration.

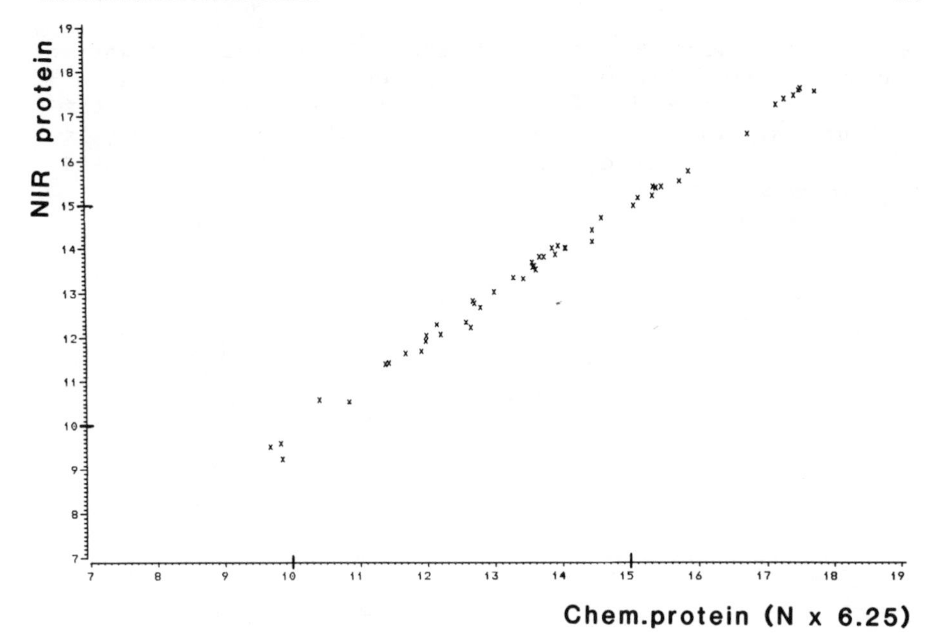

Figure 3: Protein determined in NIR reflectance instrument plotted against protein determined by a traditional method (Kjeldahl-nitrogen x 6.25) in 50 "unknown" wheat flour samples.

Advantage No. 2: Automatic error warning

In addition one gets automatic warnings from the measurement instrument if the calibration conditions are violated, e.g. by instrument drift or abnormal samples. Figure 4 shows the residual NIR-spectrum for the 10 flour samples from Figure 2:

Each residual spectrum represents the difference between the NIR input spectrum and its best fit to a calibration model, as function of wavelength. Ideally, these residual spectra should be zero except for random measurement noise. But the figure clearly shows remaining information in certain spectral ranges which correspond to the light absorption ranges of water. Note also that the maxima for the residual peaks do not appear at exactly the same wavelength for the different samples. It is known that the water light-absorbance spectrum changes with sample temperature, etc. Consequently, the figure may indicate that these samples have been analyzed under somewhat varying temperature conditions.

One could program the micro-processor to give an automatic warning if remaining information is detected in the residual spectrum, as in Figure 4. This automatic outlier method has been used for distinction of barley from wheat by NIR-analysis (7). Other outlier detection methods are available, e.g. based on leverages (10).

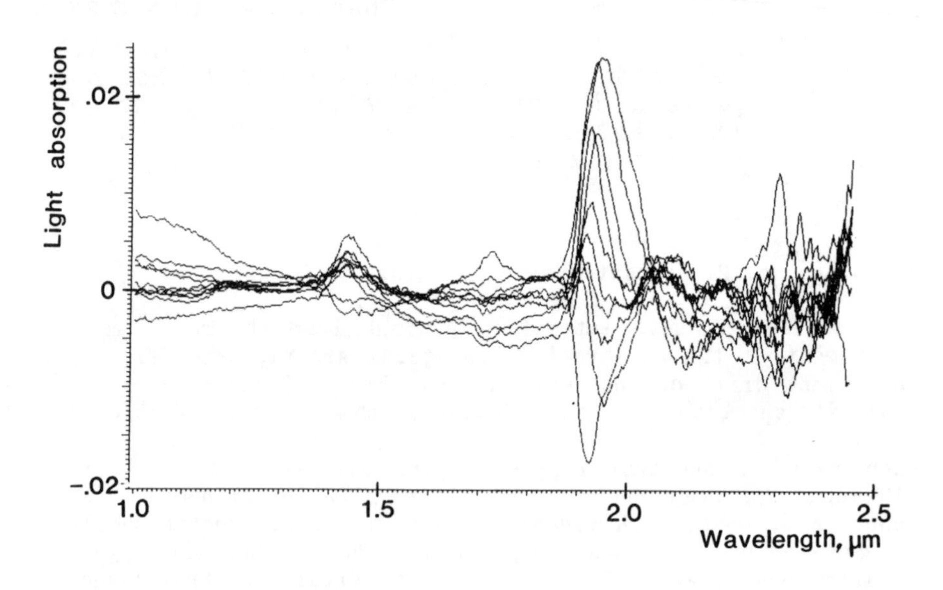

Figure 4. Residual light absorption spectra for the 10 flour samples in Figure 2 in the NIR-range, after fitting to a multivariate calibration model.

Advantage No. 3: In situ spectra

In addition to providing quantitative concentrations and automatic error warnings, multivariate calibration can also generate new information of scientific interest. Water in wheat absorbs light differently from pure water, probably due to the presence of bound water. It could be interesting to see the absorbance spectrum of water as it exists in flour. Figure 5 shows a preliminary estimate of the in situ water spectrum in flour, estimated by projecting the NIR spectra on chemical concentration data for water and protein. The figure shows a characteristic absorption peak close to 1.94 µm and a lower absorption close to 1.45 µm. The peaks coincide almost, but not completely, with the spectrum for pure water. (The negative apparent absorption at certain wavelengths is probably due to negative correlations between water and other main components in the 50 flour samples that were used in the estimation):

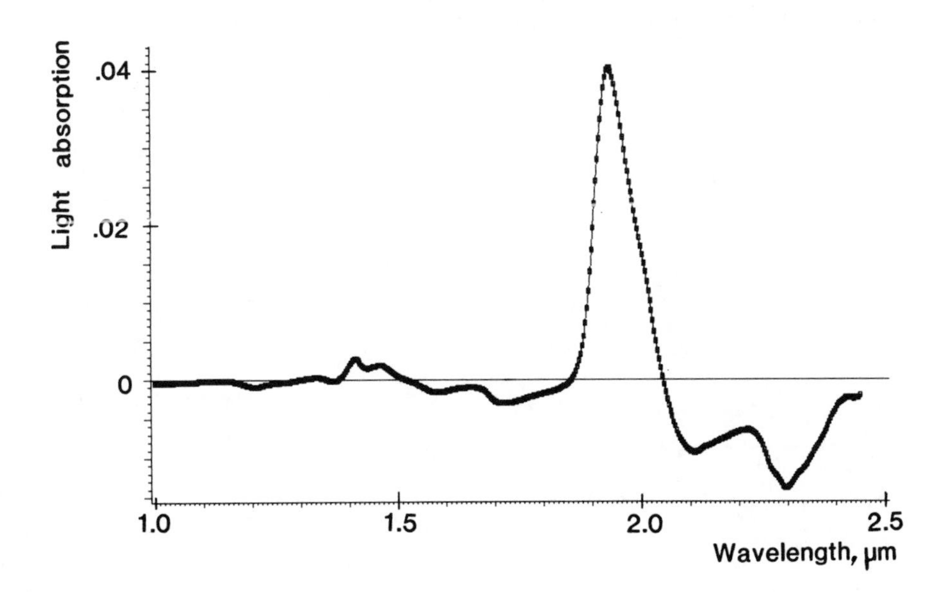

Figure 5. Light absorption spectrum of water estimated in situ in wheat flour by regression on chemically measured concentration data for water, protein, etc.

Multivariate calibration applied in chemistry

Should an analytical chemist find wheat flour somewhat too "dirty" for light spectroscopy analysis, then it may be even more surprising to realize that NIR-determined chemical concentrations can be done with higher precision than the traditional chemical reference method which has been used for calibration (e.g. Kjeldahl-N determined protein)!

We believe multivariate calibration is a very general technique with a wide potential in chemistry. In addition to the analysis of protein, water and fat by NIR reflectance (9), we have used multivariate calibration for determination of fat content in living animals by quantitative image analysis in X-ray computer tomography (10), determination of botanical components in grain by autofluorescence (11) and for predicting human taste preferences by using chemical and instrumental measurements (12).

Acknowledgments

Karl H. Norris is thanked for lending us NIR data, and Ragnhild Norang is thanked for typing and linguistic assistance.

REFERENCES

(1) Martens, H., Wold, S. and Martens, M. (1983) A layman's guide to multivariate data analysis. Proc. IUFoST Symp. "Food Research and Data Analysis", Oslo, Norway, September 1982. (Martens and Russwurm, eds.) Applied Science Publ., pp. 473-492.

(2) Martens, H. and Næs, T. (1983) Calibration as a practical problem. Proc. Nordic Symp. on Applied Statistics, Stokkand Forlag Publ., Stavanger, Norway, pp. 113-135.

(3) Næs, T. and Martens, H. (1983) Calibration as a statistical problem. Proc. Nordic Symp. on Applied Statistics, Stokkand Forlag Publ., Stavanger, Norway, pp. 137-164.

(4) Lea, P., Martens, H., Mielnik, M. and Slinde, E. (1983) Analysis of mixtures: Hemoglobin and myoglobin in various molecular states determined from visible light spectra. Proc. Nordic Symp. on Applied Statistics, Stokkand Forlag Publ., Stavanger, Norway, 165-183.

(5) Wetzel, D.L. (1983) Near-Infrared Reflectance Analysis: Sleeper among spectroscopic techniques. Analytical Chemistry 55, 12, pp. 1165A-1176A.

(6) Wold, S., Martens, H. and Wold, H. (1983) The Multivariate Calibration Problem in Chemistry solved by the PLS Method. Proc. Conf. Matrix Pencils, (A. Ruhe, B. Kågström, eds.), March 1982, Lecture Notes in Mathematics, Springer Verlag, Heidelberg, pp. 286-293.

(7) Martens, H. and Jensen, S.Å. (1983) Partial Least Squares regression: A new two-stage NIR calibration method. Proc. 7th World Cereal and Bread congress, Prague June 1982. (Holas and Kratochvil, eds.) Elsevier Publ. Amsterdam, pp. 607-647.

(8) Norris, K.H. (1983) Extracting information from spectrophotometric curves. Predicting chemical composition from visible and near-infrared spectra. Proc. IUFoST Symposium "Food Research and Data Analysis" September 1982, Oslo Norway (Martens and Russwurm, eds.) Applied Science Publ., pp. 95-113.

(9) Martens, H., Jensen, S.Å. and Geladi, P. (1983) Multivariate linearity transformation for near-infrared reflectance spectrometry. Proc. Nordic Symp. on Applied Statistics. Stokkand Forlag Publ., Stavanger, Norway, pp. 205-234.

(10) Martens, H., Vangen, O. and Sandberg, E. (1983) Multivariate calibration of an X-ray computer tomograph by smoothed PLS regression. Proc. Nordic Symp. on Applied Statistics. Stokkand Forlag Publ., Stavanger, Norway, pp. 235-268.

(11) Jensen, S.Å., Munk, L. and Martens, H. (1982) The botanical constituents of wheat and wheat milling fractions. I. Quantification by autofluorescence. Cereal Chemistry 59, pp. 477-484.

(12) Martens, M., Lea, P. and Martens, H. (1983) Predicting human response to food quality by analytical measurements. The PLS regression method. Proc. Nordic Symp. on Applied Statistics. Stokkand Forlag Publ., Stavanger, Norway, pp. 185-203.

AUTOCORRELATION AND TIME SERIES ANALYSIS

Dr. Ir. H.C. Smit

University of Amsterdam, Amsterdam, The Netherlands

INTRODUCTION

In several disciplines time series analysis is of increasing importance. It is used (1) in a number of applications:

- -Optimal forecast, i.e. the estimation of future values of the known current and past values of the series up to the present time.
- -Parameter estimation, i.e. the estimation of system parameters from time series (signals) generated during a measurement procedure.
- -Transfer function estimation. A transfer function typifies the inertial characteristics of a linear system.
- -Information extraction, i.e. the extraction of relevant information from time series containing much more but not relevant information. The separation of signal and noise (noise reduction, filtering, signal estimation) belongs to this category.
- -Optimal control. A time series of (analytical) results can be used for optimum process control.

In this paper a short overview of some basic theory of autocorrelation and time series will be given, concluding with an example of the application in a practical software package.

A time series (TS) can be deterministic, i.e. exactly determined by a known function, but of course the study of stochastic time series is in general more interesting and most theory is directed to stochastic, non-deterministic TS. In *discrete* TS the observations are made at some (fixed) interval. If the set of observations is continuous, the TS is said to be *continuous*. Discrete TS may arise by sampling a continuous TS

B. R. Kowalski (ed.), Chemometrics. Mathematics and Statistics in Chemistry, 157–176.

and by accumulating a variable over a period of time. An example is the yield of a batch process.

TIME SERIES ANALYSIS

Time Series analysis of non-deterministic TS can be done in the time domain and in the frequency domain. In the time domain the stochastic process, from which the observed TS is regarded as being generated, can be described by probability distributions with all possible sets of data points. An alternative approach is the frequency domain description, where the stochastic process is decomposed into a continuous range of frequencies (spectrum).

In many problems it is necessary to construct a model for a stochastic TS. To fit the parametric models it is only necessary to estimate a small set of parameters from the data. An example is the moving average-autoregressive model. An extensive treatment of this kind of models is given in (1).

The description of a time series by the autocorrelation function or the spectrum is non-parametric or better multiparametric. An infinite number of parameters is required to specify the process. In this paper the attention will be mainly paid to the theory of continuous time series and non-parametric models.

Examples of applications of TS analysis in analytical chemistry are the specification of the baseline noise occurring in several analytical techniques producing "dynamic" signals (peaks in chromatography, AAS, etc.), the development and calculation of (optimum) filters and the calculation of the precision of analytical methods. Furthermore, TS analysis can be used to develop (soft-ware) "tools" for testing data processing procedures. An example of this application will be given.

RANDOM TIME FUNCTIONS

Four main types of statistical functions are used to describe the basic properties of random data:

- -mean square value
- -probability density functions (PDF)
- -autocorrelation functions (ACF)
- -power spectral densities (PSD).

The most simple description of random data (2) in rudimentary terms is given by the mean and the mean square value, respectively, given by

$$\mu_x = \lim_{T\to\infty} \frac{1}{T} \int_0^T x(t)\, dt \qquad x(t) = \text{random function} \qquad (1)$$

and

$$\psi_x^2 = \lim_{T\to\infty} \frac{1}{T} \int_0^T x^2(t)\ dt \qquad \psi_x = \text{root mean square (RMS) value} \tag{2}$$

The variance, i.e. the mean square value about the mean is given by

$$\sigma_x^2 = \lim_{T\to\infty} \frac{1}{T} \int_0^T (x(t) - \mu_x)^2\ dt \tag{3}$$

The standard deviation σ_x gives the "intensity" of the random data.

The probability density function (PDF) describes the probability that the data will assume a value within some defined range at any instant of time (Fig. 1). The amplitude probability is defined by

$$\text{Prob}(x,\ x + \Delta x) = \text{Prob}(x < x(t) \leq x + \Delta x) = \lim_{T\to\infty} \frac{\Sigma \Delta t_n}{T} \tag{4}$$

If Δx approaches to zero, the amplitude probability approaches the PDF

$$p(x) = \lim_{\Delta x \to 0} \frac{\text{Prob}(x,\ x + \Delta x)}{\Delta x} \tag{5}$$

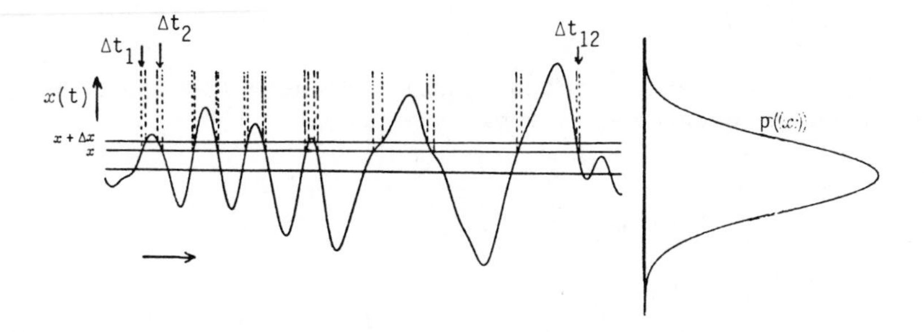

Fig. 1a Determination of a PDF

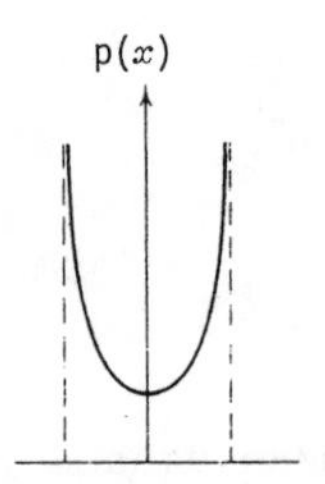

Fig. 1b PDF of a sinus.

The PDF of random noise has often a Gaussian shape. It is possible to describe a deterministic function in probabilistic terms. A random sampling of a sine wave results in the PDF shown in Fig. 1. There is a greater chance of finding the relatively slow top than to find the fast zero crossing.

A random variable $x(t)$ can be described in terms of the PDF $p(x)$. The kth moment is defined by

$$m_k = \int_{-\infty}^{+\infty} x^k \, p(x) \, dx = E\left\{x^k\right\} \tag{6}$$

where $E\{\}$ stands for the expected value.

The first moment is

$$E\left\{x\right\} = \int_{-\infty}^{+\infty} x \, p(x) \, dx = \mu_x \tag{7}$$

The central moments are

$$m_n' = \int_{-\infty}^{+\infty} (x - \mu_x)^n \, p(x) \, dx = E\left\{(x - \mu_x)^n\right\} \tag{8}$$

In case of data from two random processes the common properties can be described by the joint moments

$$m_{kr} = \int_{-\infty}^{+\infty} \int_{-\infty}^{+\infty} x^k y^r \, p(x,y) \, dx \, dy = E\left\{x^k y^r\right\} \tag{9}$$

These processes are uncorrelated if $E\{xy\} = E\{x\} \, E\{y\}$, they are orthogonal if $E\{xy\} = 0$, and they are independent if $p(x,y) = P_x(x) \cdot P_y(y)$.

Figure 2 shows an ensemble of sample functions forming a random process. The mean value (first moment) of the random process at time t_1 can be computed by summing the instantaneous values of each sample function and dividing by the number of sample functions.

In a similar way a correlation (joint moment) between values of the random process at two different times can be determined by computing the ensemble average of the product of instantaneous values at two times, t_1 and $t_1 + \tau$. For the random process the mean value $\mu_x(t_1)$ and the autocorrelation function $R_{xx}(t_1, t_1 + \tau)$ are given by

$$\mu_x(t_1) = \lim_{N\to\infty} \frac{1}{N} \sum_{k=1}^{N} x_k(t_1) \tag{10}$$

and

$$R_{xx}(t_1, t_1 + \tau) = \lim_{N\to\infty} \frac{1}{N} \sum_{k=1}^{N} x_k(t_1)x_k(t_1 + \tau) \tag{11}$$

If $\mu_x(t_1)$ and $R_{xx}(t_1, t_1 + \tau)$ do not vary with t_1, then the process is (weakly) stationary. If all higher moments are independent of t_1, then the process is strongly stationary.

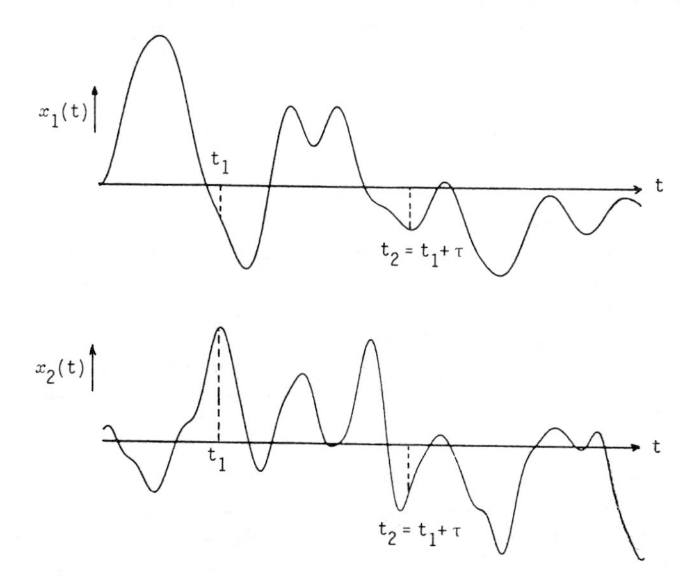

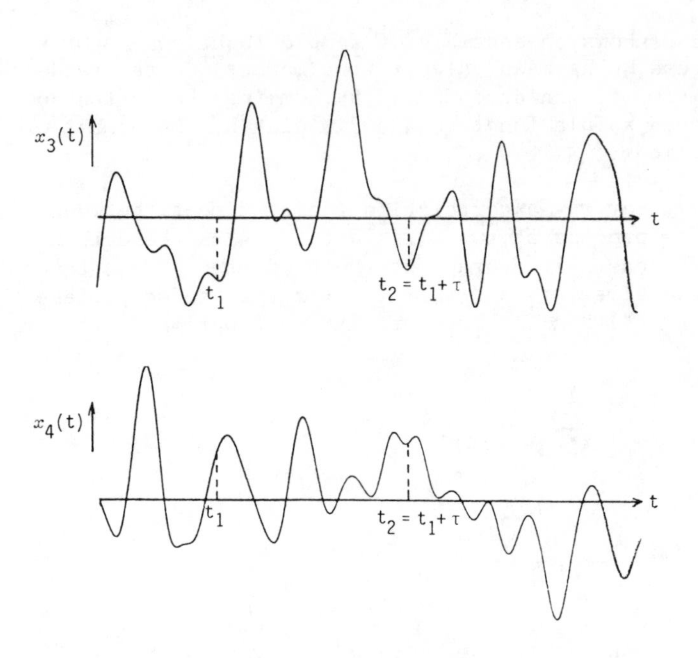

Fig. 2 Random process

For one sample function k the mean value $\mu_x(k)$ is given by eq. (1). The autocorrelation function for the kth sample function is

$$R_{xx}(\tau,k) = \lim_{T\to\infty} \frac{1}{T} \int_0^T x_k(t)\, x_k(t+\tau)\, dt \tag{12}$$

The random process is <u>ergodic</u> if $\mu_x(k)$ and $R_{xx}(\tau,k)$ are the same when computed over different sample functions; they do not differ if computed as a time average or an ensemble average, and all statistics can be determined from a single function $x(k,t)$. $R_{xx}(\tau)$ is always a real valued even function of τ ($R_{xx}(\tau) = R_{xx}(-\tau)$). Furthermore, $R_{xx}(\tau)$ has a maximum at $\tau=0$ ($R_{xx}(0) \geq |R_{xx}(\tau)|$ and $\mu_x \equiv \sqrt{R_{xx}(\infty)}$ (not in special cases like periodic signals).

The ACF is very important to establish the influence of values at any time over values at a future time. Periodic signals also have a periodic ACF; random signals diminish to zero, assuming $\mu_x = 0$. An ACF is a very powerful tool to detect periodic signals masked by noise.

The power spectral density (PSD) of a random function is defined as the Fourier transform of the ACF:

$$S(\omega) = FT\left\{R_{xx}(\tau)\right\} \tag{13}$$

For real process $R_{xx}(\tau)$ is real and even. Therefore $S(\omega)$ is also even $(S(-\omega) = S(\omega))$, and

$(S(-\omega) = S(\omega))$, and

$$S(\omega) = \int_{-\infty}^{+\infty} R_{xx}(\tau) \cos \omega\tau \, d\tau$$

$$= 2\int_{0}^{\infty} R_{xx}(\tau) \cos \omega\tau \, d\tau \tag{14}$$

The physical realisable one-sided power spectral density function $G(\omega)$, where ω only varies over $(0,\infty)$, is defined by

$$G(\omega) = 4\int_{0}^{\infty} R(\tau) \cos \omega\tau \, d\tau \tag{15}$$

RECONSTRUCTION OF TIME SERIES

Linear Mean Square Estimation

The basic problem is to estimate a random variable (r.v.)y by a linear function $ax + b$, where x is another r.v. We have to find two constants a and b that minimize the mean square error:

$$p_e = E\left\{[y - (ax+b)]^2\right\} \tag{16}$$

Assuming for simplicity

$$E\left\{x\right\} = E\left\{y\right\} = 0 \tag{17}$$

then

$$p_e = E\left\{(y - ax)^2\right\} \tag{18}$$

has to be minimized.

The constant a can be found using the orthogonality principle: The constant a is such that y - ax is orthogonal to x:

$$E\left\{(y - ax)x\right\} = 0 \tag{19}$$

The minimum mean square error is given by

$$P_{em} = E\left\{(y - ax)y\right\} \tag{20}$$

The proof will be omitted.

The proof will be omitted.

APPLICATIONS

Prediction

The ACF $R_{xx}(\tau)$ can be used for prediction, given a process s(t) and a constant λ. The problem is to estimate s(t + λ) in terms of s(t):

$$s(t+\lambda) \approx a\, s(t) \tag{21}$$

Applying the orthogonality principle gives

$$E\left\{[s(t+\lambda) - a\, s(t)]s(t)\right\} = 0$$

$$\rightarrow R_{xx}(\lambda) - a\, R_{xx}(0) = 0 \tag{22}$$

hence

$$a = \frac{R_{xx}(\lambda)}{R_{xx}(0)} \tag{23}$$

Filtering

Given a process s(t) (signal) and a process x(t), with

$$x(t) = s(t) + n(t) \tag{24}$$

where n(t) is noise. The problem is to estimate s(t) in terms of $x(t)$ with given t. The value a is determined from

$$E\left\{[s(t) - a\,x(t)]x(t)\right\} = 0$$

$$\rightarrow\ R_{sx}(0) - a\,R_{xx}(0) = 0$$

$$a = \frac{R_{sx}(0)}{R_{xx}(0)} \tag{25}$$

$R_{sx}(\tau)$ is the cross correlation of s(t) and $x(t)$, defined by

$$R_{sx}(\tau) = E\left\{s(t)\,.\,x(t+\tau)\right\} \tag{26}$$

The mean square error is

$$\begin{aligned} p_e &= E\left\{[s(t) - a\,x(t)]s(t)\right\} = \\ &= R_{ss}(0) - \frac{R^2_{sx}(0)}{R_{xx}(0)} \end{aligned} \tag{27}$$

With the assumption that $s(t)$ is orthogonal to $n(t)$, we have

$$\left.\begin{aligned} R_{sx}(\tau) &= R_{ss}(\tau) \\ R_{xx}(\tau) &= R_{ss}(\tau) + R_{nn}(\tau) \\ a &= \frac{R_{ss}(0)}{R_{ss}(0) + R_{nn}(0)} \\ p_e &= \frac{R_{ss}(0)\,.\,R_{nn}(0)}{R_{ss}(0) + R_{nn}(0)} \end{aligned}\right\} \tag{28}$$

Signal integrating (peak area determination) is important in several analytical chemical techniques. Suppose we have a random baseline noise $n(t)$. The problem is the estimation of the integral of $n(t)$ over the interval (0,T) in terms of the values of $n(t)$ at the endpoints of this interval:

$$\int_0^T n(t)\,dt \approx a_0\,n(0) + a_1\,s(T) \tag{29}$$

a_0 and a_1 must be such that

$$E\left\{\left[\int_0^T s(t)\,dt - a_0\,s(0) - a_1\,s(T)\right] s(0)\right\} = 0 \tag{30}$$

and

$$E\left\{\left[\int_0^T s(t)\,dt - a_0\,s(0) - a_1\,s(T)\right] s(T)\right\} = 0 \tag{31}$$

Hence

$$\int_0^T R_{xx}(t)\,dt = a_0\,R_{xx}(0) + a_1\,R_{xx}(T) \tag{32}$$

and

$$\int_0^T R_{xx}(T-t)\,dt = a_0\,R_{xx}(T) + a_1\,R_{xx}(0) \tag{33}$$

Both integrals are equal, so that

$$a_0 = a_1 = \frac{\int_0^T R_{xx}(t)dt}{R_{xx}(0) + R_{xx}(T)} \tag{34}$$

For $T \to 0$ we have $a_0 \approx a_1 \approx T/2$. Thus

$$\int_0^{\Delta T} s(t)\,dt \approx \frac{\Delta T}{2}\,[s(0) + s(T)] \tag{35}$$

the trapzoid approximation.

OPTIMUM ESTIMATOR

Suppose the data are available over an entire interval. Again two processes $s(t)$ and $x(t)$ are given (see eq. 24), where $x(t)$ is known. The problem is to estimate $s(t)$ for a specific t by a linear combination of the known values of $x(t)$ in the interval (a,b). Suitable weights $h(\xi)$ have to be found, such that with

$$s(t) \approx \int_a^b h(\xi)\, x(\xi)\, d\xi \tag{36}$$

the error

$$p_e = E\left\{\left[s(t) - \int_a^b h(\xi)\, x(\xi)\, d\xi\right]^2\right\} \tag{37}$$

is minimal. According to the orthogonality principle we have

$$E\left\{\left[s(t) - \int_a^b x(\alpha)\, h(\alpha)\, d\alpha\right] x(\xi)\right\} = 0 \qquad a \le \xi \le b \tag{38}$$

However

$$E\left\{s(t)\, x(\xi)\right\} = R_{sx}(t - \xi) \tag{39}$$

and

$$E\left\{x(\alpha)\, x(\xi)\right\} = R_{xx}(\alpha - \xi) \tag{40}$$

The result is

$$R_{sx}(t - \xi) = \int_a^b R_{xx}(\alpha - \xi)\, h(\alpha)\, d\alpha \tag{41}$$

The weights $h(\xi)$ have to satisfy eq. (41). This integral has to be solved. In general, the solution is difficult, but a numerical solution is always possible.

OPTIMUM RECURSIVE ESTIMATOR

The recursive optimum filter procedure is of growing importance in several disciplines including analytical chemistry. A short introduction with discrete signals will be given below [3]. In case of *optimum* estimation both the model of the process to be estimated and the *observation model*, i.e. the properties of the measurement system, have to be known.

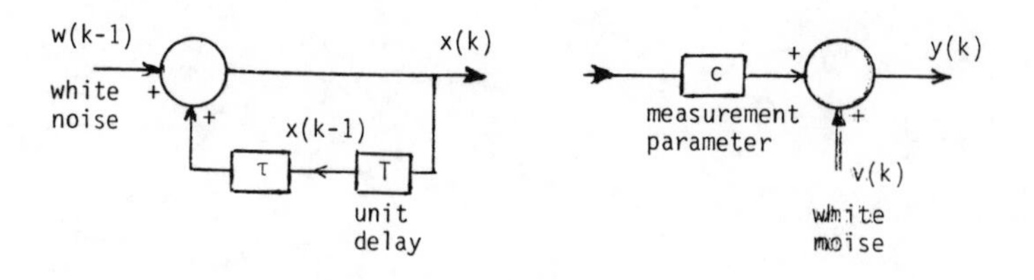

Fig. 3 Signal process model Observation model

Figure 3 shows an example of possible models. In this particular case the signal process model can be considered as an auto-regressive process of the first order. The system parameter τ plays the role of a time constant of the process. The measurement noise is assumed to be white noise.

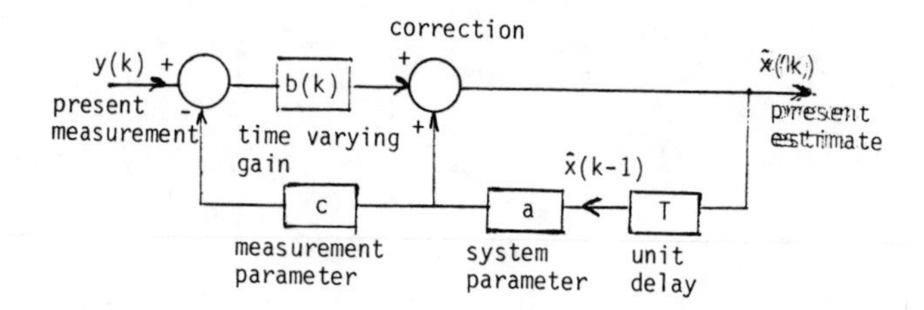

Fig. 4 Discrete Kalman filter

Figure 4 shows an optimum recursive estimator algorithm (Kalman filter). The estimation $\hat{x}(k)$ is composed of the previous estimate $\hat{x}(k-1)$ amplified by a system parameter a and correction derived from the present measurement y(k). c is an observation parameter. The estimation of x is

$$\hat{x}(k) = a\,\hat{x}(k-1) + b(k)\left[y(k) - a\,c\,\hat{x}(k-1)\right] \qquad (42)$$

The basic principle is that the mean square error p_k is minimized with respect to the time varying factor b_k. The mean square error is

$$p(k) = E\left\{e^2(k)\right\} \qquad (43)$$

and the error is

$$e(k) = \hat{x}(k) - x(k) \qquad (44)$$

Substituting eq. (42) in eq. (44) and eq. (43) and differentiating with respect to a(k) and b(k) gives the so-called orthogonality equations. The optimum values for b(k) can be found when the signal model is specified. For more details see [3].

In our example,

$$b(k) = c\, p_1(k)\, [c^2 p_1(k) + \sigma_v^2]^{-1} \tag{45}$$

where

$$p_1(k) = a^2 p(k-1) + \sigma_w^2 \tag{46}$$

The mean square error is

$$p(k) = p_1(k) - c\, b(k)\, p_1(k) \tag{47}$$

The algorithm can be extended to a vector Kalman filter, dealing with multidimensional signals.

THE GENERATION OF NOISE WITH PRE-DETERMINED CHARACTERISTICS

The theory on autocorrelation and time series analysis, treated in this paper, can be used in analytical chemistry for several purposes. The noise characteristics of detectors can be specified, (optimum) filters can be developed and the influence of noise with respect to the uncertainty in the analytical results can be determined. Another example is the use of the theory for the generation of noise with pre-determined characteristics. A realistic stationary detector noise can be generated and used for testing peak finding procedures, data processing, information retrieval methods, simulation and automation. A short description will be given; more details can be found in [4].

We start with white noise $x(t)$ with the statistical properties

$$E\left\{x(t)\right\} = 0$$

and

$$E\left\{x(t_1)\, x(t_2)\right\} = E\left\{x(t)\, x(t+\tau)\right\} = R_{xx}(\tau) = \delta(\tau) \tag{48}$$

The ACF of white noise is a delta Dirac distribution. Real white noise is not realistic, but a set of independent numbers representing a "discrete" white noise, which can easily be generated. We may assume that any stationary stochastic signal can be considered as originating from white noise, which has passed a shaping filter. A filter procedure can be described by a convolution integral

$$y(t) = \int_{-\infty}^{t} h(\tau)\, x(t-\tau)\, d\tau \tag{49}$$

where h(t) is the impulse response of the filter (Fig. 5), $x(t)$ can be considered to be built up from pulses, and $h(\tau).\ x(t-\tau)$ is the response on such an impulse at the time $\tau = t$ in the past. $y(t)$ is the sum (integral) of the impulse response caused by $x(t)$.

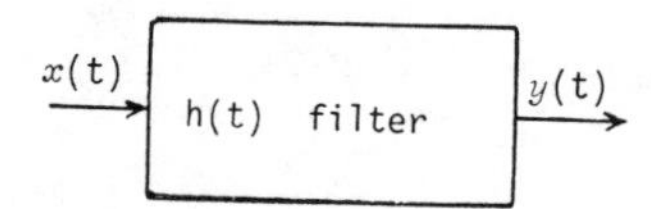

Fig. 5 Filter with impulse response h(t)

The expected value of $y(t)$ is

$$E\left\{y(t)\right\} = \int_{\infty}^{t} h(\tau)\, E\left\{x(t-\tau)\right\} d\tau = 0 \tag{50}$$

Now we can determine the autocorrelation function of $y(t)$:

$$\begin{aligned} R_{yy}(\tau) &= E\{y(t)y(t+\tau)\} \\ &= E\left\{\int_0^\infty h(\tau_1)x(t-\tau_1)d\tau_1 \int_0^\infty h(\tau_2)x(t+\tau-\tau_2)d\tau_2\right\} \\ &= \int_0^\infty\int_0^\infty h(\tau_1)h(\tau_2)\ E\{x(t-\tau_1)x(t+\tau-\tau_2)\}d\tau_1\, d\tau_2 \\ &= \int_0^\infty\int_0^\infty h(\tau_1)h(\tau_2)\ R_{xx}(\tau+\tau_1-\tau_2)d\tau_1\, d\tau_2 \end{aligned} \tag{51}$$

If R_{xx} is a δ-function, then we have

$$R_{yy}(\tau) = \int_0^\infty\int_0^\infty h(\tau_1)h(\tau_2)\delta(\tau - \tau_1 + \tau_2)d\tau_1\, d\tau_2$$

$$= \int_0^\infty h(\tau_1)h(\tau + \tau_2)d\tau_2 \qquad (52)$$

This is the system correlation function. $R_{yy}(\tau)$ is an even function of τ, hence $h(\tau)$ will be even too.

Fourier transform gives

$$FT[R_{yy}(\tau)] = H^2(\omega) \qquad (53)$$

The impulse response can easily be calculated

$$H(\omega) = \sqrt{FT[R_{yy}(\tau)]} \qquad (54)$$

$$h(\tau) = FT^{-1}\left\{\sqrt{FT[R_{yy}(\tau)]}\right\} \qquad (55)$$

Thus, the impulse response of the shaping filter with the desired characteristics can be found by Fourier transformation of the desired ACF, taking the square root and Fourier back-transformation. This procedure can be performed by a computer in a discrete form, where the impulse response h(t) is replaced by a discrete weighting function h(k).

SAMSON

A software package (SAMSON), based on the given theory and capable of generating noise with widely diverging spectral properties is developed [4]. Four standard types of noise can be generated, namely

1) first order noise $\quad R(\tau) = \sigma^2 e^{-|\tau|/T_x}$

2) Gaussian noise $\quad R(\tau) = \sigma^2 e^{-\tau^2/T_x^2}$

3) damped cosine first order noise $\quad R(\tau) = \sigma^2 e^{-|\tau|/T_x} \cos \omega_0 \tau$

4) 1/f noise $\quad R(\tau) = \sigma^2 \dfrac{E_1\left(\frac{\tau}{t_n}\right) - E_1\left(\frac{\tau}{T_m}\right)}{\ln t_n/t_m}$

$E_1(\xi)$ is an exponential integral defined by

$$E_1(\xi) = \int_{\xi}^{\infty} \frac{e^{-s}}{s}\, ds$$

Furthermore, the input to the program can be a measured or arbitrarily chosen autocorrelation function (or power spectrum).

Type 1, first order noise, can originate from shaping filters (or processes) described by a first order differential equation (low pass filters). Many amplifiers can be considered as a first order system. Gaussian filters cause the second noise type. The third type is originating from a bandpass filter as used in lock-in amplifiers. Generally it can originate from filters described by negative discriminant second order differential equations with very high $\omega_0 T_x$, allowing neglection of the sine term in the solution. The fourth type, 1/f noise, is observed in many detectors and other noise sources. In all cases, the described noise types are observed as the output of the shaping filter, if white noise is applied as input.

To determine whether the ACVF of the noise generated by SAMSON fits the model used as input ACVF, the confidence intervals for the input ACVF and for the given number of noise data generated have to be estimated. This can be done using the Bartlett formula [5]:

$$\sigma^2 \left[R_{xx}(\tau)\right] = \frac{1}{(T-\tau)^2} \int_{-T+\tau}^{T-\tau} (T-|r|-\tau)\left\{R_{xx}^2(r) + R_{xx}(r-\tau)R_{xx}(r+\tau)\right\} dr \qquad (56)$$

In SAMSON optionally the confidence intervals (99.72%) are depicted: the ACVF of the generated noise should be within the interval $R_{xx}(\tau) \pm 3\sigma\left[R_{xx}(\tau)\right]$, where $R_{xx}(\tau)$ is the input ACVF.

Figures 6, 7 and 8 show some typical examples of generated noise and corresponding ACF and PSD. SAMSON has proven to be a useful tool for many purposes, in particular in the testing of data processing and time series reconstruction procedures. It is an example of how the theory of autocorrelation and time series analysis can be applied.

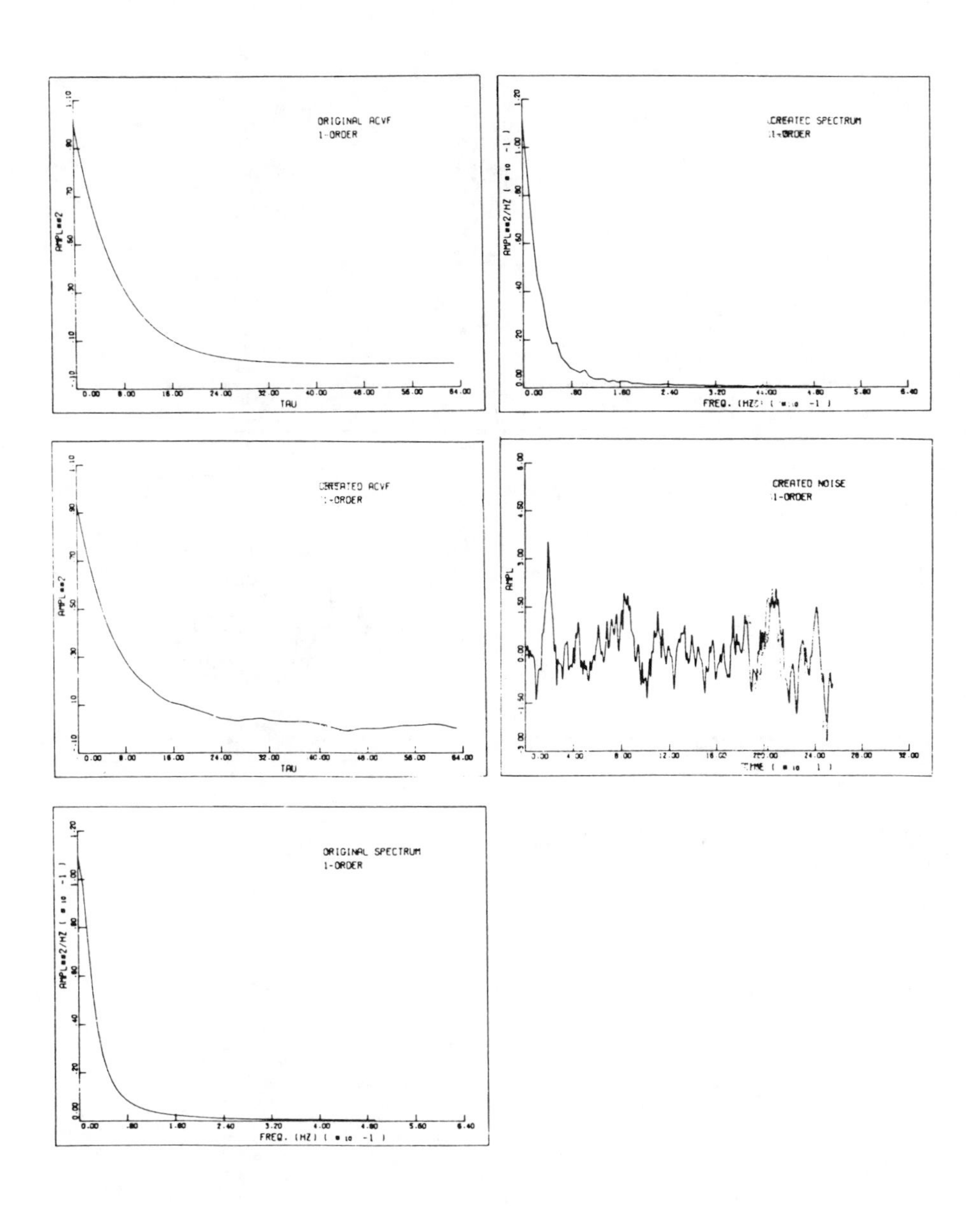

Fig.6 First order noise.

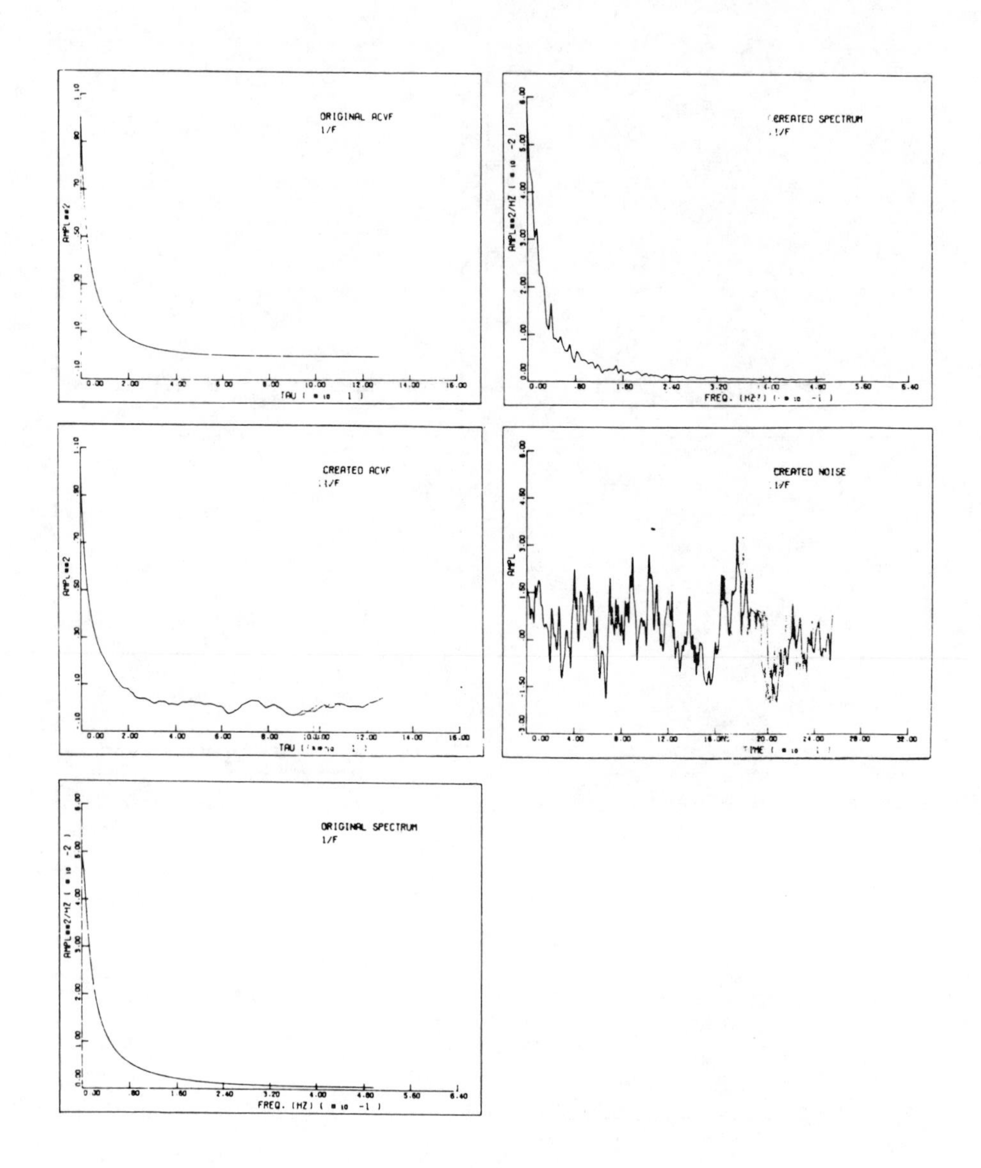

Fig.7 1/f noise.

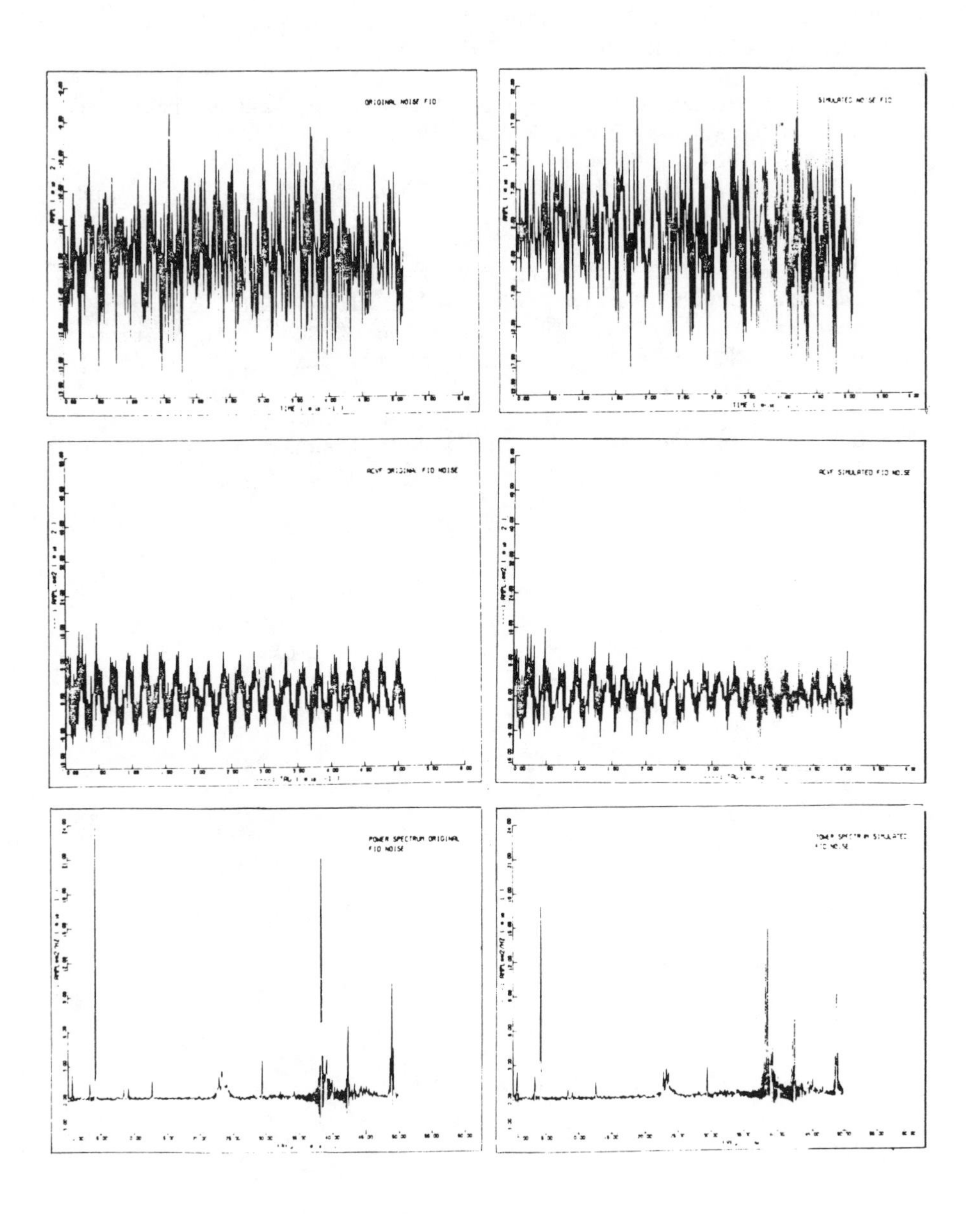

Fig.8 Simulation of flame ionisation detector noise.

REFERENCES

1) G.E.P. Box and G.M. Jenkins, "Time Series Analysis," San Francisco, Holden-Day, 1976.

2) J.S. Bendat and A.G. Piersol, "Measurement and Analysis of Random Data," New York, Wiley, 1968.

3) S.M. Bozic, "Digital and Kalman Filtering," London, Edward Arnold, 1979.

4) J.M. Laeven, H.C. Smit and J.V. Lankelma, "A Software Package for the Generation of Noise with Widely Diverging Spectral Properties. The Simulation of Realistic Stationary Detector Noise in Analytical Chemistry," Anal. Chim. Acta, to be published.

5) G.M. Jenkins and W.G. Watts, "Spectral Analysis and Its Applications," San Francisco, Holden-Day, 1969.

SAMPLING

Prof. drs. G Kateman
tel. 080-558833, ext. 3173
Department of Analytical Chemistry
University of Nijmegen
Toernooiveld
6525 ED NIJMEGEN, The Netherlands

Introduction

Chemometrics is a discipline within Chemistry, mainly within analytical chemistry, that can be described as "the software part of the tools for measuring in analytical chemistry".

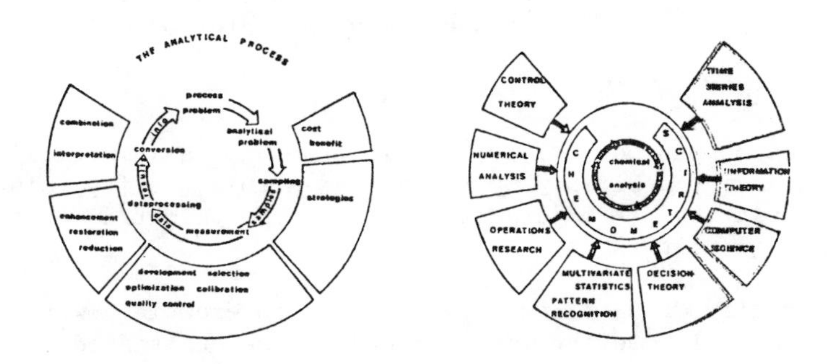

Figure 1a

Figure 1b

These tools can be used to develop formal strategies for the analytical tasks; they can be used to develop decision models and they can be used in the processing of analytical data (KA 81, MA 78, VN 83, Figure 1).

B. R. Kowalski (ed.), Chemometrics. Mathematics and Statistics in Chemistry, 177–203.

In the development of analytical strategies, chemometrics plays a crucial role. A strategy is an ideal plan to optimally reach a defined goal. The success of a strategy depends on the degree of completion of the plan. One of the yardsticks to

Laboratory strategies

Sampling for	Description Monitoring Control
Method selection by	Linear programming Decision trees Pattern Recognition
Routing by	Planning Experimental design Sequential optimisation Queueing theory Simulation

Figure 2a

Selection strategy:	which procedure produces maximal information?
Combinatory strategy:	which combination of procedures produces maximal information?
Parameter strategy:	which combination of parameter values produces maximal information?

Figure 2b

measure success can be quality, quality being the (numerical) value of a set of desired properties. As a rule quality is a composite quantity. A major difficulty is the weighting of the compounding quantities.

The quality of sampling (Figure 2) is an example how the property "optimal relevant information" can be quantified in an objective way, weighting the influences of sampling frequency, sample size, accuracy and speed of the analysis and capacity of the laboratory. In this lecture the role of chemometrics in the quality assessment of sampling, as a measure of the success of a sampling strategy, will be shown.

Quality

Quality in general is an important aspect of analytical chemistry. The results of the analytical chemist are used for (often important) decisions and must be trustworthy. This can be so only by a good standard of quality. The governments of most countries are interested in the quality of analytical results concerning drugs, clinical chemistry, the environment, or trade. Rules like Good Laboratory Practice or Quality Assessment Programs show this concern.

Quality control in chemical analysis is an old concept. As long as there has been chemical analysis, there has been the need to control the quality of its performance. There are many techniques that can be applied to improve the quality of analytical methods, sampling, and data handling. Often the results of these techniques can be used as a criterion for quality. The management of a laboratory has a need to estimate the quality of its efforts and, if this quality does not meet the standards, to improve it. Quality can nearly always be improved, but at the expense of much ingenuity, work, and money. It would be desirable if some criterion could be applied that allows the optimization of quality; that is, the sum of the yield of the object of analysis and the negative yield (cost) of the analysis itself should be maximal. This yield may be expressed in terms of money, but, though money is a good and dependable yardstick, it is not the only one.

Efforts in human ingenuity, although these are difficult to measure, could be the parameter of optimization. The amount of manpower or the level of sophistication of the instruments required can serve as a measure of ingenuity.

At this time there is no universal standard in operation. Until recently, accuracy and precision were the only quality parameters in use. These could not be optimized since it was impossible to estimate the yield and the cost corresponding to the various levels of accuracy, for example.

The first applications of quality control date back to the beginning of the century. In 1908, Gosset (GO 08), an amateur of statistics and a professional chemist, published a paper on the error of a mean under his pen name, "Student". He introduced a parameter known since then as "Student's-t". It allows analytical chemists (and others) to compare quantitatively the results of two items such as methods of analysis or samples. Here the quality parameter in use is the level of probability that the two means match.

Another statistical parameter that could be used is the standard deviation. This quality parameter combined with Fisher's "F-test" indicated the probability that the scatter in the results of a particular experiment is better or worse than the scatter found in a very reproducible standard series of analytical results. Although these quality parameters were known they were not in common use among analytical chemists.

In 1928 Baule and Benedetti-Pichler (BA 28) published a paper on the minimum size of samples of inhomogeneous lots, in which they indicated the quality parameters for sampling.

The first major impact on quality control in analytical chemistry was caused by the introduction of the control chart, familiarized by Shewhart in 1931 (SH 31). Such a chart allows the continuous supervision of the quality parameters accuracy and precision in the analytical laboratory. It enables corrective measures when a control line is crossed. A serious drawback of this method is the position of the control line that is arbitrarily chosen.

After 1940 a host of publications appeared on the application of statistical techniques in analytical chemistry. However, most of these were directed to the application of statistics in the validation of analytical results. Its main purpose is the determination of the probability that certain effects in objects or processes can be detected by analytical chemical results. In analytical chemistry the application of statistical techniques for the control of analytical procedures was restricted to the comparison of analytical procedures, without establishing more quality parameters for analytical methods than standard deviation and variance. (See, for example, AL 69, BE 75, DA 72, DI 69, DU 74, NA 63, FI 55, and YO 51.) In 1963 Van der Grinten (GR 63, GR 65, GR 66) published some simplified equations that were derived from the rules of Wiener on control theory. This allowed the quantitative evaluation of the quality of measurements with respect to the object that was measured. In the years that followed Van der Grinten togther with Kateman explored the possibilities of the quality parameter "measurability" at the chemical works of DSM in the Netherlands. To each component in a chemical process that was sampled and analyzed one could attribute a quality number when some parameters of the process (time constant and standard deviation of the uncontrolled process) were known. In 1971 Leemans (LE 71) published some of the results in the analytical literature. It became clear that, apart from accuracy and reproducibility, and also the speed of analysis, the frequency of sampling and their merit for the application of the results could be quantified and optimized.

These developments were not the only ones that caused a new view of quality control in chemical analysis. Around the year 1970 a discussion of the necessity of analytical chemistry arose - mainly in the American analytical literature - that culminated in the rhetorical question "Analytical chemistry, a fading discipline?" (FI 70). The discussion was pursued in Europe and resulted in a series of definitions. These were formulated by Kateman and Dijkstra (KA 79) within three categories of definitions:

1. Analytical chemistry* produces information by application of available analytical procedures in order to characterize matter by its chemical composition.

2. Analytical chemistry* studies the process of gathering information by using principles of several disciplines in order to characterize matter or systems.

3. Analytical chemistry* produces strategies for obtaining information by the optimal use of available procedures in order to characterize matter or systems.

In Europe the Arbeitskreis "Automation in der Analyse" (AR 73) formulated the view that analytical chemistry should be considered as a "black box", a system with known input and output and rules interconnecting these. According to this point of view, operations research, the science that studies this kind of system, could be applied to analytical chemistry. Massart (MA 75, MA 77) presented the first survey of research using this new approach. According to this approach, the application of operations research techniques to analytical chemistry indicates some new quality parameters:

- information as a measure of quality of a method
- waiting time as a measure of planning and routing in analytical chemical laboratories.

A third phenomenon, parallel to the use of control theoretical concepts and operations research, was the introduction in analytical chemistry of integrated systems of statistics and correlation analysis, often grouped under the name "pattern recognition" (KO 72). These techniques allow, for instance, the selection of those analytical results that are indispensable and sufficient for describing a certain phenomenon. The quality of analytical chemistry, expressed in relevance of measured data, seems possible now. Moreover, it might be possible to select an analytical method that, given the properties of the problem under investigation, will give an optimal solution of the problem (VA 77, JA 81, JN 81, JN 83).

The methods mentioned here can be grouped under the heading "chemometrics" (KO 75). This means that many of the techniques that are covered by chemometrics can be used in the validation of analytical actions, procedures, and organizations, that is, to establish quality in the broadest sense of the word.

The development in clinical chemistry was parallel and apparently independent. The number of analyses in this field

* The degree of optimization of a procedure, if quantitative optimization methods are used.

doubles each five years; thus clinical chemists are confronted with the need to cope with this problem. Here the first line of attack was automation, and as is clearly visible when one visits a clinical laboratory, the progress in this field is great. Clinical chemists became aware of the quality problem quite early, even before automation was begun, when confronted with the enormous differences in clinical chemical data between various laboratories. Clinical chemists came to recognize that their laboratories are in many ways similar to a manufacturing plant: reagents and samples are received and subjected to a variety of manipulations using specialized tools and instruments. The final product is the analytical data that are produced.

In 1947 Belk and Sunderman (BE 47) introduced the Shewhart chart as a means to control the product. The conclusions of Belk and Sunderman based on their pioneer survey were later confirmed by the results of similar studies (TO 63, DE 64, ST 65, BE 67). All revealed a wide range of reported values for the analysis of identical specimens among a group of laboratories. In 1950 a formal intralaboratory quality control system, based on estimation of accuracy and precision, was first described by Levey and Jenning (LE 50). Along this line many nationwide, interlaboratory control systems have been established, many under governmental supervision, others under the supervision of boards founded by clinical chemists. However, at present there is no system available that measures quality unambiguously and creates a uniform quality of clinical chemical measurements. The vast problems caused by the increasing work load also gave rise to some research on information content, cost, and waiting times (VA 74). This research seems to be independent from the analytical operations research trend described earlier.

Quality is the (numerical) value of a set of desired properties. In order to control the quality of analytical chemical methods and results, one must define the set of desired properties of the analysis as well as the way to quantify these and the standards to be set.

For practical reasons it is sufficient to consider only those properties that can be quantified and for which standards can be set. Of course it might be possible that more properties are desired, but if they cannot be quantified, it is useless to mention them. The process of measuring and varying analytical quality can be compared with the control of a technological process. In principle the same procedures can be applied and many of the statistical and control theoretical procedures described elsewhere could be of use. However, an analytical laboratory is far more difficult to control than a chemical process.

One of the criteria that seems to fulfill the requirements of measure of quality is information. If we adapt a definition of analytical chemistry that fits best the aforementioned categories, the definition formulated by Gottschalk (GO 72), we read: "Analytical chemistry produces optimal strategies for obtaining and validation of relevant information in order to characterize states and processes in matter".

It seems appropriate to focus on the words optimal, relevant and information to get an idea of how to formulate quality. Optimal is connected with planning, decisions and efficiency; relevancy is the quotient of required and produced information.

Information

Information can be defined as the decrease in uncertainty. Shannon (SH 49) defined information quantitatively:

$$I = H_{before} - H_{after} \tag{1}$$

with the entropy

$$H = \sum_{i=1}^{n} P_i \, \ell n \, P_i \tag{2}$$

Here, ℓn indicates the natural logarithm, eq to the tran 2 and P_i is the probability of occurence of the event. In this way the entropy of a histogram can be quantified as:

$$H = -\Sigma [P(x_i)\Delta x] \ell n [P(x_i)\Delta x] \tag{3a}$$

or

$$H = -\int_{-\infty}^{\infty} P(x) [\ell n P(x)] dx (-\text{constant}) \tag{3b}$$

The entropy of the normal (Gaussian) distribution is (EC 71, EC 82)

$$H = \tfrac{1}{2} \ell n (2\sigma^2 \pi e) \tag{4}$$

If it is known a priori that the concentration range of an unknown component can be found between c_o and c_e (say between 0 and 100%) and if it can be ascertained by analysis that the concentration in fact is between γ_o and γ_e the obtained information is:

$$I = \ell n(|c_e - c_o|) - \ell n(|\gamma_e - \gamma_o|) = \ell n \frac{(|c_e - c_o|)}{(|\gamma_e - \gamma_o|)} \tag{5a}$$

The unit of information is the bit, ld2.

When normal distributions are assumed, e.g. the results of analysis

$$I = \ln(\sigma_b/\sigma_a) \text{ or generally } I = \tfrac{1}{2}\ln\{|cov_b|/|cov_a|\} \quad (5b)$$

For rectular/normal distributions before and after measurement

$$I = \ln\{\frac{(|c_e - c_o|)}{\sigma\sqrt{2\pi e}}\} \text{ or } I = \ln\{\frac{|c_e - c_o| n^{\frac{1}{2}}}{2ts}\} \quad (5c)$$

The problem of formulating information unambiguously is complicated, however, when relevancy is introduced. This can be illustrated with "Couffignal's paradox". A telegram is sent, containing n bits of information. Now one binary digit is added with the following meaning:

0: all wrong; pay no attention to the telegram

1: telegram is all right; you can use it.

Now the telegram contains n+1 bit of information but the last 0 destroys all the value of the information in the telegram and a last 1 simply adds redundancy. Obviously relevancy has to do with the value of information, and information as such has to do with scarcity. In the analytical laboratory this problem is often encountered but there are ways to meet these difficulties.

Consider an analytical laboratory and its environment (KA 79), the purpose of the analysis, the object of the analysis and the analytical procedures (Figure 3).

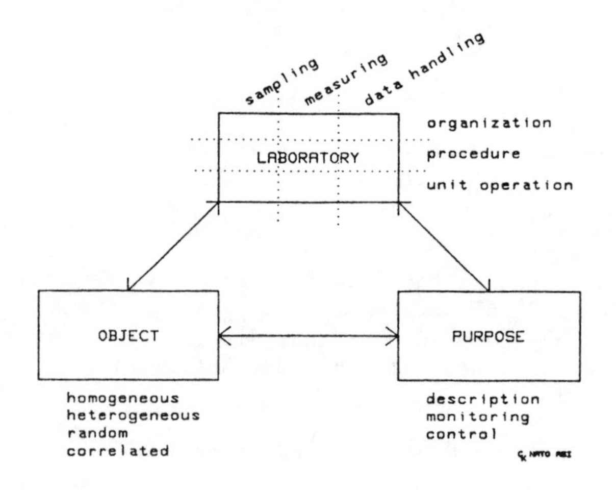

Figure 3

The need to analyze objects arises from lack of information, but as a rule much information about the object can be obtained without analyzing. The purpose defines the kind of information. Relevancy, efficiency and amount of information are measures of quality here. Sampling, as one of the aspects of analysis can be used to demonstrate these aspects.

Sampling

The quality of the sampling method is given by the similarity between the reconstruction of the composition of an object and the object itself, as far as it is influenced by the sampling strategy. It depends on the characteristics of the object and the purpose of the reconstruction. For a mere description of the object the quality can be expressed in the sample quality, that is in uncertainties concerning the means of results of samples and total object, or information.

For threshold monitoring of the object, e.g., warning for a dangerous concentration, the sampling quality can be expressed in the probability that a threshold crossing will be detected, or the information required to keep that probability at a certain level.

For object control or real time process reconstruction the quality can be expressed as measurability or "dynamic information". One type is the internally correlated object (GY 79, MS 78, KA 78).

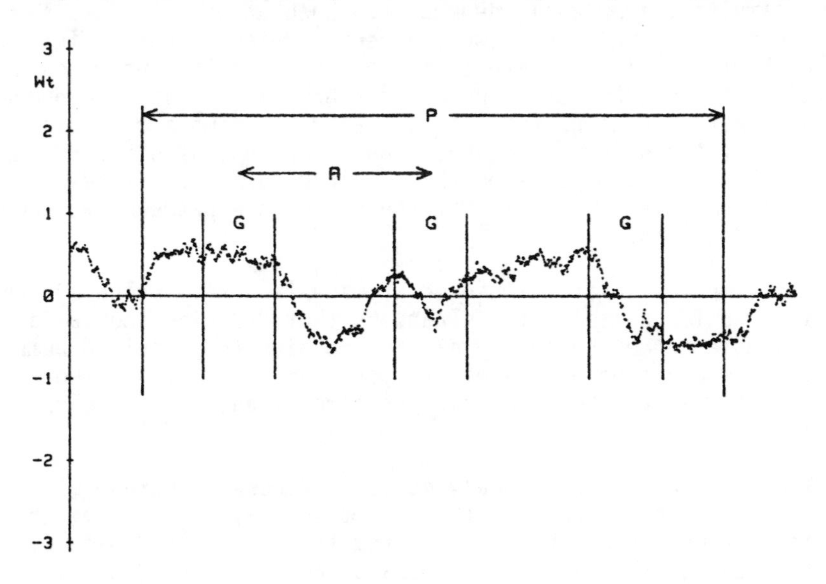

Figure 4

Since the sampling procedure affects the sample, the so-called sampling parameters that define this procedure must be considered. The object or process is represented by the distance 0-t. It is an ergodic time series which means a process that can be represented mathematically by a time series, consecutive numbers representing, say, composition as a function of time, of such a length that the properties of the time series - mean standard deviation, and autocorrelation - are independent of the length. In this series a lot can be represented by the distance P, a non-ergodic time series (Figure 4).

A sample is the sum of n sample increments or grabs, taken during a time span G with a frequency 1/A. In the rest of this paper the influence of the parameters, G, A, and n on the estimation of the composition of the lot P or the process is considered for various types of lots.

Objects can be internally correlated in time or space; for example the composition of a fluid emerging from a tank does not show random fluctuations, but is correlated to the composition in earlier or later sampled product.

Factors causing the internal correlation of object include 1) diffusion or mixing within the object, for example, in mixing tanks, buffer hoppers, and rivers and 2) varying properties of the producer of the object, reactors or emitters, for instance. In both situations samples are mutually dependent. When an object shows a large internal correlation, two adjacent samples do not differ much from each other. The difference between two samples increases with greater distance, however. One may ask how to estimate the number of grabs that have to be taken to get a "real" sample or gross sample. As stated for the case of the heterogeneous objects, the number of grabs, n, depends on the required variance of the sample, this variance being so small that the difference between two samples cannot be detected.

The sample size is influenced by many factors, including the lot size. Unfortunately, the equations that describe the variance of the sample cannot be simply altered to give the required number of grabs, but must be estimated by iterative methods. Another method is to derive them from a graphic representation of the equations.

When the object to be analyzed is a process, a stream of material of infinite length with properties varying in time, the sampling parameters can be derived from the process parameters. When the object is a finite part of a process, however, usually called a lot, the description of the real composition of the lot depends not only on the parameters of the process the lot is derived from, but also on the length of the lot. The sample now must represent the lot, not the process.

Lots derived from Gaussian, stationary, stochastic processes of the first order allow a theoretical approach. In practice most lots seem to fulfill the above requirements with sufficient accuracy to justify the following equations (MS 78, KA 78). An estimate of the mean m of a lot with size P can be obtained by taking n samples of size G, equally spaced with a distance A. In this case P = nA and the size of the gross sample S = nG. Here it is assumed that it is not permissible to have overlapping samples and that A > G. The relevant parameter is the variance in the composition of the gross sample compared to the variance of the lot. This variance σ_*^2 is thought of as being composed of the variance in the composition of the gross sample itself σ_m^2, the variance in the composition of the whole lot σ_μ^2, and the covariance between m and μ:

$$\sigma_*^2 = \sigma_m^2 + \sigma_\mu^2 - 2\sigma_{m\mu} \tag{6}$$

as shown in MU 78. The terms in eq. (6) are given by

$$\sigma_m^2 = \frac{2\sigma_x^2}{ng}\{g-1+\exp(-g)+[\exp(-g)+\exp(g)-2]x$$

$$\cdot\ x(\frac{\exp(-a)}{1-\exp(-a)} - \frac{\exp(-a)[1-\exp(-p)]}{n\ 1-\exp(-a)^2})\} \tag{7}$$

$$\sigma_\mu^2 = \frac{2\sigma^2}{p^2}[p-1+\exp(-p)] \tag{8}$$

$$\sigma_{m\mu}^2 = \frac{\sigma_x^2}{npg}\{2ng+[1-\exp(-p)](\frac{\exp(-g)-1}{1-\exp(-a)} + \frac{\exp(g)-1}{1-\exp(a)})\} \tag{9}$$

where σ_x^2 = variance of process

$p = P/T_x \quad g = G/T_x \quad a = A/T_x$

T_x = correlation factor of the process

As can be seen inequations 6-9, σ_*^2 depends on a number of factors. The properties of the process from which the lot stems are described by σ_x, the standard deviation of the process, and T_x, the correlation factor of the process. The only relevant property of the lot here is its size p, expressed in the same units as T_x. These units may be time units such as when T_x is measured in hours. In this case T_x is usually called the time constant of the process. The lot size is expressed in hours as well. When describing a river, the lot size can be the mass of water that flows by in one day or year.

The correlation factor T_x can be called a space constant when the unit is used in length, however. Here the lot size is expressed in length units. This is the case, for example, when a lot of manufactured products contained in a conveyor belt or stored in a pile of material in a warehouse is considered. The correlation factor also can be expressed in dimensionless units as when bags of products are produced. In this case the lot size is also expressed in terms of items (number of bags, drums, tablets). The properties of the sample are the grab size G, expressed in the same units as T_x and P, the distance between the middle of adjacent grabs A (also in the units of T_x, P and G), and the number of grabs n that form the sample. If the sample size is chosen such that the subsequent samples are taken without interruption, it can be shown that $\sigma_*^2 = 0$. This result is not surprising because here the whole lot has been sampled: $A = P/n$, and the sample size $nG = P$. For an uncorrelated lot, with $T_x' = 0$, it can be shown that σ_m, σ_μ, and $\sigma_{m\mu}$ and consequently σ_* are all 0. If, however, the sample size $G = 0$, $\sigma_m^2 = \sigma_x^2/n$ and $\sigma_*^2 = \sigma_x^2/n$.

In other words, if the lot size is very large in time constant units, one sample of finite size suffices, but if the grab size is near zero, a number of samples must be taken (see Figures 5 and 6). Note that here the sample can be obtained from one part of the object (e.g., when G≠0) or from several, equally spaced places (e.g. when G=0).

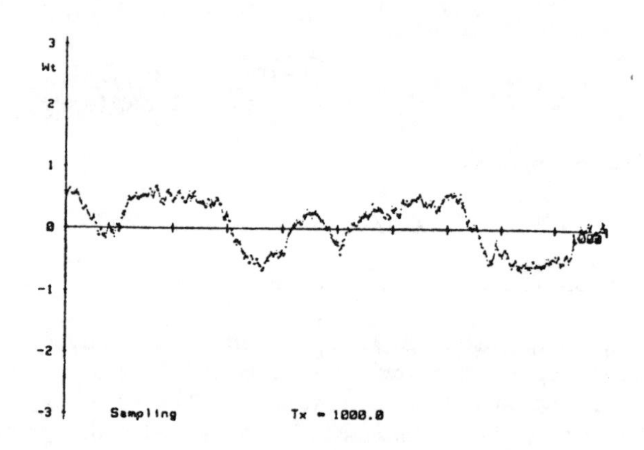

Figure 5

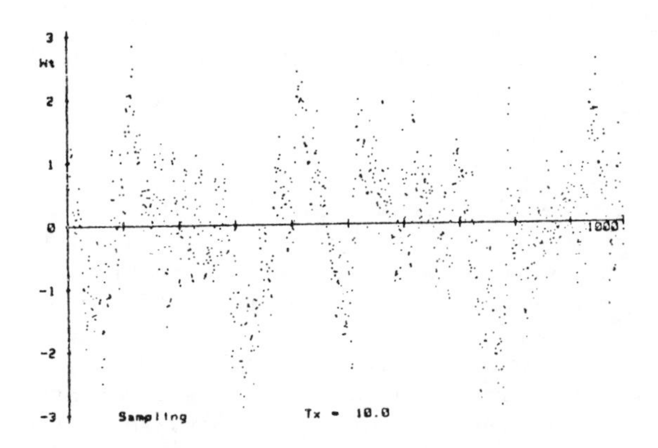

Figure 6

When $T_x \to \infty$ the situation of a homogeneous object is approached. In this case it can be shown that σ_m, σ_μ, and $\sigma_{m\mu}$ all approach the value of σ_x; therefore σ_* will be 0 and one sample of any desired size suffices. In highly autocorrelated lots only one small sample is needed to describe the lot with sufficient accuracy. For medium lot sizes P, or values of T_x between 0 and ∞, the smallest sample needed to obtain a certain value is one made up from many small grabs. When fewer but larger grabs are taken, the size of the sample increases (Figure 7).

The information, I, is given by

$$I = \tfrac{1}{2}\ln(\sigma_x^2/\sigma_*^2) \tag{10}$$

The minimum number of grabs to obtain information cannot be derived directly but can be obtained from graphical representations. The optimal number of grabs to obtain the relevent information for this purpose can also be obtained from the graphs (Figure 7).

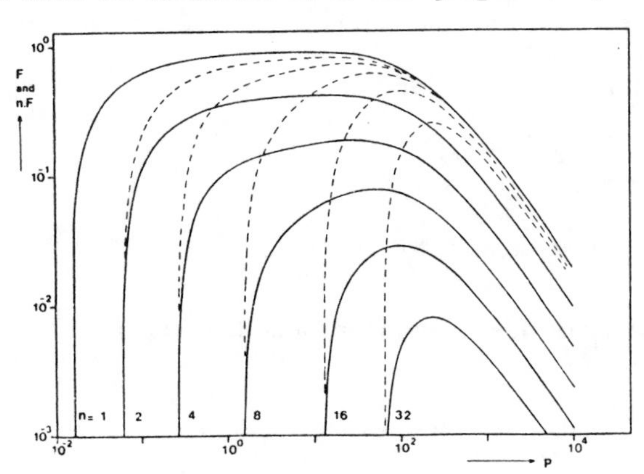

Figure 7

Relationship between the relative lot size P/T_x and the relative sample size F (= G/P) for different values of the number of samples n.

Monitoring

If the purpose of the sampling method is threshold monitoring an a posteriori determination of information is worthless as is a total reconstruction of the process values in time. The only relevant information is given by comparison of the process values with a given threshold value. If a first order autoregressive process - a very common type of process in industry and nature - is analyzed without error and the result is immediately available, then at that moment, there is no uncertainty about the process value. After a while, the process value has changed and the entropy has increased to

$$H(\tau) = -P(\tau)\ \ell n\ P(\tau) - (1-P(\tau))\ \ell n\ (1-P(\tau))$$

where $P(\tau)$ = pobability that the process value exceeds the threshold value. When $H(\tau)$ becomes too large, a new analysis is required. Depending on the required reliability, a limit should be set to permissible uncertainties. This means that every time this limit is reached, an analysis must be made with an information yield equal to this uncertainty limit.

Müskens (MU 78) derived that one should sample at a time τ after an analytical result x_t according to

$$\tau = -T_x \cdot \ell n \left\{ \frac{T_r \cdot X_t + Z[X_t^2 - q(T_r^2 - Z^2)]^{\frac{1}{2}}}{X_t^2 + qZ^2} \right\} \tag{11}$$

where T_r = threshold value (normalized to zero mean and unit standard deviation)

X_t = process value (normalized to zero mean and unit standard deviation)

Z = reliability factor

q = $(\sigma_x^2 + \sigma_a^2)/\sigma_x^2$

σ_x^2 = variance process

σ_a^2 = variance method of analysis

In this way optimal use is made of each analysis (see Figures 8 and 9).

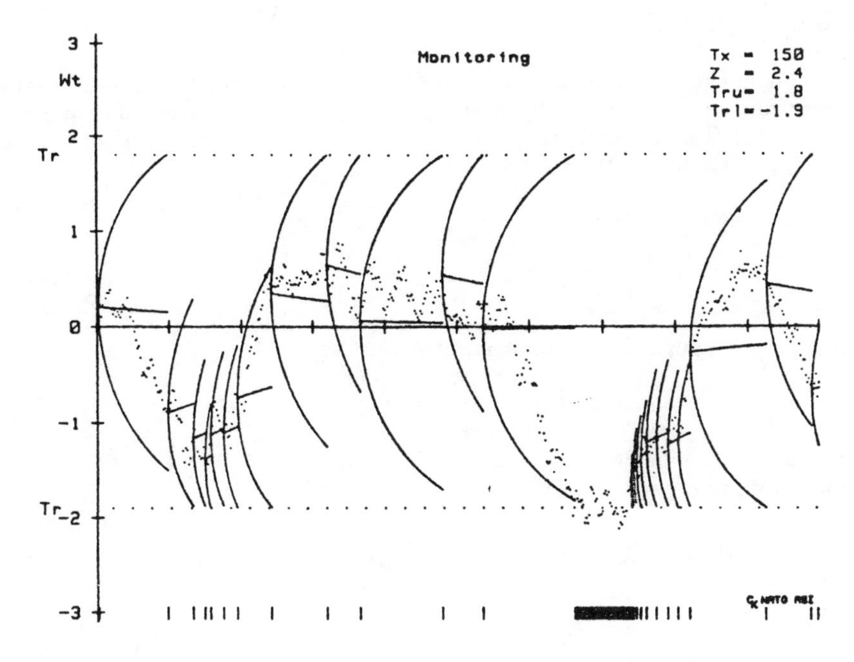

Figure 8

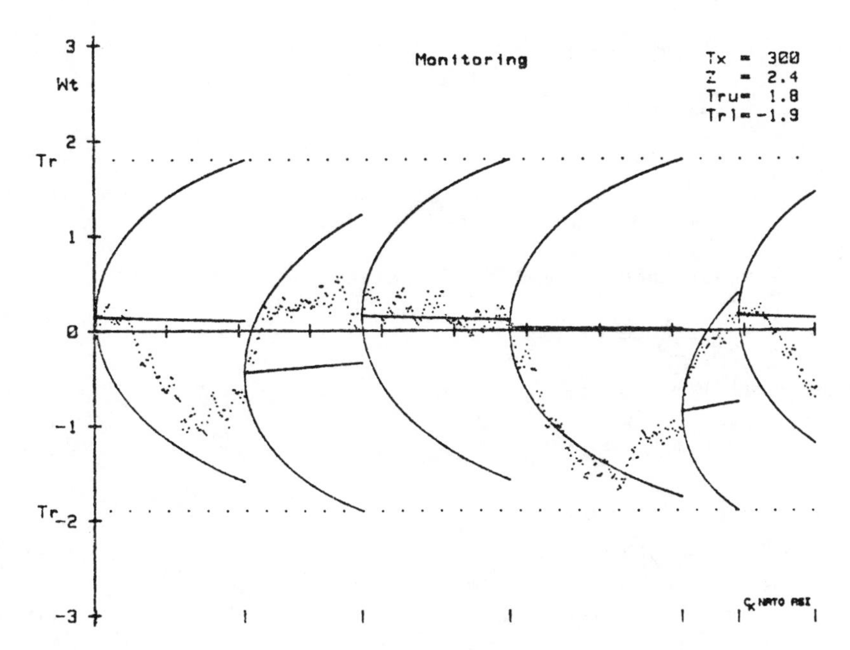

Figure 9

Control

When the purpose of the sampling is real-time process reconstruction or control, "dynamic" information is needed and a process reconstruction with a minimal or optimal reconstruction error is the goal (Figure 10).

A measurability factor m is defined by

$$m^2 = (\sigma_x^2 - \sigma_e^2)/\sigma_x^2 = 1 - \sigma_e^2/\sigma_x^2 \tag{12}$$

where σ_x^2 = variance of process deviation

σ_e^2 = reconstruction error

According to Van der Grinten (GI 64, GR 63, GR 65, GR 66) the measurability factor m is given by

$$m = [\exp-(d+\tfrac{1}{2}a+1/3y)](1-Sa\cdot t_e^{\frac{1}{2}}) \tag{13}$$

where $d = D/T_x$ $a = A/T_x$ $g = G/T_x$ $t_e = T_e/T_x$ and $s_a = \sigma_a/\sigma_x$

D = analysis time

A = $(\text{sampling frequency})^{-1}$

G = sample size

T_x = time constant (correlation factor) of process

T_e = time constant of measuring device

σ_a^2 = variance of method of analysis

σ_x^2 = variance of process

Equation (13) can be rewritten as

$$m = m_D \cdot m_A \cdot m_G \cdot m_N \tag{14}$$

where

$$m_D = \exp(-d) \tag{15a}$$

$$m_A = \exp(-a/2) \tag{15b}$$

$$m_G = \exp(-g/3) \tag{15c}$$

$$m_N = 1 - s_a t_e^{\frac{1}{2}} \tag{15d}$$

From equation 13 or 14 it follows that the maximum obtainable controllability factor will never exceed the smallest of the composing factors. This implies that all factors should be considered in order to eliminate the restricting one. It also means that a trade-off is possible between high and low values of the various factors. If, or example, m_N. the factor influenced by the reproducibility of the measuring device or the method of analysis, is high and m_A is low, the frequency of sampling can be increased causing m_A to increase. The reproducibility of the analysis - in order to cope with the higher rate of sampling is the product $m_A m_N$ is higher than before.

From $m_A = \exp(-a)$ it follows that decreasing a, the distance between samples or the reciprocal of the sampling frequency, causes m_A to increase. The higher the sampling rate, the better the possibility of controlling the process. This does not mean, however, that the highest obtainable frequency is the best, for sampling costs rise about linearly with 1/a. Moreover the optimal value of a depends on balancing the costs of analysis agent costs of an eventual process failure.

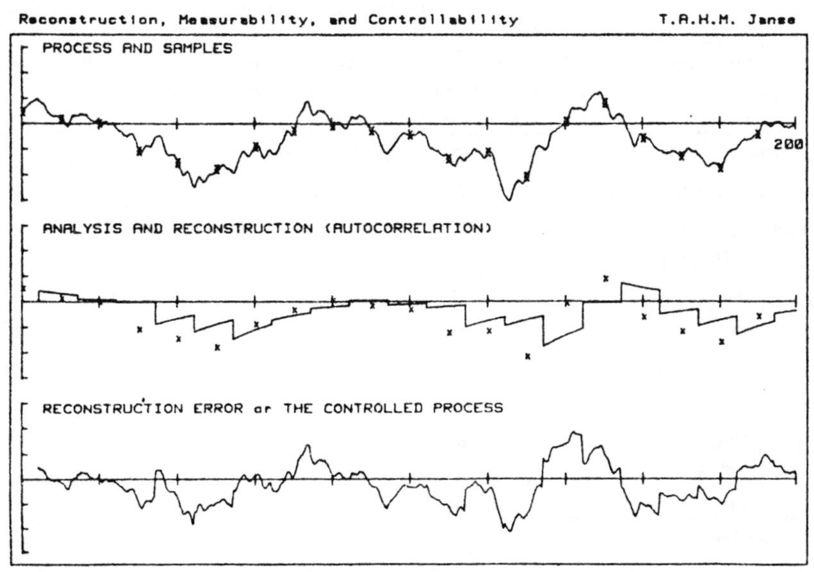

Figure 10

From $m_G = \exp(-g/3)$ it is clear that decreasing g, the sample size increases m_G. The result is that, when sampling for control, it is always better to use samples that have the minimum size required by analysis. This is contrary to the situation when describing a lot. A special case is encountered when the time constant of the instrument, for example, an automatic analyzer, sets the sampling rate. Here T_e a, so

$$m_N = 1 - s \cdot a^{\frac{1}{2}} \tag{15b}$$

The information yield of the analysis is expressed as the difference in uncertainty or entropy of the object with and without analysis (JA 83).

$$I_{dyn} = \tfrac{1}{2} \ln \sigma_x^2/\sigma_e^2 \tag{16a}$$

or with equation (12)

$$I_{dyn} = - \ln (1-m^2) \tag{16b}$$

This information measure can be used as quality measure as it describes the effectiveness of the analytical method, as well as the effectiveness of the sampling scheme. Some experimental method, for nitrogen assay in fertilizer are compared in Table 1.

Criterion and Analytical Technique	Dead Time of analysis (minutes)	Std. Dev. of Analysis (%N)	m_D	m_N	m
Total N, classical	75	0.17	0.32	0.99	0.24
distillation	75	0.17	0.32	0.99	0.24
Total N, DSM automated analyzer	12	0.25	0.84	0.97	0.65
NO_3-N, Technicon Auto analyzer	15½	0.51	0.79	0.92	0.58
NO_3-N, specific electrode	10	0.76	0.86	0.85	0.58
$NH_4NO_3/CaCO_3$ ratio X-ray diffraction	8	0.8	0.89	0.83	0.59
Total N, fast neutron activation analysis	5	0.17	0.93	0.99	0.74
Specific gravity γ-ray absorption	1	0.64	0.98	0.88	0.69

Table 1 Methods for analysis of nitrogen in fertilizer (LE 71) A sampling frequency of 2 samples/hour is assumed, which means that m_A = 0.80; m_D = measurability caused by analysis time, m_N = measurability caused by accuracy.

The Laboratory

The sampling rate, and therefore the information yield is not only set by the sampling scheme, but also by the laboratory, a rather flexible part of the scheme. If the frequency of sampling is too high queueing of samples occurs. For dynamic information delivery this means loss of information.

As Janse showed, the theory of queueing can be applied to study the effects on information yield of such limited facilities (JA 83, JE 83). Basically, limited facilities result in a limited number of practicable analytical measurements. If this number is occasionally exceeded, waiting times occur, and a situation may arise in which more samples enter the system than leave it. Besides this quantity, the ways in which a certain number of samples are offered and processed are important. In this respect, the quality of organizational aspects can be quantified. In queueing theory, such systems have been extensively studied (KL 76, GR 74). One of the simplest queueing models is the M/M/1 model, based on an exponential distribution of the inter-arrival times, as well as an exponential distribution of the analytis times, with one analyst. Such a model could represent a one-analyst laboratory with no organization at all. A perfectly organized system could be modelled by the D/D/1 model (D for deterministic), in which samples arrive at constant intervals and where every sample takes exactly the same amount of measurement time with an analyst who is always present. Of course, such models are purely theoretical. Conclusions related to more realistic situations may be obtained by simulation experiments (VA 79, VA 80, VO 81, JE 83).

The way that samples are taken from the process fixes the sample input for the laboratory; the sampling interval time equals the inter-arrival time for samples to the laboratory. The mean sampling frequency is λ ($\lambda=1/\bar{A}$). The laboratory capacity is fixed by one analyst; his average analysis rate is μ ($\mu=1/\bar{D}$).

The absence of a plan can be modelled by a M/M/1 system: the sampling scheme is characterized by a (memory-less) Poisson process resulting in an exponential distribution for the inter-arrival times. The time required to analyze a sample is also unpredictable and also given an exponential distribution. As long as the utilization factor (the ratio λ/μ) is below 1, the system functions. However, waiting times occur, and when the utilization factor is increased the time spent by samples in the system increases exponentially. This "dead time" T_d (waiting + service time) has a negative effect on the process reconstruction, and thus on the information delivered by the chemist.

With some approximations the following general solution is applicable:

$$m^2 = (\lambda\ T_x/2)[(L(f(a))-1] \cdot L(f(T_d))]\ s = 2/T_x \tag{17}$$

where L is the Laplace transform, s the Laplace operator, T_x the time constant of the process, A the sample interval time, T_d the time spent in the system (dead time), and λ the mean sample frequency.

Back-transformation to the time domain is often very complicated, if possible at all. For the M/M/1 system, the result is

$$m^2 = [T_x/(T_x+2\overline{A})]\{(1-\rho)T_x/[(1-\rho)T_x+\overline{D}]\} \tag{18a}$$

For the D/M/1 system (fixed arrival rates), the result is

$$m^2 = \exp(-2\overline{A}/T_x)\{(1-\rho)T_x/[(1-\rho)T_x+2D]\} \tag{18b}$$

For the M/D/1 system (fixed analysis times), the equation is

$$m^2 = [T_x/(T_x+2\overline{A})]\{2(1-\rho)\overline{A}/[(2\overline{A}-T_x)\exp(2\overline{D}/T_x)+T_x]\}$$

In these equations $\overline{D}$ is the mean analysis time, $\overline{A}$ the mean sample interval time, and ρ the utilization factor.

The information yield (related to m by equation 16b) is shown as a function of the utilization factor ρ for the M/M/1 system (Figure 11). There is an optimal mean sample frequency: however, at the utilization factor 0.5, it is surprisingly low. Only half of the analyst's capacity is used in obtaining the maximal information yield, or in having the best process control by such system. At lower utilization factors, the information yield decreases because of the lower sampling frequency. At higher utilization factors, the information decreases because of the increasing waiting time in the system. The information-decreasing effect of the waiting times overrides the information-increasing effect of the higher sampling frequency.

In a D/D/1 system there are no waiting times. Here it is obvious that a utilization factor of 1 gives maximal information. With the theory outlined above, it is possible to develop a theory for analytical planning. First, it is important to distinguish between the influence of a stochastic input (M/D/1 system) and a stochastic processing time (D/M/1 system). The latter is clearly more favorable (Figure 12).

Various plans to improve the performance of the laboratory or to increase the information yield are possible. One conclusion drawn from the above is that smoothing the input is advantageous (from M/M/1 to D/M/1).

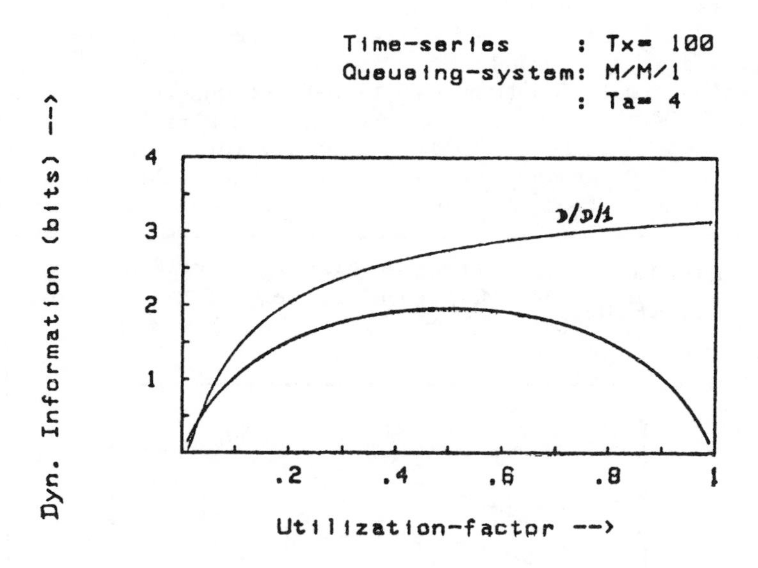

Figure 11

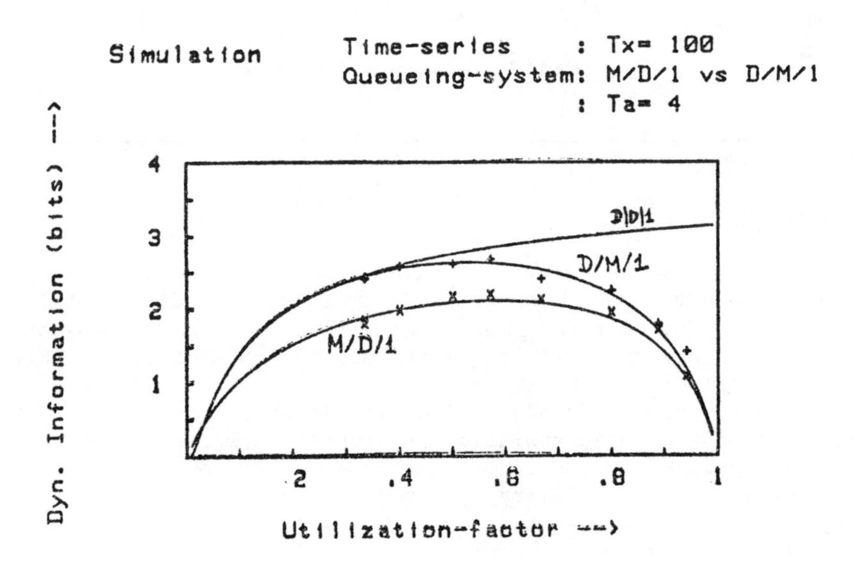

Figure 12

A totally different kind of plan is to fix priorities for the various samples entering the laboratory. Two possibilities are considered. In one, the priorities influence the sequence in which the samples are processed. With respect to process control, it can be shown by simulation that a Last-In-First-Out (LIFO) scheme gives a higher information yield than the standard First-In-First-Out IFIFO) scheme (Figure 13). This is hardly surprising, as for a first-order A.R. process only the most recent samples gives relevant information. In fact all the samples that queue could be discarded.

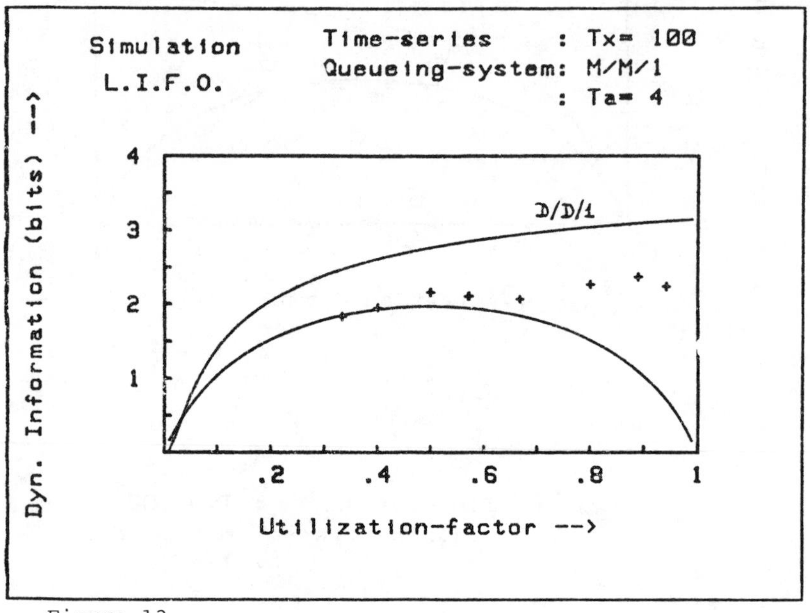

Figure 13

Conclusion

It is shown that information as a measure of quality can be used through the whole analytical process from sampling until laboratory organization. The methods that are used to evaluate quality are all chemometrical methods. The optimal way to obtain information seems to be to produce only the required information (relevancy) and make use of all available prior knowledge. In fact correlation and autocorrelation are important means, as they can be used for prediction. Further examples of the use of chemometrics in quality assessment, quality improvement and quality control can be found in pattern recognition, state- and parameter estimation by Kalman filtering and, of course, in the classical statistical approach.

In pattern recognition in essence all correlation between patterns and features is eliminated and discrimination between clusters is favoured, increasing the information (Ja 81, JN 81, JN 83). Kalman filtering uses optimally all available information from the past. It allows prediction of state and parameters, thereby reducing the amount of measurements required to obtain a predetermined level of information. Examples can be found in multicomponent analysis (PO 79, DI 80), drift correction (PO 80), calibation (VA 83), variance reduction (PO 83, JS 83), etc.

Department of Analytical Chemistry
University of Nijmegen
Toernooiveld
6525 ED NIJMEGEN
The Netherlands

This lecture is based on parts of the work of our research group.

Staff members:

G. Kateman	1972-
B.G.M. Bandeginste	1973-
P.J.W.M. Müskens	1973-1978
C.B.G. Limonard	1974-1978
P.F.A. van der Wiel	1977-
F.W. Pijpers	1977-
H.N.J. Poulisse	1978-1981
T.A.H.M. Janse	1979-
P.C. Thijssen	1981-

Assistants:

H.C.G. Debets	1972-1977
W.J.M. Philipse	1979-

and more than 75 students.

REFERENCES

AL 69 P.L. Alger, Mathematics for Science and Engineering, McGraw-Hill, New York, 1969.

AR 73 Arbeitskreis, Automation in der Analyse, Talanta, 20, 811 (1973).

BA 28 B. Baule, A.A. Benedetti-Pichler, Z. Anal. Chem. 74, 442 (1928).

BE 47 W.P. Belk, F.W. Sunderman, Am. J. Clin. Pathol. 17, 854 (1947)

BE 67 R.E. Berry, Am. J. Clin. Pathol. 47, 337 (1967).

BE 75 R.J. Behtea, B.S. Duran, T.L. Boullion, Statistical Methods for Engineers and Scientists, Marcel Dekker, New York, 1975.

DA 72 O.L. Davies, P. L. Goldsmith, Statistical Methods in Research and Production, Oliver and Boyd, Edinburgh, 1972.

DE 64 F.B. Desmond, N. Z. Med. J. 63, 716 (1964).

DI 69 W.J. Dixon, F.J. Massey, Introduction to Statistical Analysis, 3rd Ed., McGraw-Hill, New York, 1969.

DI 80 C.B.M. Didden, H.N.J. Poulisse, Anal. Letters, 13 (All), 921 (1980)

DU 74 O.J. Dunn, V.A. Clark, Applied Statistics: Analysis of Variance and Regression, Wiley, New York, 1974.

EC 71 K. Eckschlager, Coll. Czech. Chem. Commun., 36, 3016 (1971).

EC 82 K. Eckschlager, V. Štěpánek, Anal. Chem. 54 (11), 1115A (1982).

FI 55 D.J. Finney, Experimental Design and its Statistical Basis, Cambridge University Press, 1955.

FI 70 A.F. Findeis, M.K. Wilson, W.W. Meinke, Anal. Chem. 42 (7), 26A (1970).

GI 63 P.M.E.M. van der Grinten, Control Eng. 10, (12), 51 (1963).

GO 08 W.S. Gosset, Biometrika, 6, 1 (1908).

GO 72 G. Gottschalk, Z. Anal. Chem. 258, 1 (1972).

GR 63 P.M.E.M. van der Grinten, Control Eng. 10 (10), 87 (1963).

GR 65 P.M.E.M. van der Grinten, J. Instrum. Soc. Am. 12 (1), 48 (1965).

GR 66 P.M.E.M. van der Grinten, J. Instrum. Soc. Am. 13 (2), 58 (1966).

GR 74 D. Gross, C.M. Harris, Fundamental of Queueing Theory, Wiley, New York, 1974.

GY 79 M. Gy, Sampling of Particulate Materials: Theory and Practice, Elsevier, Amsterdam, 1979.

JA 81 R.T.P. Jansen, F.W. Pijpers, G.A.J.M. de Valk, Anal. Chim. Acta 133, 1 (1981).

JA 83 T.A.H.M. Janse, G. Kateman, Anal. Chim. Acta, 150, 219 (1983).

JE 83 T.A.H.M. Janse, G. Kateman, to be published.

JN 81 R.T.P. Jansen, F.W. Pijpers, G.A.J.M. de Valk, Ann. Clin. Biochem. 18, 218 (1981).

JN 83 R.T.P. Jansen, Ann. Clin. Biochem. 20, 41 (1983).

JS 83 R.T.P. Jansen, H.N.J. Poulisse, Anal. Chim. Acta, 151, 441 (1983).

KA 78 G. Kateman, P.J.W.M. Müskens, Anal. Chim. Acta 103, 11 (1978).

KA 79 G. Kateman, A. Dijkstra, Z. Anal. Chem. 247, 249 (1979).

KA 81 G. Kateman, F.W. Pijpers, Quality Control in Analytical Chemistry, Wiley, New York, 1981.

KL 76 L. Kleinrock, Queueing Systems, Vol. I; Theory, Wiley, New York, 1976.

KO 72 B.R. Kowalski, C.F. Bender, Anal. Chem. 44, 1406 (1972).

KO 75 B.R. Kowalski, J. Chem. Inf. Comput. Sci. 15, 201 (1975).

LE 50 S. Levey, E.R. Jenning, Am. J. Clin. Pathol. 20, 1059 (1950).

LE 71 F.A. Leemans, Anal. Chem. 43 (11), 36A (1971).

MA 75 D.L. Massart, L. Kaufman, Anal. Chem. 47 (14), 134A (1975).

MA 77 D.L. Massart, H. de Clerq, R. Smits, Reviews on Analytical Chemistry, lecture presented at Euroanalysis Conference II, Masson, Paris, 1977, 119.

MA 78 D.L. Massart, A. Dijkstra, L. Kaufman, Evaluation and Optimization of Laboratory Methods and Analytical Procedures, Elsevier, Amsterdam, 1978.

MS 78 P.J.W.M. Müskens, G. Kateman, Anal. Chim. Acta 103, 1 (1978).

MU 78 P.J.W.M. Müskens, Anal. Chim. Acta 103, 445 (1978).

NA 63 M.G. Natrella, Experimental Statistics, National Bureau of Standards Handbook 91, U.S. Govt. Printing Office, Washington DC, 1963.

PO 79 H.N.J. Poulisse, Anal. Chim. Acta 112, 361 (1979)

PO 80 H.N.J. Poulisse, P. Engelen, Anal. Lett. 13 (A14), 1211 (1980).

PO 83 H.N.J. Poulisse, R.T.P. Jansen, Anal. Chim. Acta 151, 433 (1983).

SH 31 W.A. Shewhart, Economic Control of the Quality of Manufactured Product, Van Nostrand, New York, 1931.

SH 49 C.E. Shannon, W. Weaver, The Mathematical Theory of Communication, University Press of Illinois, Urbana, 1949.

ST 65 J.V. Straumfjord, B.E. Copeland, Am. J. Clin. Pathol. 44, 242 (1965).

TO 63 D.B. Tonks, Clin. Chem. 9, 217 (1963).

VA 74 I.R. Vaananen, S. Kivirikko, J. Koskenniemi, J. Koskimies, A. Relander, Methods Inf. Med. 13 (3), 158 (1974).

VA 77 B.G.M. Vandeginste, Anal. Lett. 10, 661 (1977).

VA 79 B.G.M. Vandeginste, Anal. Chim. Acta 112, 253 (1979).

VA 80 B.G.M. Vandeginste, Anal. Chim. Acta 122, 435 (1980).

VA 83 B.G.M. Vandeginste, J. Klaessens, G. Kateman, Anal. Chim. Acta 150, 71 (1983).

VN 83 B.G.M. Vandeginste, Anal. Chim. Acta, 150, 199 (1983).

VO 81 J.G. Vollenbroek, B.G.M. Vandeginste, Anal. Chim. Acta 133, 85 (1981)

YO 51 W.J. Youden, Statistical Methods for Chemists, Chapman and Hall, London,1951.

RECOMMENDED READING

General

KA 81 G. Kateman, F.W. Pijpers, Quality Control in Analytical Chemistry, Wiley, New York, 1981.

MA 78 D.L. Massart, A. Dijkstra, L. Kaufman, Evaluation and Optimization of Laboratory Methods and Analytical Procedures, Elsevier, Amsterdam, 1978.

Information

EC 82 K. Eckschlager, V. Štěpánek, Anal. Chem. 54 (11), 115A (1982).

KA 81 G. Kateman, F.W. Pijpers, Quality Control in Analytical Chemistry, chapter 4.2, Wiley, New York, 1981.

MA 78 D.L. Massart, A. Dijkstra, L. Kaufman, Evaluation and Optimization of Laboratory Methods and Analytical Procedures, chapter 8, Elsevier, Amsterdam, 1978.

Sampling

JA 83 T.A.H.M Janse, G. Kateman, Anal. Chim. Acta 150, 219 (1983).

GY 79 M. Gy, Sampling of Particulate Materials: Theory and Practice, Elsevier, Amsterdam, 1979.

KA 78 G. Kateman, P.J.W.M. Müskens, Anal. Chim. Acta 103, 11 (1978).

KA 81 G. Kateman, F.W. Pijpers, Quality Control in Analytical Chemistry, chapter 2, Wiley, New York, 1981.

LE 71 F.A. Leemans, Anal. Chem. 43 (11), 36A (1971).

MA 78 D.L. Massart, A. Dijkstra, L. Kaufman, Evaluation and Optimization of Laboratory Methods and Analytical Procedures, chapters 26 and 27, Elsevier, Amsterdam, 1978.

MS 78 P.J.W.M. Müskens, G. Kateman, Anal. Chim. Acta 103, 1 (1978).

MU 78 P.J.W.M. Müskens, Anal. Chim. Acta 103, 445 (1978).

AUTOMATIC CONTROL OF CHEMICAL PROCESSES

N. Lawrence Ricker

University of Washington,
Seattle, Washington, 98195

Modern strategies for the automation of chemical processes are surveyed with an emphasis on control algorithms that incorporate a mathematical model of the system to be controlled. This approach can be advantageous when simpler classical feedback methods fail to provide adequate control.

INTRODUCTION

Chemical processes are becoming increasingly complex as designers attempt to maximize the efficiency of energy and raw-material consumption. Consequently, it is more important than ever to augment the decision-making abilities of process operators with automated data-gathering and process-control systems.

The design of such systems has been the domain of instrument specialists and process engineers. It is clear, however, that other specialists, e.g., chemometricians, could play an important role if they had an appreciation of typical system design problems.

System Design Objectives

The overall tasks of a process automation system are: 1) to minimize the effect of disturbances in the process inputs (e.g., variations in the composition of raw materials), and 2) to maintain "optimal" operation in the face of changes in the desired process throughput and other economic and physical constraints within which the plant must operate. Implicit in

B. R. Kowalski (ed.), Chemometrics. Mathematics and Statistics in Chemistry, 205–223.

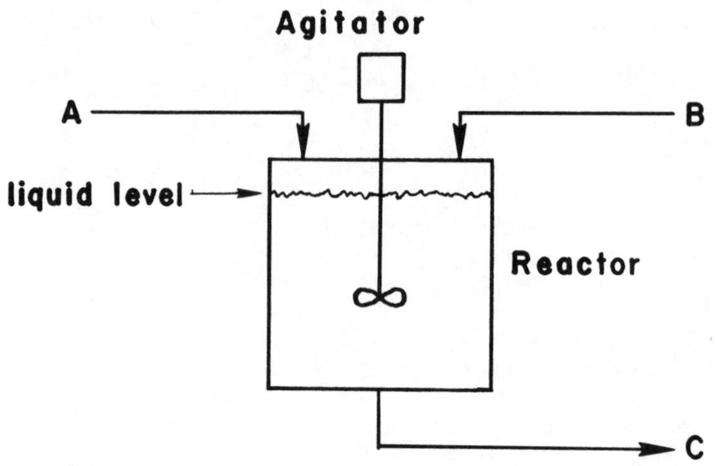

FIGURE 1: Schematic of a Chemical Reaction Process

these overall goals are other considerations such as the need to assure safety of operation and system reliability, satisfy environmental regulations, meet standards on product quality, provide information needed by management to assess the efficiency of the operation, etc.

An Example Problem

As an illustration of design objectives and potential operating problems, let us consider the automation of a simple isothermal liquid-phase chemical reaction process. The process, shown in Figure 1, consists of an agitated cylindrical tank into which there is a continuous feed of two raw material streams, A and B, and a continuous withdrawal of a product stream, C. Raw material A is a byproduct of another operation in the plant; its flowrate is dictated by the throughput of that process and cannot be regulated by our reactor system. We can, however, regulate the flowrates of streams B and C.

Let us assume that the primary objective of the regulation system is to hold the product composition at a specified optimal value. This optimal value would be a function of current process economics and other factors and will, in general, vary with time; its calculation would be the job of an independent part of the process automation system.

The secondary objective of the regulation system is to control the level of liquid in the reactor. Clearly, we must prevent the reactor from draining or overflowing, but we require more precise regulation than this because a change in level changes the fluid residence time and, in general, the reactor conversion. We will assume that there is an optimal level that varies with throughput.

Conventional Regulation Strategy

The conventional approach is to simplify such regulation problems through a strategy of "divide-and-conquer," i.e., that one can satisfy a given control objective by adjusting a single process variable, and that adjustments in other process variables will have only a minimal effect on this control objective. The problem is then to determine the best pairings between individual control objectives (the controlled variables) and the available process adjustments (the manipulated variables).

Two possible pairings for the reactor control system are shown in Figures 2a and 2b. The scheme shown in Figure 2a compares the measured composition of stream C to that specified by the process optimizer (not shown). If there is a discrepancy, the composition controller, CC-B, changes the setpoint of the flowrate controller, FC-B. This, in turn, compares the measured flowrate signal from FT-B with the setpoint and adjusts the valve, FCV-B until the flowrate of B is at the setpoint. Similarly, an error in the liquid level causes an adjustment in the flowrate of stream C. In Figure 2b, on the other hand, an error in the composition of C causes an adjustment in the flowrate of C, and an error in the level causes an adjustment in the flowrate of B.

Both cases are examples of feedback control. We take corrective action if and only if we observe an error in a controlled variable. This has the important practical advantage that we need not know what has caused the error. Moreover, the four controllers shown in each scheme can be implemented in a standardized way. In chemical processes the PID (proportional-integral-derivative) controller is the standard (1,2,3). "Tuning" such a controller to fit the process requires only a rudimentary knowledge of the process dynamics -- another advantage.

Given that the equipment required to implement the two schemes is essentially identical, is one preferable over the other? An experienced process engineer would immediately select scheme 2a because of its favorable causality, i.e., a change in the flowrate of B would directly affect the composition of C and a change in the flowrate of C would have a strong effect on the level.

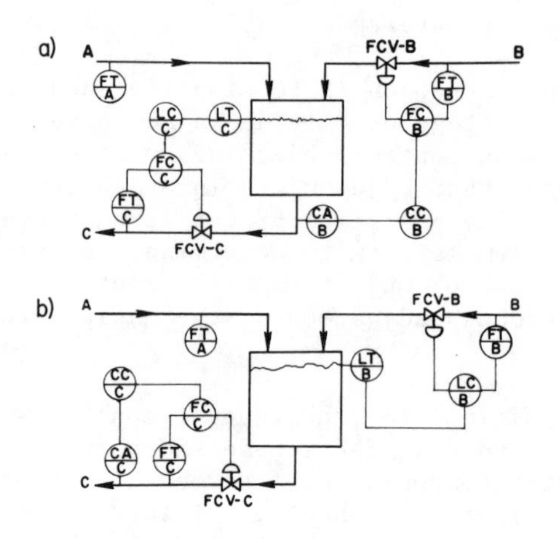

KEY TO SYMBOLS

FT = Flowrate Transmitter
FC = Flowrate Controller
LT = Level Transmitter
CA = Composition Analyzer

FCV = Flow Control Valve
LC = Level Controller
CC = Composition Controller

FIGURE 2: Two Possible Control Schemes for the Reactor

In Figure 2b, on the other hand, the couplings between the manipulated and controlled variables are less direct. Suppose, for example, that B is a concentrated raw material and only a small amount is required relative to the flowrate of A. Clearly, scheme 2a would provide better level control. In either case, an adjustment in the flowrate of C will affect the level more directly than the product composition. (In less obvious situations, calculation of the relative gain array (1) will suggest likely pairings).

Interaction Between Control Objectives

Even in scheme 2a, however, it is possible that an adjustment in the flowrate of B would strongly affect both the product composition and the level. This violates the basic tenet of the divide-and-conquer strategy and would result in "interaction" between the controller responsible for holding the level and that responsible for the composition. Note that there is no direct communication between LC-C and CC-B -- they are assumed to be independent. Consequently, it is possible for these controllers to work at cross purposes.

In general, if it is impossible to reduce interactions to an acceptable level by changing the pairings of controlled and manipulated variables or by changing the fundamental design of the process, a more sophisticated regulation strategy might solve the problem. Such strategies will be considered in more detail below.

Other Problems with Simple Feedback Regulation

Other possible shortcomings of the schemes shown in Figures 2a and 2b have to do with the basic nature of feedback control, namely that <u>an adjustment can only be made in response to a measured error</u>. Success of the feedback strategy thus requires that 1) the error is directly measurable and, 2) that the measured error is an accurate indication of the <u>current</u> process state. These requirements are often violated, especially in the case of composition control.

It is rare, for example, to be able to provide a reliable, accurate, on-line device for the measurement of the composition of a mixture. The current practice in many cases is to provide an indirect measurement of composition -- density, refractive index, viscosity, etc. One occasionally finds true on-line analyzers such as gas chromatographs in industry, but current instruments are relatively expensive and difficult to calibrate and maintain. Furthermore, they are often slow. One must minimize all time delays in the feedback system or suffer the consequences: sluggish and inaccurate control (1,2,3).

Delays inherent in the process itself are equally detrimental. For example, if the composition of stream A were to change, this might take a significant time to show up as an error in the product composition. Meanwhile, the composition of stream A might have changed again. In other words, error measurements alone, even if accurate, can be a poor indicator of the current state of the process.

USE OF A MATHEMATICAL MODEL IN PROCESS REGULATION

In order to progress beyond the regulatory capability of the simple divide-and-conquer feedback strategy outlined above, we need a way to build more knowledge of the process characteristics into the regulation system. If a model of the process were available we could exploit its predictive capabilities in the following ways:

1. Measure potential system disturbances such as a change in the flowrate of stream A in Figure 2 and compensate for such disturbances before they cause an error in a controlled variable. In its simplest form this is called "feedforward" control (1,2).

2. Infer the state of an unmeasured variable (such as a composition) from the value of secondary measurements. This is "inferential control" (1) or "state estimation" (4).

3. Prevent interaction between control objectives, sometimes referred to as "decoupling" (2,4).

4. Compensate for large time delays in a feedback structure so as to provide more responsive regulation, usually termed "deadtime compensation" (1,4).

Note that such a model might also be useful for optimization of the processing conditions. There are, in general, many ways to implement the above ideas; details can be found in the cited references. Here we will consider specific methods that are in current favor both in academic research and industrial applications.

Static vs Dynamic Models

It is first necessary to review the different types of mathematical models that are used in control applications. An overall classification can be made between static models, which depend only on the current state of the process, and dynamic models, which depend not only on the current state but also on the cumulative effect of the past history of the process operation.

Static models are strictly valid only in the limit of continuous, steady-state process operation. It can happen, however, that all elements of a process respond so rapidly to changing conditions that in effect, the process is always at a quasi-steady-state for the current operating conditions. This greatly simplifies the modeling problem; the elimination of time as an independent variable reduces dimensionality of the model and often results in a model that is a set of algebraic, rather than differential, equations. For this reason, a static model is sometimes used as an approximation in control applications.

Here, however, we will consider only dynamic models. The substitution of an approximate static model is a straightforward simplification of this more general case.

Lumped vs Distributed-Parameter Models

A lumped-parameter model is one in which the system is idealized to consist of interconnected, discrete elements, each of which has uniform properties. In a distributed-parameter system, we remove the restriction of uniform properties and allow the properties of an element to vary with spatial dimension. In Figure 1, for example, if we consider the reactor to be perfectly mixed so that at any instant of time the temperature and composition are uniform throughout, the reactor can be modeled as a lumped element. If, on the other hand, it appears that there would be significant variations with respect to the horizontal and/or vertical dimensions, this could be modeled using a distributed-parameter representation.

It should be mentioned that it is common practice to approximate a distributed-parameter system by breaking it up into interconnected lumps. The reactor might, for example be divided into two or three zones with characterized convection flows circulating material between zones.

Mechanistic vs Empirical Models

In either case, if our goal is a mechanistic model of the system, we begin by writing the applicable conservation equations. For example, we could equate the rate of accumulation of component A in the reactor in Figure 1 to the net influx of that component less its rate of consumption in the reaction. The result would be an ordinary differential equation (ODE) in the case of a lumped-parameter system and a partial differential equation (PDE) in the case of a distributed-parameter system. We would then add the applicable constitutive equations, e.g., reaction kinetics, thermodynamic relationships, etc. The model of the entire system would consist of coupled ODEs, algebraic equations, and/or PDEs. In general, these would be non-linear. Good reviews of mechanistic modeling in the context of chemical processes are given in (1) and (2).

Unfortunately, the effort required to derive an accurate mechanistic model is often more than can be justified if the goal is merely to implement a good process control system. An alternative is some form of empirical model identified directly from process data. Many such models and related identification techniques are available. Such models usually have the following properties:

- linearity with respect to the independent and dependent variables.
- a time-invariant structure, e.g, constant parameters in a parametric model.

Of particular importance are the "time series" methods. See, e.g., reference (5) and the discussions by H. Smit and G. Kateman elsewhere in this volume.

The more general non-linear and mechanistic models can always be linearized so as to fit into the framework of any of the empirical models. Since the calculational requirements are thereby greatly reduced and since one can bring many more theoretical weapons to bear on a linear model, this is often done for control system analysis.

Continuous-Time vs Sampled-Data Models

Most chemical processes are continuous systems rather than a series of discrete events. With the increasing use of computer techniques, however, it has become common practice to represent a process as a sampled-data system, i.e., even though each variable in a process varies continuously with time, we only sample each variable at discrete intervals and have no knowledge of what happens between these intervals. See, e.g., the discussion by G. Kateman elsewhere in this volume. One advantage of this approach is that the usual empirical sampled-data models are coupled linear difference equations, which are convenient for computations.

INTERNAL MODEL CONTROL

Let us now consider Internal Model Control (IMC) (6,7,8) as an example of how an empirical model of the process dynamics might be used. The usual form of the process model is an "pulse-response" input-output representation. Figure 3 shows hypothetical pulse-responses for a change in the flowrate setpoint for FC-B in Figure 2a.

Experimental Measurement of the Pulse Response

It is helpful to understand how these might be obtained experimentally. First, note that in this example the sampling interval for the two output variables, y_1 (concentration of product) and y_2 (liquid level in reactor) is 15 seconds. Choice of the length of the sampling interval is an important decision -- see the chapter by G. Kateman and references (1) and (2).

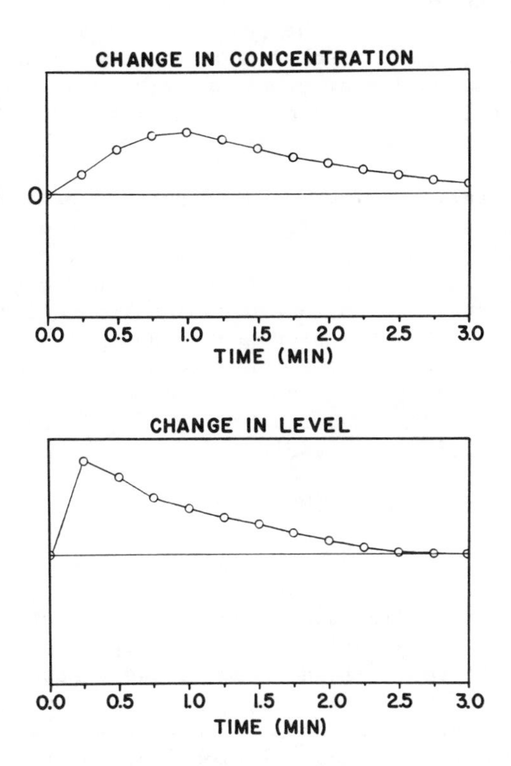

FIGURE 3: Pulse Responses for Increase in FC-B Setpoint

The outputs and the manipulated variable, u_1 (the FC-B setpoint) are assumed to be at constant, steady-state values prior to time t = 0. At t = 0, we increase u_1 by one "unit" from its nominal value and hold it there for one sampling interval, then change it back to the nominal value. We then measure the effect of this square pulse in u_1 on the two outputs. The circles shown could be actual data points taken at each sampling time or smoothed and normalized data from a number of replications. In any case, the sequence of deviations from the nominal conditions comprise the pulse response and can be used directly in the IMC strategy. The same type of experiment could be performed for the other manipulated variable shown in Figure 2a, the setpoint of FC-C.

The responses shown in Figure 3 show that the level responds relatively quickly to a change in inlet flowrate whereas because of the mixing capacitance of the liquid held up in the reactor, the response of the product concentration is more drawn out. In general, the pulse response goes on to infinite time, but in the case of most chemical processes (and with an appropriate choice of sampling interval), the variable is nearly back at its original condition within 20-40 sampling intervals and the pulse response is truncated at that point.

In a real plant, it might be difficult to generate a perfect square pulse. Fortunately, that is not really necessary -- any shape will do as long as the pulse duration is relatively short and the magnitude is large enough to cause a significant change in the outputs (so as to avoid signal-to-noise problems) but small enough so that the process response is approximately linear (2,9). An alternative to purposeful introduction of pulses is to infer the pulse response from data recorded during periods of measured, random disturbances (5) but this is less likely to give satisfactory results.

Use of the Pulse Response in IMC

Once the pulse responses are known for each pair of manipulated variable, u_i, and output variable, y_k, we can use this information to predict the response of the outputs to a sequence of arbitrary variations in the manipulated variables. It is also possible to model and predict the response of the outputs to "disturbance" variables, i.e., things we can measure but over which we have no control, such as the flowrate of stream A in Figure 2a. The calculational procedure is very simple. See, e.g., references (5) and (9).

In the IMC formulation, the typical approach (6,7,8) is to use matrix algegra; the relationship between the manipulated and output variables is given by

$$y = G\,u + b \qquad (1)$$

where y is a vector sequence of output variables, u is a sequence of manipulated variables, G is a rectangular matrix containing the pulse-response data in a certain format, and b is a bias vector. If, for example, we specify a sequence of future values of each manipulated variable over a prediction horizon of N sampling intervals, we could use equation 1 to calculate the corresponding sequence of future y values. The sequence of bias values, b, is to account for the effect of past variations in u. If unknown, this would be set to zero. Details are available in reference (8).

The Control Objective

Let us assume that we are running the reactor process continuously and are able to change the settings of the manipulated variables at the beginning of each sampling interval. Each time such a change is made, the new value is maintained by a "zero-order hold" (a computer hardware device) until the beginning of the next sampling interval. The control algorithm's job is: given a sequence of measurements indicating the past and current state of the process, determine the values at which to set each manipulated variable for the next sampling interval.

We must first define what we mean by "good" control. IMC is a form of optimal control in which the main goal is accurate regulation of the controlled variables. The mathematical objective is then to minimize the difference between the actual value of the controlled variables and a specified (reference) value. In reference (8) the objective is to maximize

$$J = -\frac{1}{2}(y_r - y)^T H (y_r - y) \qquad (2)$$

where J is a weighted-least-squares measure of the accuracy of regulation, y_r is a vector sequence of reference values corresponding to y, the sequence of output values, H is a positive-definite weighting matrix (usually diagonal), and the superscript T denotes the transpose.

Consider, for example, the situation shown in Figure 4. Imagine that we are currently at time t = 0 (in arbitrary time units). The actual measured values of the output variable y_1 for the current time and 4 previous sampling intervals are indicated by open circles (we would have similar information for all other outputs). The desired (reference) value of y_1 is shown as a dashed line at $y_1 = 0$. Regulation has been good in the past but the current value of y_1 deviates from the reference value.

Based on measurements made in the past and on our knowledge of the pulse responses for the process, we predict what the current value of y_1 should be. The difference between this prediction and the current measurement is the prediction error, assumed to be constant in the future. This determines the value of the bias vector, b, in equation 1.

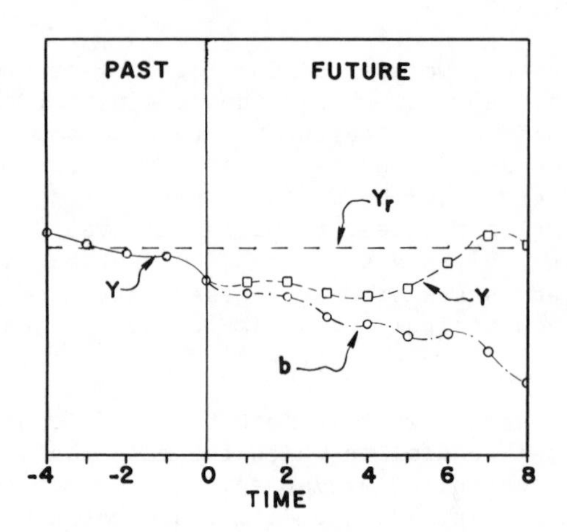

FIGURE 4: Example of IMC Prediction Scheme

We next predict what would happen if the manipulated variables were to remain at their current settings. Suppose that y_1 would then behave as shown by the open circles for positive times in Figure 4. In this example, the status quo is an unattractive policy and we thus predict what the process would do if we used some other sequence of values for the manipulated variables. Suppose that for an arbitrary choice the prediction for y_1 is the sequence of open squares in Figure 4, which is clearly better than the status quo.

Optimization by Quadratic Programming

The optimization problem, then, is to determine the sequence of u values that minimizes the deviations from the reference sequence in the sense defined in equation 2, above. Note that this involves a weighting of all outputs, not just y_1. If, for example, we want to control y_1 more accurately than y_2, we could do this by proper specification of the elements of the weighting matrix, H. Details are available in (7) and (8).

In general, there are constraints on the permitted values of u. Reference (8) assumes that these are in the form of upper and lower bounds:

$$0 \leq u \leq d \tag{3}$$

where d is a vector of upper bounds and the manipulated variables are normalized such that their lower bound is zero. This combination of a quadratic objective function and linear inequality constraints is a quadratic programming problem.

Ricker (8) discusses how the optimal sequence of future values of the manipulated variables can be determined in an efficient manner such that the necessary calculations can be done in real time (to keep pace with the process).

As described above, we predict the optimal value of each manipulated variable for N sampling intervals into the future. It is important to note, however, that in IMC only the first value in each manipulated-variable sequence is actually implemented; the entire optimization procedure is then repeated at the beginning of the next sampling interval. If we were instead to use the entire sequence of N values it is likely that an unexpected disturbance would enter the system. The best way to detect and correct for this is to check the prediction error and update the control strategy as frequently as possible. (Of course, frequent updating increases the computational requirements).

Since only the first element in the sequence is used one might then ask why it is necessary to carry out the predictions beyond a point one sampling interval into the future. The reason is that the full impact of a change in a manipulated variable might not be felt until two, three, or many sampling intervals in the future. If the control algorithm neglects to look far enough ahead, it tends to make relatively large changes in manipulated variables in order to minimize the current output errors, not realizing that this will cause even greater problems in the future. Examples of control algorithms that only look at one sampling interval in the future and hence turn out to work poorly in process applications are given, e.g., in reference (5) and in many journal articles.

Example of IMC Performance

Ricker (8) describes the application of IMC to a computer-simulated process. An overview of the control problem is shown in Figure 5. The simulation represents a modern, full-scale 6-effect evaporation system in a kraft pulp mill. The purpose of the evaporation system is to remove water from "weak black liquor" (WBL), an aqueous effluent (about 85% water) from the pulping process that contains dissolved non-volatile organic and inorganic compounds. The product from the evaporator is "strong black liquor" (SBL), in which the concentration of the dissolved solids has been raised to about 55% by weight.

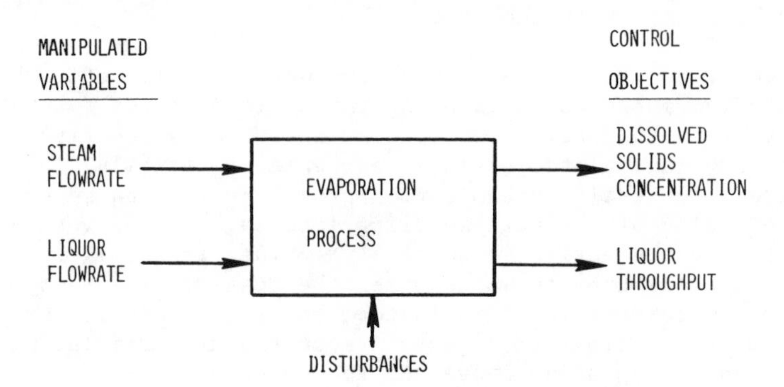

FIGURE 5: The Evaporator Control Problem

The primary objective of the evaporator control system is to hold the concentration of SBL solids close to a specified target value (which can be measured continuously using a refractive-index sensor). If insufficient water is evaporated, the overall energy efficiency of the kraft process decreases and there is a safety hazard in subsequent processing. If too much water is evaporated, the SBL becomes hard to transport and can deposit undesirable scale on the heat-transfer surfaces.

The secondary objective is to meet a target value for the average throughput of WBL. This target may change several times a day in order to accommodate demand in other parts of the kraft process. Tight control of throughput is unnecessary as long as the average over an hour or two of operation is close to the target.

Manipulated variables that can be used to achieve the two objectives are the flowrate of the steam used to supply heat to the system and the flowrate of incoming WBL, as shown in Figure 5. A change in the WBL flowrate has a large impact on both control objectives (i.e., there is a strong interaction) whereas a change in the steam flowrate affects only the SBL solids concentration.

An approximate, linear, pulse-response model was derived from the non-linear process simulation. This was used as the internal model for IMC. Tests included the addition of random process noise and other efforts to identify the effects of mismatch between the internal model and the behavior of the actual process (represented by the non-linear simulation). The IMC strategy was then tested for a period of simulated plant operation and its performance was compared to that for a conventional PID-controller-based, divide-and-conquer strategy. Typical results are shown in Figures 6, 7 and 8.

Figure 6 shows a sequence of 500 minutes including an initial period in which the SBL target concentration (setpoint) is constant, followed by a 10% pulse in the setpoint. Note that between times designated 100 and 200 there was a 10% pulse in the setpoint for the throughput (see Figure 7). This caused large disturbances in the SBL concentration in the case of the PID control method whereas the IMC method eliminated the undesirable interaction between the two control objectives. Also, IMC tracks the pulse change in the SBL setpoint much more accurately than the PID controller.

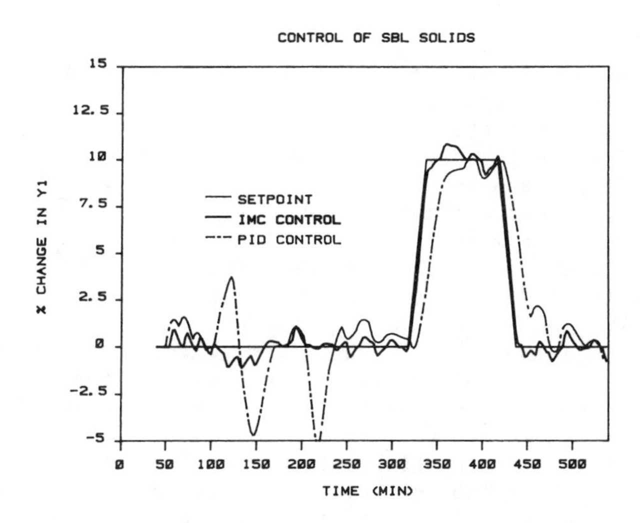

Figure 6: Comparison of IMC and PID Control of SBL Solids

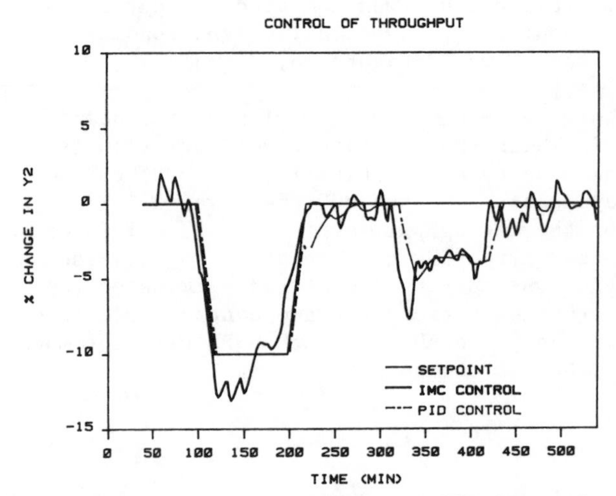

FIGURE 7: Comparison of IMC and PID Control of Throughput

Figure 7 shows control of the throughput (equivalent to manipulation of WBL flowrate) during the same period shown in Figure 6. Since the primary objective was control of SBL solids concentration, the weighting matrix, H, in the IMC objective function was defined to force IMC to minimize errors in SBL solids, sacrificing short-term control of the throughput when necessary to achieve the primary objective. This is the reason for the deviations from the setpoint between times 100 and 200 in the case of IMC. Note, however, that the average of the throughput in the IMC case was reasonably close to the target.

Between times 300 and 450, the steam flowrate was at its upper limit (see Figure 8) and it was necessary to reduce the throughput in order to meet the target on the SBL solids (which was increasing during that period). This was done automatically for both the IMC and the PID schemes, although it was necessary to resort to a rather elaborate control structure to achieve this in the case of the PID scheme (8). Again, proper initial specification of the H matrix was all that was necessary in IMC.

Figure 8 shows the changes in the steam flowrate that were needed to achieve the performance shown in Figures 6 and 7. It has been a common criticism of optimal control schemes that they tend to request extreme changes in the manipulated variables in order to achieve tight control. In Figure 8, it is clear that, if anything, the IMC method requires less extreme variations than the PID method.

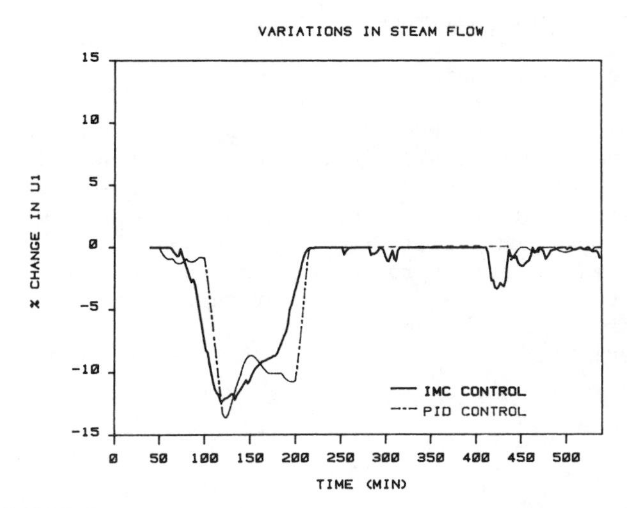

FIGURE 8: Comparison of IMC and PID Manipulation of Steam Flowrate

Another common criticism of optimal controllers is that they fail to compensate for abnormal conditions such as a manipulated variable being at its upper or lower limit. When this happens, the control system loses a degree of freedom and performance can easily deteriorate. In these examples, the upper limit on the steam flowrate is equivalent to $u_1=0$ in Figure 8. Note that even though possible variations in the steam flowrate are severely restricted between times 250 and 550, the performance of IMC is maintained. This is because the on-line quadratic programming method proposed by Ricker (8) allows the effect of such constraints to be evaluated explicitly and efficiently.

ADAPTIVE CONTROL

Although the IMC scheme described above is extremely flexible and has worked admirably in recent industrial applications, it has an important disadvantage: it assumes that each pulse-response in the G matrix is constant. In reality, a process evolves with time and its fundamental dynamic characteristics change. The utility of a predictive scheme like IMC hinges on whether the algorithm can maintain adequate control in spite of the inevitable mismatches between the predictions of the internal model and the behavior of the real process. Tolerance of such mismatches is termed "robustness."

It has been shown (7,8) that IMC has favorable robustness properties but that control performance degrades if the mismatch between the model and the process is large. This suggests the need to adjust the internal model to track changes in the process characteristics. Controllers of this type are said to be "adaptive" or "self-tuning."

The current status of the field of adaptive control has been reviewed by Seborg *et al*. (10). A detailed discussion is beyond the scope of this paper. It is clear, however, that if model-based methods are to be widely accepted, the determination and updating of the model will have to be automated. Excellent tutorial articles on this approach are available (11,12). Briefly, successful applications of adaptive control exist for simple problems (e.g., one controlled variable and one manipulated variable) but a solution of the general problem remains elusive.

SUMMARY

The majority of process control problems can be solved by a simple divide-and conquer strategy that requires little knowledge of the process dynamics. For more difficult problems, one can increase the accuracy of control by making a model of the subject process an integral part of the control algorithm. The IMC formulation is a particularly convenient way to do this and it is rapidly gaining acceptance. Quadratic programming is an accurate and efficient method for the required on-line optimization of the manipulated variables.

The model used in IMC is usually identified using data from a series of process tests. There is, however, an incentive to increase the intelligence of the controller by allowing it to determine its own process model and to adjust the model to track changes in the process characteristics.

In any case, chemometricians can help in solving control problems in several ways. First, since control improves as the data available from the process become more current and more accurate, it would be beneficial if chemometric techniques could be used to allow on-line analyses to be done more rapidly and/or with better accuracy. This would include, for example, methods to resolve overlapping peaks in a chromatogram. A second area in which research is needed is in the use of secondary measurements to accurately estimate a variable of primary interest, e.g., the use of several temperature measurements to estimate an unknown composition. A third area is in the development of on-line process model identification techniques that are reliable enough to be used in adaptive control systems. A fourth possibility is the use of chemometrics to detect and diagnose abnormal process conditions which could then be dealt with by process operators or by the automatic regulation system.

REFERENCES

1. Stephanopoulos, G., 1984, "Chemical Process Control," Prentice-Hall, New Jersey.

2. Luyben, W.L., 1973, "Process Modeling, Simulation, and Control for Chemical Engineers," McGraw-Hill, New York.

3. Shinskey, F.G., 1967, "Process Control Systems," McGraw-Hill.

4. Ray, W.H., 1981, "Advanced Process Control," McGraw-Hill.

5. Box, G.P.; Jenkins, G.M., 1976, "Time Series Analysis" (Revised Ed.), Holden-Day, San Francisco.

6. Garcia, C.E.; Morari, M., 1982, Ind. Eng. Chem. Process Des. Dev., 21, pp. 308-323.

7. Garcia, C.E.; Morari, M., 1982, "Internal Model Control II: Design Procedure for Multivariable Systems," submitted to Ind. Eng. Chem. Process Des. Dev.

8. Ricker, N.L., 1983, "Use of Quadratic Programming for Constrained Internal Model Control," submitted to Ind. Eng. Chem. Process Des. Dev.

9. Kouvaritakis, B.; Kleftouris, D., 1980, Int. J. Control, 31(1), pp. 127.

10. Seborg, D.E.; Shah, S.L.; Edgar, T.F., 1983, "Adaptive Control Strategies for Process Control: A Survey," paper from the AIChE Diamond Jubilee Meeting, Washington, D.C.

11. Harris, C.J.; Billings, S.A. (Ed.), 1981, "Self-Tuning and Adaptive Control: Theory and Applications", Peter Peregrinus Ltd., London.

12. Astrom, K.J., (Guest Editor), 1982, Optimal Control Applic. and Methods, 3(4).

DATA ANALYSIS IN CHROMATOGRAPHY

Dr. Ir. H.C. Smit

University of Amsterdam, Amsterdam, The Netherlands

INTRODUCTION

An ideal chromatogram shows a number of well-separated Gaussian peaks and a flat noise-free baseline. The required analytical information can be obtained by determining the height or the area of the peaks, the peak position and the peak width.

However, in reality a chromatogram is never ideal; the peaks are in general more or less asymmetrical (tailing), the separation is not always sufficient (low resolution), a non-stationary (drifting) baseline may be present as well as noise including spikes. Data processing and data analysis in quantitative chromatographic analysis involve not only simple peak area determination by integration, but may include peak deconvolution, noise reduction by (optimum) filtering, baseline drift correction, spike suppression, and peak finding procedures. The data processing has to be completed with an estimation of the systematic error and the uncertainty in the results.

In this paper some possible procedures in chromatographic data processing are discussed, including a treatment of the determination of the uncertainty in signal integrating analytical methods. The theoretical base is emphasized.

NOISE IN SIGNAL INTEGRATING METHODS

Let us consider the following problem. Suppose the systematic error in a chromatographic process, including conventional data processing (peak area determination), is negligible. The question is to determine the precision (uncertainty) of a

B. R. Kowalski (ed.), Chemometrics. Mathematics and Statistics in Chemistry, 225–250.

peak area determination, if the signal (peak) and the baseline noise are well-defined. The noise is supposed to be ergodic and therefore, stationary. Furthermore, it is assumed that the noise is independent of the signal. The dynamics of a linear time-independent system are specified by the impulse response, i.e. response h(t) to a δ function (Fig. 1).

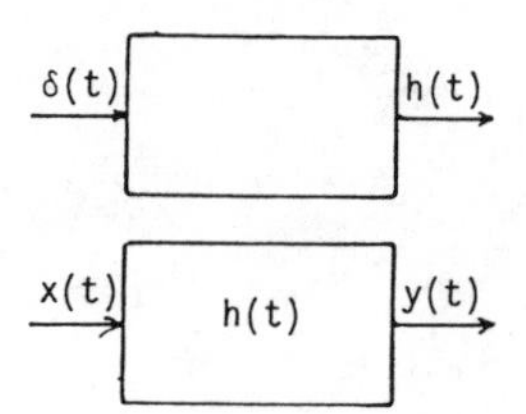

Fig. 1 Impulse response

The impulse response h(t) can be used to determine the response y(t) on an arbitrary time function x(t):

$$y(t) = \int_0^\infty h(\tau)\; x(t-\tau)\; d\tau = h(t) * x(t) \tag{1}$$

Equation (1) shows a convolution integral.
The generalised function $\delta(t)$ is defined by

$$\left.\begin{aligned} &\int_{-\infty}^{\infty} \delta(t)\; f(t)\; dt = f(0) \rightarrow \int_{-\infty}^{\infty} \delta(t)\; dt = 1 \\ &\delta(t) = 0 \qquad\qquad (t \neq 0) \end{aligned}\right\} \tag{2}$$

An integrator is a linear time-independent system with the output

$$y(t) = \int_{-\infty}^{t} x(t)\; dt \tag{3}$$

where x(t) is the input function.
The impulse response can be calculated by substituting $\delta(t)$ for x(t)

$$h(t) = \int_0^\infty \delta(t)\; dt = 1 \tag{4}$$

The complex frequence response $H(j\omega)$ is defined as the Fourier Transform of $h(t)$

$$H(j\omega) = FT\{h(t)\} \tag{5}$$

The power spectral density (PSD) of the $y(t)$ (see these proceedings, Time Series Analysis and Autocorrelation) can be calculated from $H(j\omega)$ and the PSD $G_x(\omega)$ of $x(t)$

$$G_y(\omega) = |H(j\omega)|^2 \, G_x(\omega) \tag{6}$$

$H(j\omega)$ of an ideal integrator is

$$H(j\omega) = FT\{h(t)\} = \frac{1}{j\omega} \tag{7}$$

Combining eq. (6) and eq. (7) gives

$$G_y(\omega) = \frac{1}{\omega^2} G_x(\omega) \tag{8}$$

The variance σ^2 of noise can be calculated by integrating the power spectrum

$$\sigma^2 = \int_0^\infty G(\omega)\, d\omega \tag{9}$$

The problem of determing the variance of integrated noise is apparently solved by integrating the $G_y(\omega)$ in eq. (9).

However, an infinite integration time is not realistic; in practice, a signal (peak) is integrated in an interval with time duration T. The practical impulse response is

$$\left.\begin{array}{ll} h(t) = 1 & (0 \le t < T) \\ h(t) = 0 & (\text{else}) \end{array}\right\} \tag{10}$$

The complex frequency response is

$$H(j\omega) = \int_{-\infty}^{+\infty} h(t)\, e^{-j\omega t} = \frac{1 - e^{-j\omega T}}{j\omega} \tag{11}$$

and

$$G_y(\omega) = |H(j\omega)|^2\, G_x(\omega) = \frac{\sin^2 \omega T/2}{(\omega/2)^2}\, G_x(\omega) \tag{12}$$

The variance is

$$k\overline{\sigma_I^2} = \int_0^{\infty} \frac{\sin^2(\omega T/2)}{(\omega/2)^2}\, G_x(\omega)\, d\omega =$$

$$= 2T \int_0^{\infty} G_x(\omega)\, \frac{\sin^2 \omega T/2}{(\omega T/2)^2}\, d\,\frac{\omega T}{2} \tag{13}$$

The bar and the index k are usual in annotating an ensemble average.

As an example eq. (13) will be applied to white noise, bandlimited by an ideal low pass filter with rectangular bandpass

$$\begin{aligned} G_x(\omega) &= K \qquad (0 < \omega < \omega_0) \\ G_x(\omega) &= 0 \qquad (\omega > \omega_0) \end{aligned} \tag{14}$$

The variance of the non-integrated baseline noise is

$$\sigma_n^2 = \int_0^{\infty} G(\omega)\, d\omega = \int_0^{\omega_0} G(\omega)\, d\omega \tag{15}$$

Hence

$$K = \frac{\sigma_n^2}{\omega_0} \tag{16}$$

Substituting $G_x(\omega)$ in the derived equation (13) gives

$$k\overline{\sigma_I^2} = \sigma_n^2 \frac{2T}{\omega_0} \int_0^{\omega_0 T/2} \frac{\sin^2(\omega T/2)}{(\omega T/2)} \, d\omega T/2 \tag{17}$$

$\omega_0 T/2$ has a large value in chromatography and the upper limit $\omega_0 T/2$ can be replaced by ∞ without much error. The value of the resulting integral is known ($= \frac{\pi}{2}$). The final result is

$$\left.\begin{aligned} k\overline{\sigma_I^2} &= \sigma_n^2 \frac{2T}{\omega_0} \cdot \frac{\pi}{2} = \pi \frac{T}{\omega_0} \\ k\overline{\sigma_I} &= \sigma_n \sqrt{\frac{\pi T}{\omega_0}} \end{aligned}\right\} \tag{18}$$

A more general treatment can be given in the time domain; primarily the demand of ergodicity is not maintained. Consider the random variable

$$I = \int_a^b n(t) \, dt \tag{19}$$

The problem is to determine the expected value $E\{I^2\}$ of I^2. $E\{I^2\} = \sigma_I^2$ can be used to calculate the uncertainty in integration procedures due to the integrated baseline noise, where $\sigma_I^2 = k\overline{\sigma_I^2}$, determined in the frequency domain I^2, is written as

$$I^2 = \int_a^b n(t_1) \, dt_1 \int_a^b n(t_2) \, dt_2 \tag{20}$$

Because of the independence of the integration limits eq. (20) can be written as

$$I^2 = \int_a^b \int_a^b n(t_1) \, n(t_2) \, dt_1 \, dt_2$$

Hence

$$E\{I^2\} = \int_a^b\int_a^b E\left\{n(t_1)\ n(t_2)\right\} dt_1\ dt_2$$

$$= \int_a^b\int_a^b R(t_1,t_2)\ dt_1\ dt_2 = \sigma_I^2 \tag{21}$$

The autocorrelation function $R(t_1,t_2)$ appears more or less naturally. If the integration interval lies between $-\frac{T}{2}$ and $\frac{T}{2}$, eq. (21) becomes

$$\sigma_I^2 = \int_{-T/2}^{T/2}\int_{-T/2}^{T/2} R(t_1,t_2)\ dt_1\ dt_2$$

If the baseline noise is assumed to be stationary, then the result is

$$\sigma_I^2 = \int_{-T/2}^{T/2}\int_{-T/2}^{T/2} R(t_1-t_2)\ dt_1\ dt_2 \tag{22}$$

This equation can be simplified [1]

$$\int_{-T/2}^{T/2}\int_{-T/2}^{T/2} R(t_1-t_2)\ dt_1\ dt_2 = \int_{-T}^{T} (T-|\tau|)\ R(\tau)d\tau = \sigma_I^2 \tag{23}$$

Equation (23) can be used for the determination of the uncertainty if peaks are integrated in the presence of (stationary) baseline noise with autocorrelation function $R(\tau)$ and integration time T.

The autocorrelation function (ACF) of first order noise, i.e. white noise bandlimited by a first order low pass filter with time constant T_1, is

$$R(\tau) = \sigma_n^2 \exp\left(\frac{-|\tau|}{T_1}\right) \tag{24}$$

where σ_n^2 is the variance of the first order noise. Substituting eq. (24) in eq. (23) gives

$$\sigma_I^2 = \sigma_n^2 \left[2\, TT_1 + 2\, T_1^2 \left\{ \exp\left(- \frac{T}{T_1} \right) - 1 \right\} \right] \tag{25}$$

In chromatographic practice ($T >> T_1$) eq. (25) can be reduced to

$$\sigma_I^2 = \sigma_n^2 \cdot 2\, TT_1 \tag{26}$$

Also, a practical ACF of a detector can be determined. Substitution in eq. (23) gives σ_I for each value of the integration time T: Figure 2 shows an example from practice. A detailed description of the treated problem is given in [1].

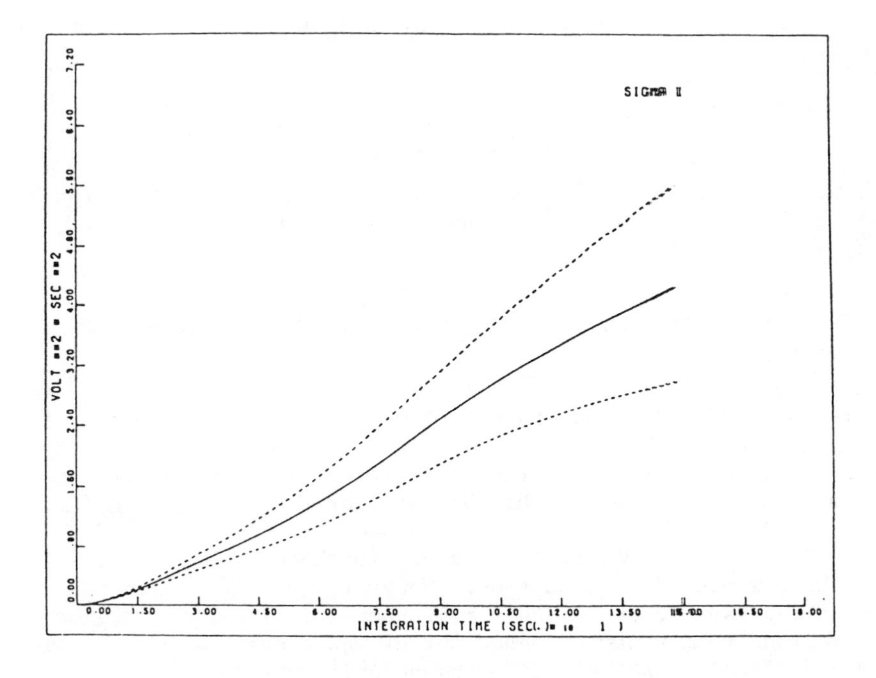

Fig. 2 The variance of a piece of integrated baseline with a 90% confidence interval. The integration was performed after application of a 2nd order correction. Noise source: FID detector.

SIGNAL ESTIMATION

For the case of conventional peak area determination by integration, the uncertainty can be calculated if stationary noise is present. The question arises, if it is possible to determine signal characteristics in a more optimal way with the intention to lower the uncertainty in the analytical results. Filtering and statistical signal estimation procedures are possibilities to enhance the signal to noise ratio.

However, it can be proved that low pass filtering before peak integration is not profitable in this respect [1]. This is not necessarily true in other cases; for instance, peak finding procedures generally require filtering.

Many filter- and signal estimation methods are known. In this paper some attention will be paid to non-linear filtering and approximation of signals by orthogonal polynomials. A short introduction will be given.

A signal f(t) can be developed in a series of functions g(t)

$$f(t) = a_1 g_1(t) + a_2 g_2(t) + \ldots\ldots + a_n g_n(t) \tag{27}$$

To determine each of the constant coefficients a_n, for instance a_k, both sides are multiplied by $g_n(t)$ and integrated.

$$\int_{-\infty}^{+\infty} f(t)\, g_n(t)\, dt = a_1 \int_{-\infty}^{+\infty} g_1(t)\, g_n(t)\, dt + \ldots\ldots + a_N \int_{-\infty}^{+\infty} g_N(t)\, g_n(t)\, dt \tag{28}$$

This can be done for all values of n and we obtain a number of results that may be used to calculate the coefficients a_n. To provide good accuracy and fast convergence (minimum number of terms), a careful choice must be made for the functions $g_n(t)$.

If the series 1, t, t^2, t^3 etc. are chosen, the result will be the moments of the function. However, quite a number of equations have to be solved simultaneously. Addition of another term means that a new extended set of equations has to be solved. This difficulty can be avoided by choosing the functions to be orthonormal, i.e. the integral of the product of two different functions g_n and g_m is zero; the integrated product $g_n . g_m$ is 1 if n = m. If this product is not 1, but some other constant (not zero), then the functions are orthogonal. Orthonormality resp. orthogonality means that addition of an extra term requires only one extra calculation, the other coefficients remaining unchanged. Several kinds of orthogonal polynomials are known. The classical polynomials are orthogonal with respect to a weighting function (orthogonality condition)

$$\int_a^b w(t)\, p_m(t)\, p_n(t)\, dt = 0 \qquad m \neq n \tag{29}$$

where $w(t)$ is the weighting function and $p_m(t)$ and $p_n(t)$ are polynomials of degree m and n [2]. An example of one of the classical orthogonal polynomials is the Hermite polynomial. The domain is $-\infty$ to $+\infty$, and the weighting function is e^{-t^2}. From the definitions, a general expression can be determined (a Rodrigues formula),

$$H_n(t) = (-1)^n e^{t^2} \frac{d^n}{dt^n} e^{-t^2}$$

and a recurrence relation can be derived, usable for computer calculations and generation of the polynomial

$$H_{n+1}(t) = 2t\,H_n(t) - 2n\,H_{n-1}(t) \tag{30}$$

The classical Hermite polynomial can be modified by multiplying each function with $\exp\left(-\frac{t^2}{2}\right)$

$$He_n(t) = H_n(t)\exp\left(-\frac{t^2}{2}\right) \tag{31}$$

orthogonal with respect to $w(t) = 1$.

The scalar product is

$$\langle He_m, He_n \rangle = \int_{-\infty}^{+\infty} He_m\, He_n\, dt = \begin{cases} 0 & n \neq m \\ \sqrt{\pi}\, 2^n n! & n = m \end{cases} \tag{32}$$

The recurrence relation is

$$He_{n+1}(t) = 2\,t\,He_n(t) - 2\,n\,He_{n-1}(t) \qquad (33)$$

The resulting polynomial is extremely useful to describe bell-shaped curves with a minimum of terms. Bell-shaped curves are very usual in analytical chemistry.

If only a few terms are selected, the polynomial approximation works like an efficient filter. If the number of peaks is increasing, then the Hermite approach will fail. A better solution is the use of the Chebyshev polynomial, with the domain $-1 \to +1$ and $w(t) = (1 - t^2)^{-\frac{1}{2}}$. The terms can be calculated from

$$T_n = \cos\,(n \arccos t) \qquad (34)$$

and the recurrence relation is

$$T_{n+1}(t) = 2\,t\,T_n(t) - T_{n-1}(t) \qquad (35)$$

The Chebyshev polynomial is very useful because the maxima and minima are of comparable magnitude and the error is spread reasonably uniform over the range. Fitting with this polynomial is not the same as a least square fit. In the least square fit one attempts to minimize the average of the square of the error and this does not preclude occasional large errors. Here the maximum value of the error is minimized.

Figure 3 shows an example of the Chebyshev polynomial approximation (filtering) of a noisy chromatogram.

NON-LINEAR REGRESSION

Non-linear regression methods can be applied in chromatography for the estimation of noisy signals, signals with baseline drift and peaks with low resolution. A least square fit to a chromatogram with functions, non-linear in the parameters, has to be made, see [3]. The parameters cannot be separated into different terms of a sum. The goodness of fit χ^2 is defined as

$$\chi^2 = \sum_{i=1}^{m} \left\{ \frac{1}{\sigma_I^2} \left[y_i - y(x_i)\right]^2 \right\} \qquad (36)$$

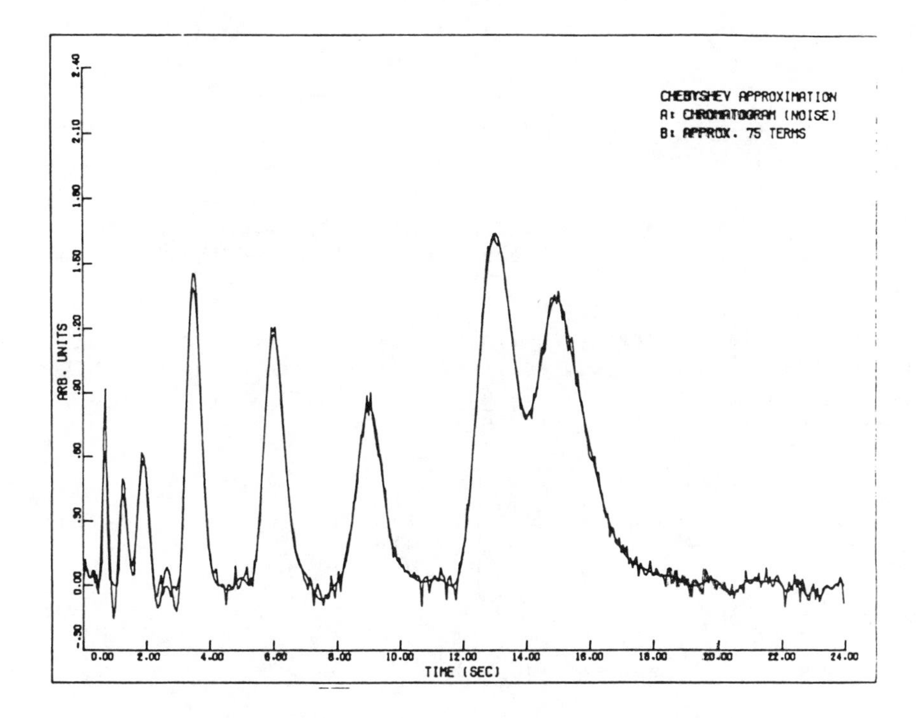

Fig. 3 Chebyshev approximation of a simulated noisy chromatogram.

where the terms are defined as follows:

y_i = data

$y(x_i)$ = fitting function

σ_I^2 = uncertainty in data points

m = number of data

For example, y(x) would be a Gaussian peak plus quadratic baseline [3]:

$$y(x) = a_1 \exp\left[-\frac{1}{2}\left(\frac{x-a_2}{a_3}\right)^2\right] + a_4 + a_5 x + a_6 x^2 \quad (37)$$

The non-linear regression procedure means minimizing χ^2 with respect to each of the parameters a_j

$$\frac{\partial}{\partial a_j} \chi^2 = \frac{\partial}{\partial a_j} \sum \left\{ \frac{1}{\sigma_i^2} \left[y_i - y(x_i) \right]^2 \right\} = 0 \tag{38}$$

Here, χ must be considered as a continuous function of the n parameters a_j describing a higher surface in a n-dimensional space.

In our laboratory a non-linear regression software package (ROAD) has been developed, and is usable in chromatography. The fitting function can be freely chosen; in the examples the Fraser-Suzuki function is used [4]

$$f(t) = H \exp \left[\frac{-\ln 2}{A^2} \left\{ \ln \left(1 + 2A \frac{t-P}{W} \right) \right\}^2 \right] \tag{39}$$

where H = height

P = place (retention time)

$W = \sqrt{8 \ln 2} \quad \sigma$ = peak width

A = asymmetry factor

The stop criterion is the relative change in χ^2 and the Marquardt algorithm is applied [3]. The program is interactive; starting parameters estimated by the operator have to be introduced. A higher order baseline is estimated simultaneously. The program contains several safety provisions. Figures 4, 5 and 6 show some examples of the applications of the program.

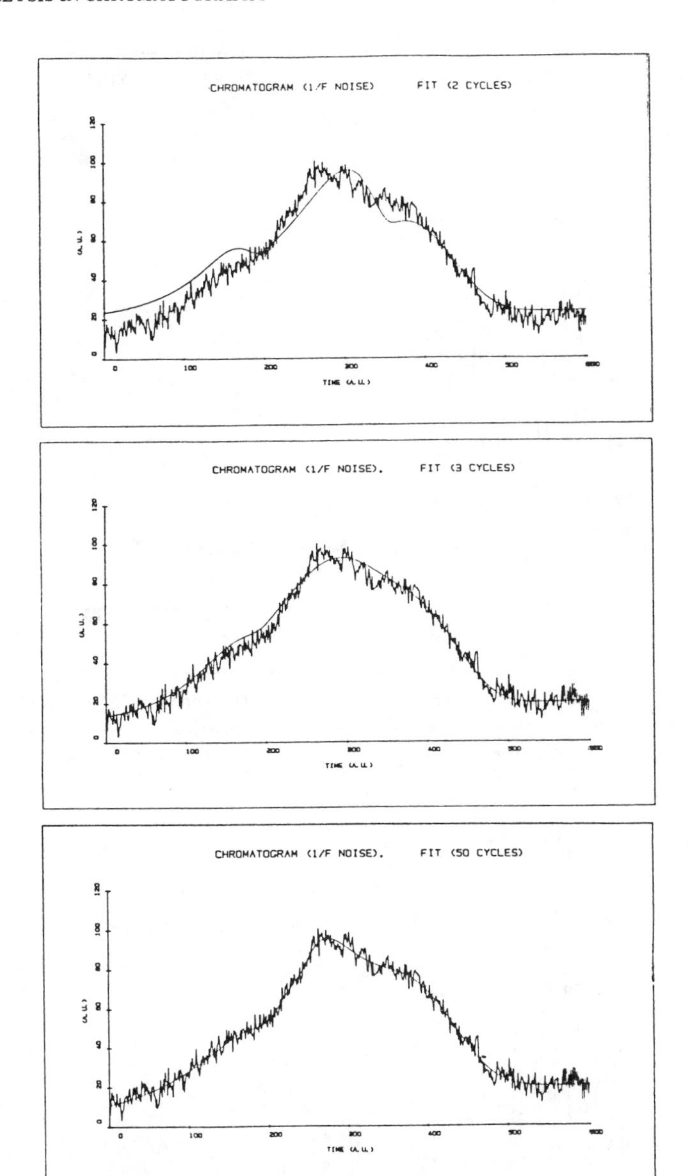

Fig. 4 Simulated chromatogram with 3 peaks. Quadratic baseline with superimposed 1/f noise. X^2 after 50 cycles: 13. Error in estimated parameter values: < 1%. Time: 5 minutes.

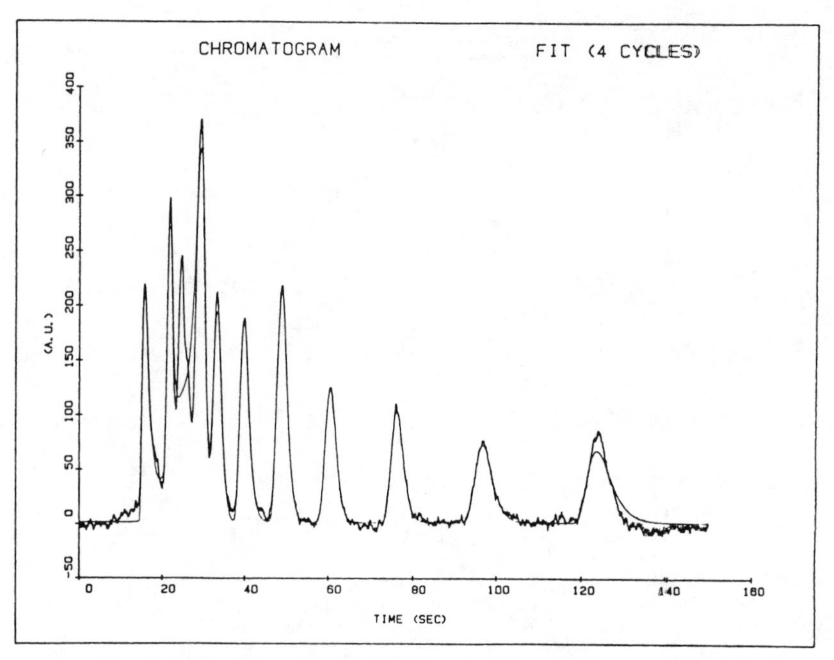

Fig. 5 HPLC chromatogram. Phenols and nitrophenols. 11 peaks, 4 cycles.

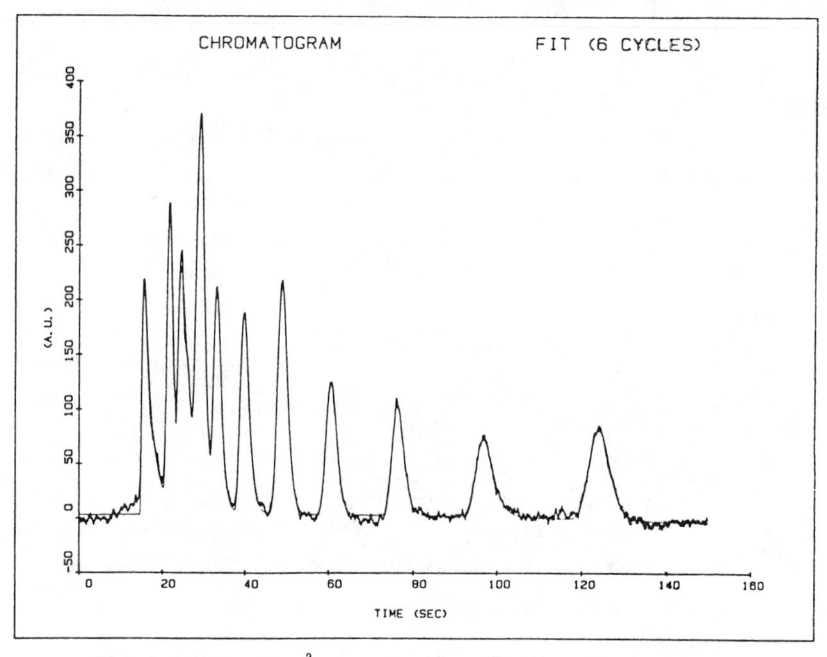

Fig. 6 See Fig. 5. χ^2 after 6 cycles: 25. Time: 38 minutes.

CORRELATION TECHNIQUES

Correlation techniques can be applied in chromatography in several ways. The usual peak area or peak height determination in quantitative chromatographic analysis does not take advantage of preknowledge about the shape of the peak. An alternative way to obtain a measure for the "intensity" of a peak, directly related to the concentration of the relating component, is to cross-correlate a peak in the chromatogram having a known peak shape with a "standard" peak. This procedure is known as correlation reception and is closely related to matched filtering. In order to establish the increase in signal to noise ratio, a part of a noisy chromatogram with two Gaussian peaks will be considered. The "chromatogram" can be described by

$$f(t) = K_1 \exp\left(-\frac{t^2}{\sigma^2}\right) + K_2 \exp - \frac{(t-\mu)^2}{2\sigma^2} \tag{40}$$

Cross-correlation of f(t) with a "standard" function $\check{f}(t)$ means the determination of the integrated product of f(t) and time τ shifted version $f_s(t + \tau)$ of $f_x(t)$. The cross-correlation function is

$$F(\tau) = \int_{-\infty}^{+\infty} f(t)\, f_s(t+\tau)dt = K_1 \sigma\sqrt{\pi} \exp - \frac{\tau^2}{2(\sigma\sqrt{2})^2} + K_2 \sigma\sqrt{\pi} \exp - \frac{(\tau-\mu)^2}{2(\sigma\sqrt{2})^2} \tag{41}$$

The peak heights $K_1\sigma\sqrt{\pi}$ and $K_2\sigma\sqrt{\pi}$ can be used as a quantitative measure for the concentration ot the two components. Of course, in practice the integration time T is finite. The noise n(t) is also multiplied by $f_s(t)$ and integrated (Fig. 7)

$$I_n = \int_{-T/2}^{T/2} f_s(t)\, n(t)\, dt \tag{42}$$

with integration interval $-T/2 \rightarrow T/2$.

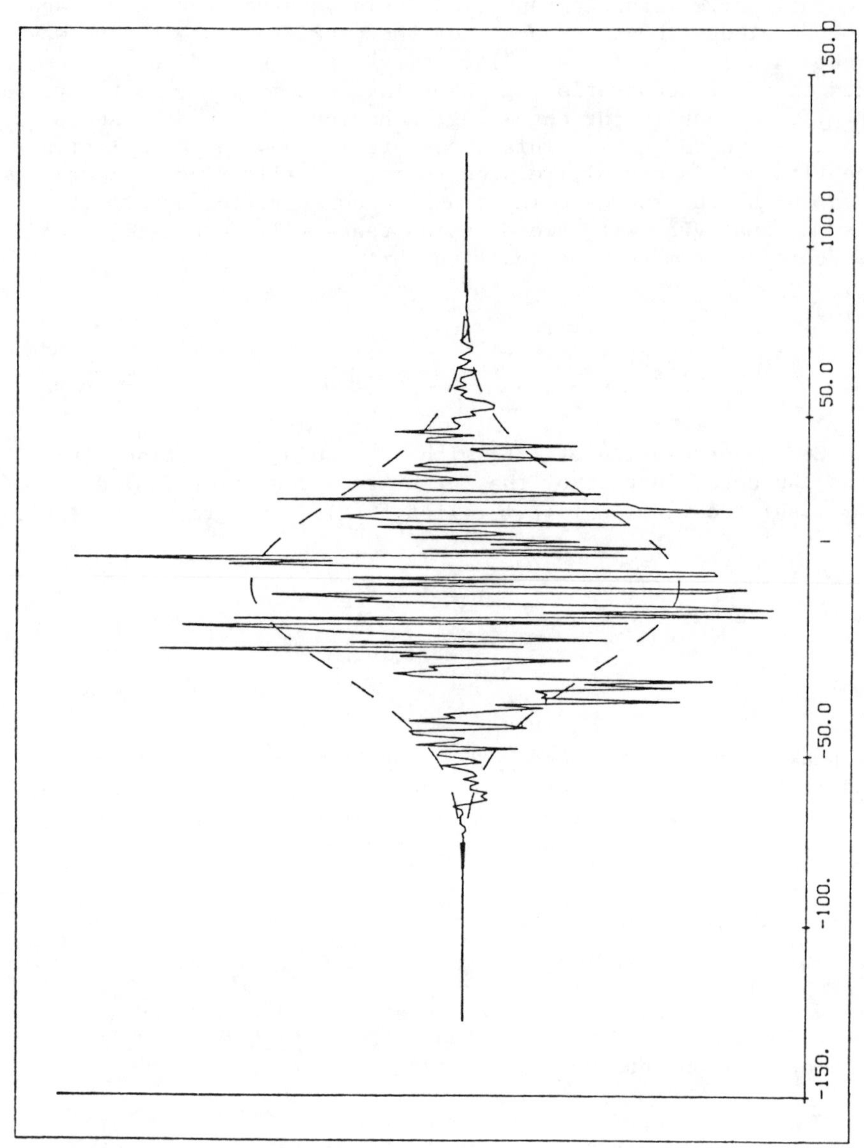

Fig. 7 Noise multiplied by function $f_s(t)$.

In order to estimate the standard deviation of I_n, we form

$$I_n^2 = \int_{-T/2}^{T/2} \int_{-T/2}^{T/2} f_s(t_1)\, f_s(t_2)\, n(t_1)\, n(t_2)\, dt_1\, dt_2 \tag{43}$$

The expected value is

$$E\{I_n^2\} = \int_{-T/2}^{T/2} \int_{-T/2}^{T/2} E\{f_s(t_1)\, f_s(t_2)\, n(t_1)\, n(t_2)\}\, dt_1\, dt_2$$

$$= \int_{-T/2}^{T/2} \int_{-T/2}^{T/2} f_s(t_1)\, f_s(t_2)\, R(t_1,t_2)\, dt_1\, dt_2 \tag{44}$$

The ratio of the peak height $K_1 \sigma \sqrt{\pi}$ with respect to $K_2 \sigma \sqrt{\pi}$ in eq. (41) and the square root of $E\{I^2\}$ determines the uncertainty in the result and has to be compared with the ratio of the peak area and the standard deviation σ_I of the integrated noise. In case of stationary first noise with an ACF

$$R(\tau) = \sigma_r^2 \exp - \frac{|\tau|}{T_1} \tag{45}$$

we obtain

$$E\{I^2\} = \int_{-T/2}^{T/2} \int_{-T/2}^{T/2} \exp - \frac{t_1}{2\sigma^2} \cdot \exp - \frac{t_2}{2\sigma^2} \cdot \sigma_n^2 \exp - \frac{|t_1-t_2|}{T_1}\, dt_1\, dt_2 \tag{46}$$

This integral can be solved and the increase in signal to noise compared with conventional area determination can be calculated. The signal to noise ratio is increased about a factor of two in this case.

CORRELATION CHROMATOGRAPHY

A quite different approach is correlation chromatography (CC). In conventional chromatography a chromatogram is obtained by a pulse-shaped injection. A chromatogram can be considered as the impulse response of the chromatographic system. However, the impulse response of a system can be determined by cross-correlation of a suitable input signal and the resulting output signal. The input signal must be a stochastic signal ("noise"). This noise must be "white", i.e. the power spectral density has to be level up to relatively high frequencies, as described in the following derivation.

The cross-correlation function of two signals, in this case the input signal x(t) and the output signal y(t) of a linear process, is by definition

$$R_{xy}(t_1,t_2) = E\left[x(t_1)\ y(t_2)\right] \tag{47}$$

$E\left[\ \right]$ denotes the expected value of the expression between the brackets. The output signal y(t) of a linear system can be calculated as a convolution of the input signal x(t) and the impulse response h(t) of the system:

$$y(t) = x(t) * h(t) = \int_0^\infty h(\tau)\ x(t-\tau)\,d\tau \tag{48}$$

Combining eq. (47) and eq. (48) gives

$$R_{xy}(t_1,t_2) = E\left[x(t_1)\int_0^\infty h(\tau)\ x(t_2-\tau)\,d\tau\right] \tag{49}$$

In this case integration and averaging can be interchanged, hence

$$R_{xy}(t_1,t_2) = \int_0^\infty h(\tau)\ E\left[x(t_1)\ x(t_2-\tau)\right]d\tau \tag{50}$$

However, the autocorrelation function of x(t) is defined as

$$R_{xx}(t_1,t_2) = E\left[x(t_1).x(t_2)\right] \tag{51}$$

and eq. (50) and (51) can be combined to

$$R_{xy}(t_1,t_2) = \int_0^\infty h(\tau)\, R_{xx}(t_1,t_2-\tau)\, d\tau \qquad (\tau > 0) \qquad (52)$$

If x(t) is stationary, then

$$R_{xx}(t_1,t_2-\tau) = R(t_2-\tau-t_1) = R(t-\tau) \qquad (53)$$

and eq. (6) becomes

$$R_{xy}(t_1,t_2) = R_{xy}(t) = \int_0^\infty h(\tau)\, R_{xx}(t-\tau)\, d\tau \qquad (54)$$

Comparing eq. (48) and eq. (54) shows that the output signal y(t) of a linear system with an input signal equal to the autocorrelation function $R_{xx}(t)$ of a signal x(t) is similar to the cross-correlation function $R_{xy}(t)$ of the input signal x(t) and the output signal y(t) resulting from x(t). A white noise, that is, white with respect to the bandpass of the system, has an impulse-shaped autocorrelation function and can be used as input function x(t) to determine the impulse response.

On further consideration a chromatographic procedure can be regarded as the determination of an impulse response; a chromatogram shows the response on the impulse-shaped injection of the sample. The prime objective of correlation chromatography is to determine the chromatogram by stochastically injecting the sample into the column and cross-correlating the input pattern and the resulting output. The noise of the chromatographic system is not correlated with the input, its contribution to the overall cross-correlation function converges to zero with increasing correlation time. A considerable improvement of the signal to noise ratio can be achieved in a relatively short time.

The separating power of the column is used very efficiently; the average dilution of the sample is only a factor of 2, which is very low compared with a factor of 500 or 600 as in conventional chromatography.

APPLICATION IN PRACTICE

The most suitable random input function, controlling the input flow of the sample, is the pseudo random binary sequence (PRBS). This function is to be preferred to other random inputs with approximately impulse-shaped autocorrelation functions for the following reasons:

1) It is a binary noise, the only two levels being +1 and -1 or +1 and 0. The levels can be used to control simple on/off valves and correspond with the injection of sample and eluent, respectively.

2) This function can easily be generated and reproduced.

3) Its special properties offer the possibility of reducing the so-called correlation noise, which is caused by a limited correlation time.

The PRBS is a logical function combining the properties of a true binary random signal with those of a reproducible deterministic signal. After a certain time (a sequence) the pattern is repeated. The PRBS generator is controlled by an internal clock, a PRBS is considered with a sequence length N and a clockperiod Δt. It is important that the estimated autocorrelation function of a PRBS, if computed over an integral number of sequences, is at all times exactly equal to the autocorrelation function. The autocorrelation function is a triangle whose base is twice as long as the clockperiod. If the clockperiod is small compared with the retention time and the width of the peaks, the triangle can be considered to be an impulse.

Figure 8 shows a simplified diagram of a correlation chromatography system. Apart from the necessary modifications of the injecting system, essential extensions to a chromatographic setup to allow CC are:

a) The pattern generator, necessary for generating either a single pulse (normal chromatography) or a PRBS (CC), and used for stimulation of the column (valve switching) and for calculation of the cross-correlation function (CCF) from the detector output and the pattern.
b) The data sampler, which is used to sample and digitize the filtered electric detector signal.
c) An arithmetic unit to calculate the CCF.
d) A display for the results.

Some extensions are not essential for CC, but greatly improve its capabilities. Thus, an interface to a data storage medium like a digital cassette or a mini-computer is useful. An interface

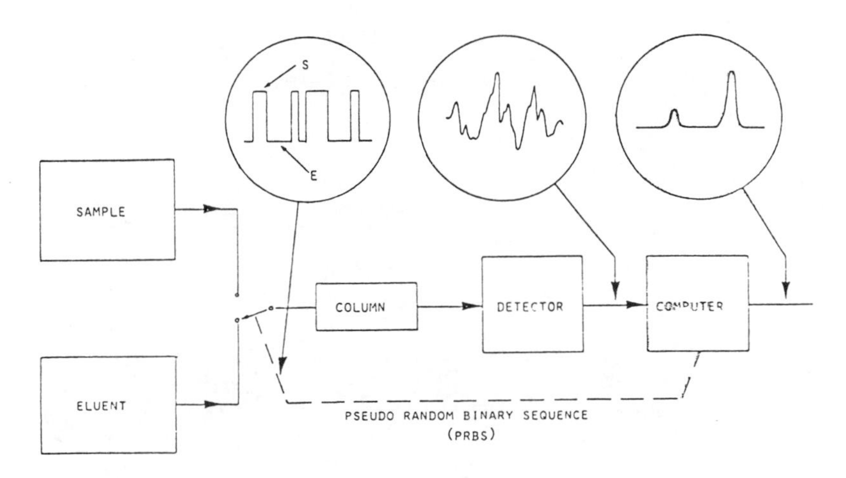

Fig. 8 Basic diagram of a correlation chromatograph.

to a hard copy unit is valuable and a facility for applying base-line corrections and for integrating (individual) peaks is desirable. In general, baseline drift correction with a polygon subtracted from the chromatogram (correlogram) is sufficient. Finally, the standard deviation of the baseline noise of the correlogram and the standard deviation of the integral should be determined.

A microprocessor is ideal in CC. In our laboratory we have developed a micro-processor-based correlator, which meets all the requirements mentioned for CC [5]. It has until now been mainly intended for use in trace analysis; other potential applications of CC (process control) require modification.

RESULTS IN TRACE ANALYSIS

The correlator has been used to obtain chromatograms (correlograms) of mixtures of polychlorophenols [6] (pentachlorophenol (PCP) and isomers of tetra-, tri- and dichlorophenol) at very low concentrations. PCP is used in wood preservatives; the less chlorinated compounds are present as impurities or are formed in the environment. PCP pollution has been reported as an environmental problem; it may contaminate soils as well as surface water.

The modified chromatographic set-up including the injection system used in the experiments is shown in Fig. 9.

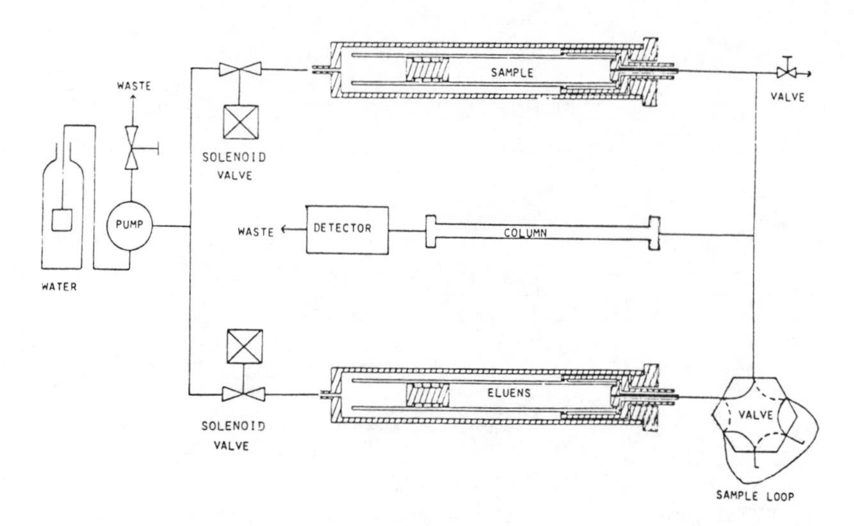

Fig. 9 Complete set-up of a correlation HPLC system. The constant water flow is, depending on a PRBS pattern, directed to either the sample or the eluent reservoir, driving a plunger forward. At the outlet of the eluent reservoir is a 6-way rotary valve allowing single injection experiments.

Figure 10 shows a plot of a conventional chromatogram representing the HPLC separation of twelve chlorinated phenols. The corresponding HPLC correlogram, the detector output during the correlation, and the origin of the correlogram are shown in Fig. 11 and Fig. 12 respectively. The concentrations shown in the chromatogram and in the correlogram differ by a factor of 50 (10 ppm and 200 ppb, respectively). During the experiments all correlograms showed large differential peaks near t = 150 s, probably caused by oxidation products of a component of the eluent, THF (tetrahydrofuran), and oxidation/degradation products of parts of the injection system.

CC is essentially a differential method and negative peaks caused by components in the eluent and positive peaks caused by components in the sample can be present. Apart from some minor differences in resolution, due to slightly changing separation conditions (temperature), the presence of extra peaks in the correlogram compared with the chromatogram should be noted. The small peak at 138 s in the correlogram is presumably an isomer and the large peak at 35 s appears in the wrong place. Probably, an unknown contaminant has given rise to a peak somewhere after the last "regular" peak. The chosen duration of the chromatogram was too short and so the peak was folded back in the correlogram, fortunately in an empty part.

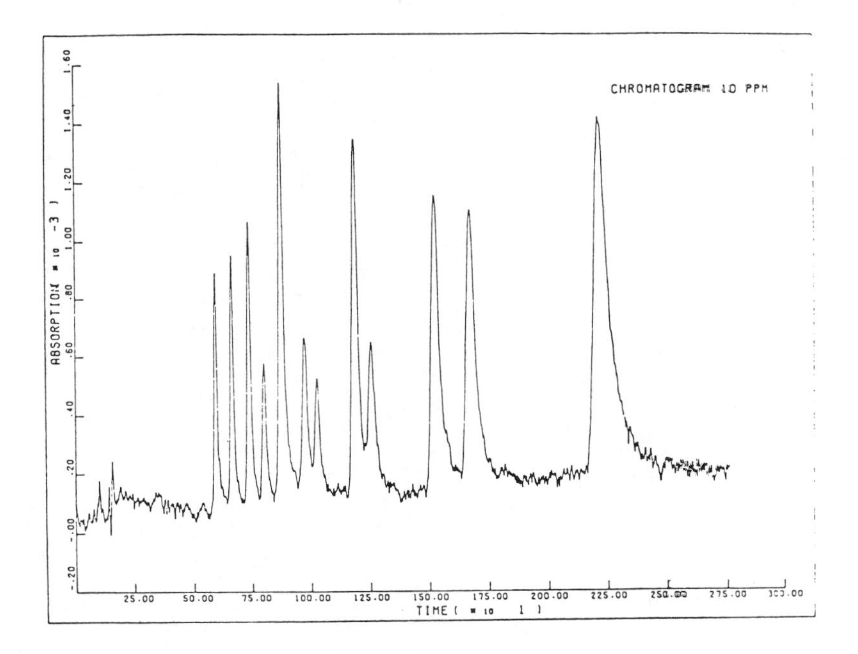

Fig. 10 Separation of twelve different chlorinated phenols by conventional HPLC. The concentration of each component is 10 ppm.

The analytical performance of CC is demonstrated in Fig. 13. A calibration curve of phenol was determined over five orders of magnitude of concentration: 0.01 - 100 $\mu g\ l^{-1}$. Conventional HPLC equipment with fluorimetric detection and a newly developed injection device for correlation HPLC [7] were used. The two higher concentrations (10 - 100 $\mu g\ l^{-1}$) were determined by conventional (reverse phase) HPLC and the two lower concentrations (0.01 - 0.1 $\mu g\ l^{-1}$) by correlation HPLC with 16 and 3 sequences of correlation time, respectively. Measurements at the 1 $\mu g\ l^{-1}$ level were performed both by conventional and correlation HPLC (1 sequence).

The bars indicated on the calibration graph represent the peak area ± 3 σ_I (arbitrary units), when σ_I is the standard deviation of the integrated noise [8]. The inner bars at the 1 $\mu g\ l^{-1}$ level represent the correlation results and the outer bars the single injection results.

The detection limit for the single injection experiments, defined as 3 σ_I, was about 0.5 $\mu g\ l^{-1}$. The detection limit with the 10 ng l^{-1} concentration was estimated to be 3 ng l^{-1} (3 ppt = 3 parts per trillion) (Fig. 14).

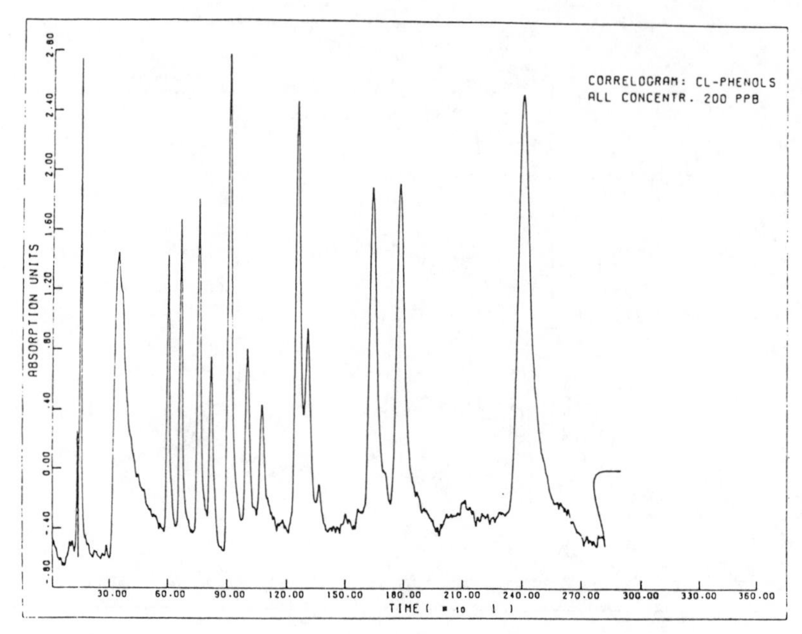

Fig. 11 Correlogram according to Fig. 10 with slightly different separation conditions. The concentration of each component is 0.2 ppm.

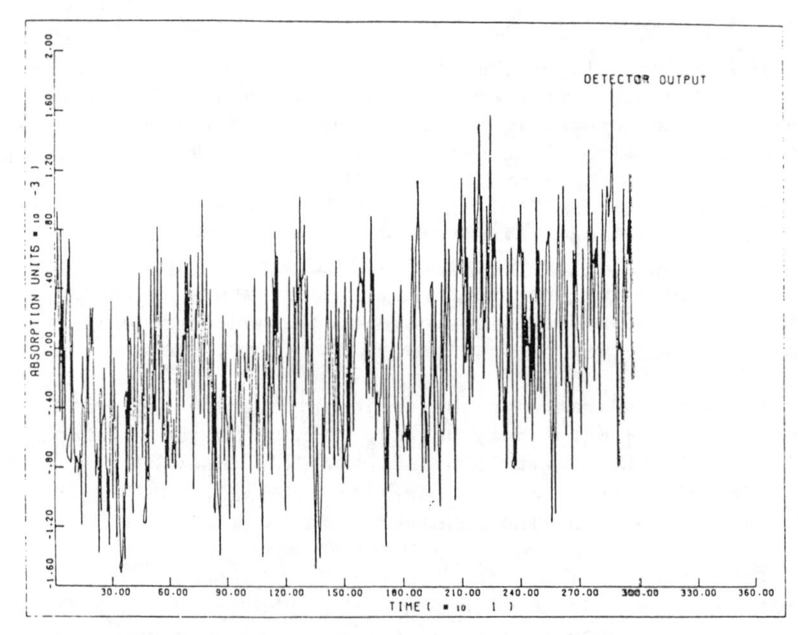

Fig. 12 Detector output during correlation, leading to the correlogram of Fig. 10.

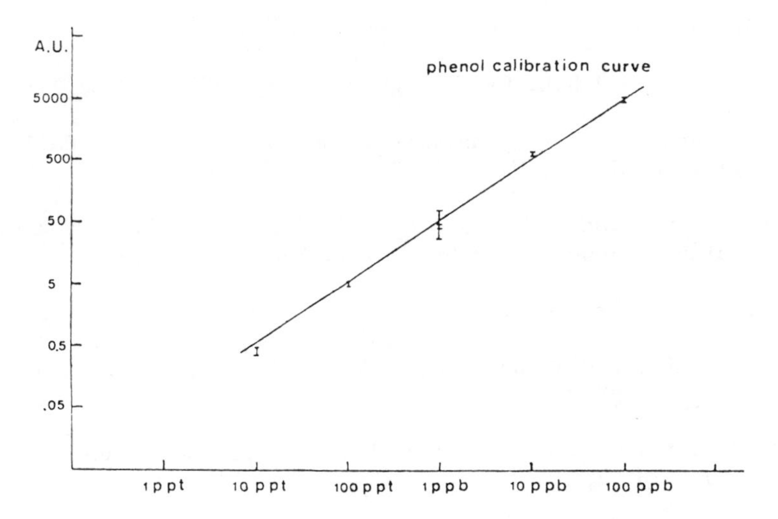

Fig. 13 Calibration graph of phenol with fluorimetric detection.

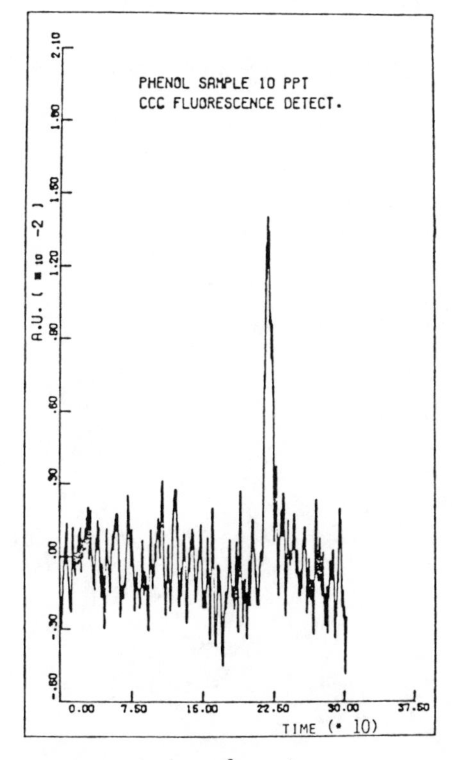

Fig. 14 Correlogram of a 10^{-8} g l^{-1} phenol sample. Detection limit for 80 minutes correlation time is approximately 3 ppt (3.10^{-9}.g l^{-1}).

REFERENCES

1) H.C. Smit and H.L. Walg, "Chromatographia" 8 (1975) 311.

2) R. Deutsch, "System Analysis Techniques." Prentice Hall, Englewood Cliffs, N.Y., 1969.

3) P.R. Bevington, "Data Reduction and Error Analysis for the Physical Sciences." McGraw-Hill, New York, 1969.

4) R.D.B. Fraser and E. Suzuki, Anal. Chem. 41 (1979) 37.

5) H.C. Smit, R.P.J. Duursma and H. Steigstra, Anal. Chim. Acta 133 (1981) 283.

6) H.C. Smit, T.T. Lub and W.J. Vloon, Anal. Chim. Acta 122 (1980) 276.

7) J.M. Laeven, H.C. Smit and J.C. Kraak, Anal. Chim. Acta 150 (1983) 253.

8) R.P.J. Duursma and H.C. Smit, Anal. Chim. Acta 133 (1981) 67.

EXPERIMENTAL DESIGN: RESPONSE SURFACES

Stanley N. Deming

Department of Chemistry, Emory University
Atlanta, Georgia 30322

I: SIMPLEX OPTIMIZATION OF VARIABLES IN ANALYTICAL CHEMISTRY

Experimental optimization by the sequential simplex method or its modifications involves trivial calculations readily done on a small digital computer or even by hand. Applications in analytical chemistry are almost unlimited. Consider the following common activities of analytical chemists:

Methods Development

In the development of new analytical methods and in the improvement or adaptation of established ones, factorial experiments and the analysis of variance are often used to investigate the effects of many continuous variables (such as reagent concentration, pH, temperature) upon the results of a give method. Those variables that exhibit a significant effect can then be adjusted to improve the results of the method, e.g., to increase sensitivity, to decrease side reactions, to improve separation.

Instrumentation

To perform at their best, complex analytical instruments must be "tuned up," a procedure that involves adjusting several instrumental parameters until optimum response is obtained.

Data Treatment

When fitting experimental data to a known mathematical model, variables in the theoretical equation are adjusted until calculated values are in close agreement with the experimental values.

B. R. Kowalski (ed.), Chemometrics. Mathematics and Statistics in Chemistry, 251–266.

All of the above activities are concerned with optimization, "the collective process of finding the set of conditions required to achieve the best result from a given situation" (1). If the variables do not interact with each other, each variable can be optimized independently of the other variables. In general, however, variables do interact with each other, and the one-factor-at-a-time approach will not always result in an optimum set of conditions. This Report presents a simple, straightforward approach to the optimization of a system in which the variables might interact.

Experimental optimization of chemical reactions has always been of strong economic concern to chemical engineers and industrial mathematicians (1). The general problem of experimentally optimizing a function of several variables was stated by Hotelling (2) and discussed by Friedman and Savage (3) who presented a sequential one-factor-at-a-time optimization procedure. Box and Wilson (4) showed that many factors can be varied at the same time to arrive at and track the optimum, thus establishing the foundations of the "evolutionary operation" (EVOP) procedure (5). EVOP techniques were designed to analyze the results of small variations in the operating conditions of industrial processes. Because of the accompanying small changes in response, many measurements are required to obtain a statistically valid decision as to the best direction to move to reach the optimum.

A more efficient method of optimization, the "sequential simplex" method, was first presented by Spendley et al. (6) and later applied to analytical chemistry by Long (7). This method does not use traditional testing of significance and is therefore faster and simpler than previous methods. It rapidly attains the experimental optimum, guided by calculations and decisions that are rigidly specified, yet almost trivially simple, which make it particularly attractive for automated optimization. The simplex method is efficient and can be easily justified for many applications.

SIMPLEX METHOD

A simplex is a geometric figure defined by a number of points equal to one more than the number of dimensions of the space. A simplex in two dimensions is a triangle. A simplex in three dimensions is a tetrahedron. The series can be extended to higher dimensions, but the simplexes are not easily visualized. Figure 1 shows a two-dimensional simplex superimposed on a contour map which represents the various iso-response lines of a function of two variables. An example of such a function might be the initial reaction rate of an enzyme-catalyzed reaction as a function of temperature and of pH. The parameters to be varied are chosen by initial factorial experiments. Single experiments are carried out to determine the responses at points A, B, and C, the vertices of the simplex.

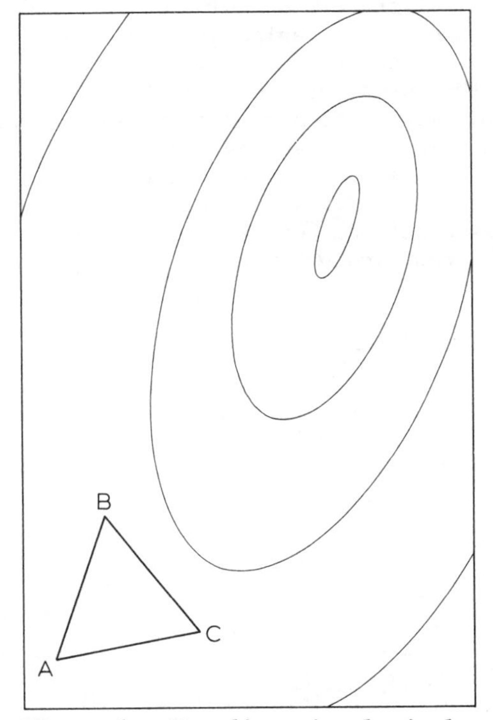

Figure 1. Two-dimensional simplex.

For purposes of illustration, two-dimensional simplexes will be used. The method is applicable to any number of dimensions. Optimization will be taken to mean maximization of response, but it could apply equally well to the process of finding a minimum.

The objective of the sequential simplex method is to force the simplex to move to the region of optimum response. The decisions required to accomplish this constitute the so-called "rules" of the simplex procedure.

Rule 1

A move is made after each observation of response. Once the responses at all vertices of a simplex have been evaluated, a decision can be made as to which direction to move. As seen in rule 2, a new simplex can be completed by carrying out only one additional observation. Thus, a move can be made after each observation of response.

Rule 2

A move is made into that adjacent simplex which is obtained by discarding the point of the current simplex corresponding to the least desirable response and replacing it with its mirror image across the (hyper) face of the remaining points. Figure 2 shows three moves in the normal progression of a two-dimensional simplex toward an optimum. Point A of the original simplex has the lowest response and is discarded, leaving points B and C. Reflection of point A across the face BC generates point D which, together with points B and C, forms the second simplex. The response at point D is observed. Only one new observation is required to complete a new simplex.

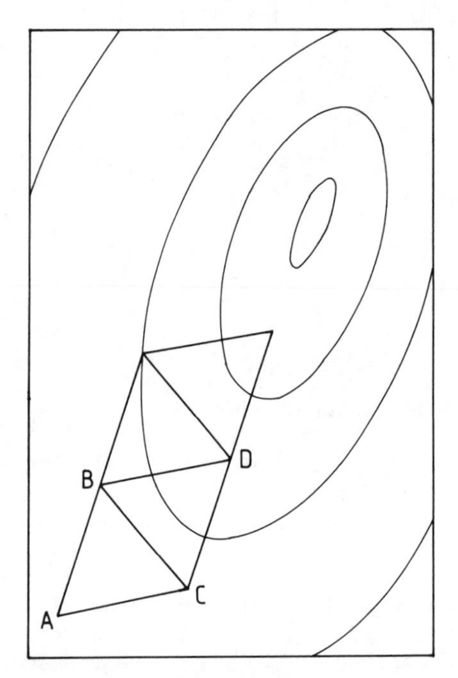

Figure 2. Progress of two-dimensional simplex toward optimum.

Discarding point C and reflecting give the simplex BDE. Finally, simplex DEF is formed after eliminating point B.

If the vertices of a k-dimensional simplex are represented by the coordinate vectors $\underline{\underline{P}}_2, \ldots, \underline{\underline{P}}_j, \ldots, \underline{\underline{P}}_k, \underline{\underline{P}}_{k+1}$, elimination of the undesirable response $\underline{\underline{P}}_j$ leaves the hyperface $\underline{\underline{P}}_1, \underline{\underline{P}}_2, \ldots,$

$\underline{\underline{P}}_{j-1}, \underline{\underline{P}}_{j+1}, \ldots, \underline{\underline{P}}_k, \underline{\underline{P}}_{k+1}$ with centroid $\underline{\underline{\bar{P}}}$:

$$\underline{\underline{\bar{P}}} = (1/k)(\underline{\underline{P}}_1 + \underline{\underline{P}}_2 + \cdots + \underline{\underline{P}}_{j-1} + \underline{\underline{P}}_{j+1} + \cdots + \underline{\underline{P}}_k + \underline{\underline{P}}_{k+1}) \quad (1)$$

The new simplex is defined by this face and a new vertex, $\underline{\underline{P}}_j^*$, which is the reflection of rejected vertex $\underline{\underline{P}}_j$ across the face through $\underline{\underline{\bar{P}}}$:

$$\underline{\underline{P}}_j^* = \underline{\underline{\bar{P}}} + (\underline{\underline{\bar{P}}} - \underline{\underline{P}}_j) \quad (2)$$

If the reflected point is lowest in the new simplex, rule 2 would reflect the simplex back to the previous one. The simplex would then oscillate and become stranded. This is shown in Figure 3 where points E and G are both less desirable than either points F or D. An exception to rule 2 is necessary.

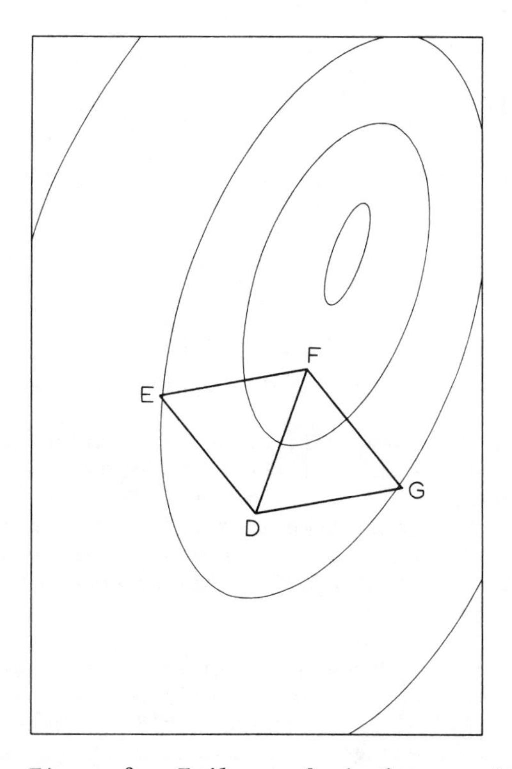

Figure 3. Failure of simplex on ridge.

Rule 3

If the reflected point has the least desirable response in the new simplex, do not reapply rule 2, but instead reject the second lowest response in the new simplex and continue. Figure 4 indicates the movement of a simplex on a ridge. Rule 3 was employed between simplexes DFG and FGH and between simplexes FIJ and IJK.

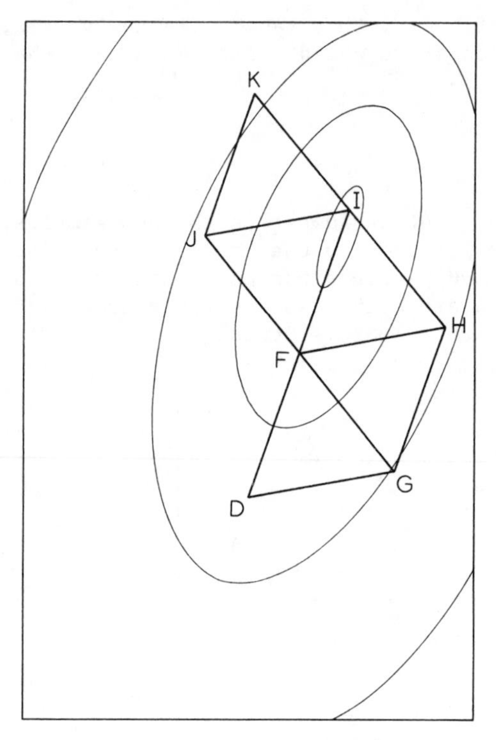

Figure 4. Progress of simplex on ridge.

It is possible to perform replicate determinations at a given vertex and to use the mean of these replicates as the response for that vertex. Further, the variance of the replicates could be used to assign significance to the differences in response at the vertices of a simplex. This is, however, unnecessary. The abandonment of traditional testing of significance can be justified for two reasons.

If the differences in response are large compared to the size of the indeterminate errors, the simplex will move in the proper direction, and repetition of observations would be wasteful. If the differences are small enough to be affected by indeterminate errors, the simplex may move in the wrong direction. However, a move in the wrong direction will probably (in a truly statistical sense) yield a lower response which would be quickly corrected by rules 2 and 3, and the simplex, though momentarily thrown off course, would proceed again toward the optimum.

A special case is that of a large positive deviation. As the simplex moves, the low responses are discarded, whereas the high responses are retained. Thus, it is possible that the simplex will become fastened to a false high result and mistake it for the true optimum. To help distinguish between an anomalously large response and a valid optimum, the following exception to rule 1 is used.

Rule 4

If a vertex has been retained in k+1 simplexes, before applying rule 2 reobserve the resonse at the persistent vertex. If the vertex is truly near the optimum, it is probable that the repeated evaluation of response will be consistently high, and the maximum will be retained. If the response at the vertex was high because of an error in observation, it is improbable that the repeat observation will also be high, and the vertex will eventually be eliminated.

Occasionally, the simplex may try to move beyond the boundaries of the independent variables.

Rule 5

If a new vertex lies outside the boundaries of the independent variables, do not make an experimental observation, but instead assign to it a very undesirable response. Application of rules 2 and 3 will then force the simplex back inside its boundaries, and it will continue to seek the optimum response.

Appropriate modifications in the chemical system or in the instrumentation allowing removal of a boundary condition may produce better optimum conditions. A formal optimization method such as the simplex is useful in identifying such situations.

When an optimum has been located, the rules force the simplex to circle (Figure 5). The simplex in the initial phase of the experiments is purposely made large to move rapidly over the response surface and locate a broad region of optimum response. A more precise definition of the optimum can be obtained by starting a new, smaller simplex near the crude optimum and letting it proceed until it circles. The optimum can be defined even more precisely by using a still smaller simplex. A practical lower limit to the size of the simplex exists. If the simplex is too small, the indeterminate errors will mask the true effects, and the simplex will wander erratically within a small area near the optimum.

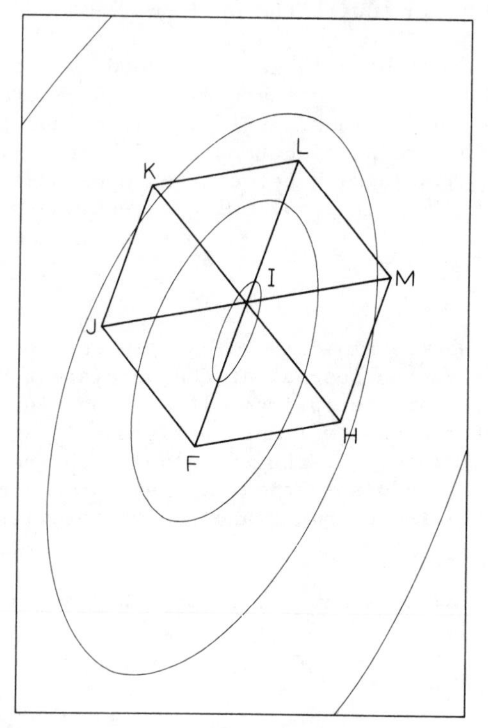

Figure 5. Circling simplex.

Table I shows numerical values for the movement of a three-dimensional simplex (8). The values represent volume percent of concentrated sulfuric acid, mg sodium sulfate, color development time, and the resulting absorbance in the optimization of a Liebermann-Burchard method for a fixed amount of cholesterol (9). Vertices 11-17 indicate the behavior of a circling tetrahedron. A boundary violation occurred at vertex 9.

There is always the possibility that a local optimum has been located, not the global (overall) optimum. At present, for the simplex method and for all other methods of optimization, experimental and numerical, it is impossible to be certain that the global optimum has been found (1). Confidence can be increased if the same optimum is reached when the optimization procedure is started from widely differing regions of the variables' domain.

The sequential simplex technique of Spendley et al. (6) suffers three limitations when it is used to locate a stationary optimum. In two dimensions there is no difficulty in determining when the optimum has been located. The simplexes will become superimposed because of the ability of triangles to close pack (Figure 5). Tetrahedra and higher dimensional simplexes will not close pack. It is not always clear when an optimum has been reached. In Table I vertex 17 is close to vertex 11 but is not superimposed upon it.

The original simplex technique has no provision for acceleration. This is usually overcome by the previously mentioned technique of running several phases, the first at low resolution to roughly define the region of optimum response, the others to more precisely "home in" on the optimum.

Third, one orientation of a simplex will cause it to attain a false optimum (Figure 6). This problem can be overcome if another phase is run in which the new simplex is rotated with respect to the previous simplex.

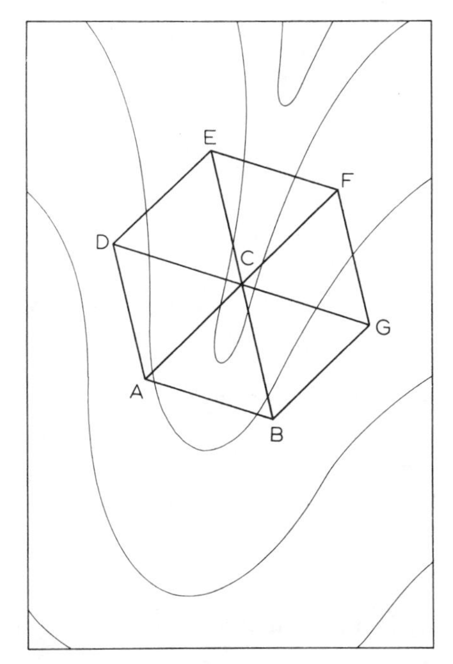

Figure 6. Second type of failure of simplex on ridge.

There are applications in which none of these "limitations" causes problems, applications that are truly "evolutionary operations" in which the simplex method is used to attain and follow an optimum around a response surface that changes with time. Such applications are not rare in analytical chemistry. Keeping instruments in their best operating conditions as various of their components change characteristics with age is one example.

II: LINEAR MODELS AND MATRIX LEAST SQUARES IN CLINICAL CHEMISTRY

Modification by Nelder and Mead (10) of the original simplex procedure gives not only a clear indication of when a sufficiently precise stationary optimum has been attained but also has the advantages of acceleration and adaptation to fit the particular response surface being studied. In addition to the operation of reflection, two new operations are used, expansion and contraction.

Consider the initial simplex represented by BNW in Figure 7.

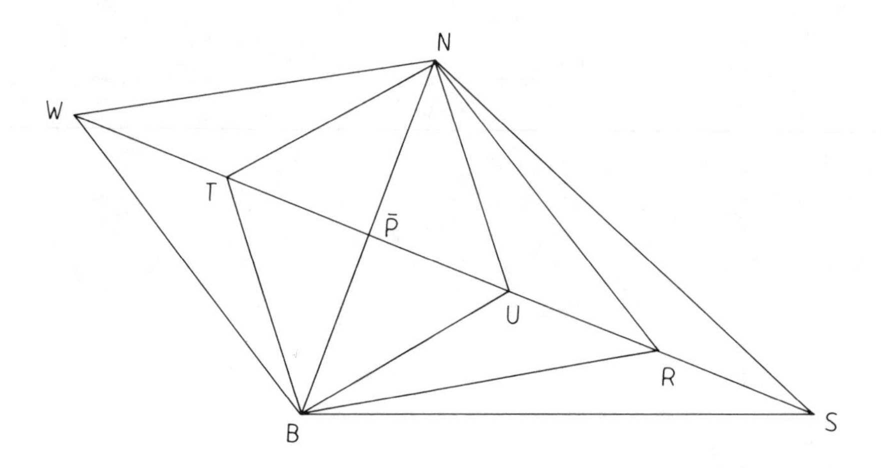

Figure 7. Possible moves in modified simplex method.

B is the best response, W is the worst response, and N is the next-to-the-worst response. $\bar{\underline{\underline{P}}}$ is the centroid of the hyperface $\overline{BN}$. Reflection of W across $\overline{BN}$ generates point R.

$$\underline{\underline{P}}_r = \bar{\underline{\underline{P}}} + (\bar{\underline{\underline{P}}} - \underline{\underline{P}}_w) \tag{3}$$

There are three possibilities to be considered for the response at point R:

The response at R is more desirable than the response at B. This indicates movement in the proper direction and suggests further investigation. Segment WR is expanded, and the response is evaluated at point S.

$$\underline{\underline{P}}_s = \underline{\underline{\bar{P}}} + \gamma(\underline{\underline{\bar{P}}} - \underline{\underline{P}}_w) \qquad (4)$$

where γ is the expansion coefficient ($\gamma>1$). If S is more desirable than B, the new simplex is BNS. If S is not more desirable than B, the new simplex is BNR.

The response at R is neither better than the response at B nor worse than the response at N. Since neither expansion nor contraction is clearly indicated, the new simplex is BNR.

The response at R is less desirable than the response at N. This indicates unsatisfactory movement and suggests contraction in this particular direction of investigation.

If the response at R is less desirable than the response at the worst previous vertex (point W), then the new, contracted simplex should lie at T which is closer to W than to R.

$$\underline{\underline{P}}_r = \underline{\underline{\bar{P}}} - \beta(\underline{\underline{\bar{P}}} - \underline{\underline{P}}_w) \qquad (5)$$

where β is the contraction coefficient ($0<\beta<1$).

If the response at R is not less desirable than the response at the worst previous vertex, then the new, contracted simplex should lie at U which is closer to R than to W.

$$\underline{\underline{P}}_u = \underline{\underline{\bar{P}}} + \beta(\underline{\underline{\bar{P}}} - \underline{\underline{P}}_w) \qquad (6)$$

If T or U should prove to be the least desirable vertex in the new simplex, corrective action can be taken to assure that the simplex does not become stranded. This action can be taken by further diminution (10) or by translation (11) of the simplex.

The simplex is halted when the step size becomes less than some predetermined value (e.g., 1% of the domain of each variable), or when the differences in response approach the value of the indeterminate error.

An example of the use of the modified simplex method in analytical chemistry is the experimental optimization of NMR magnetic field homogeneity (11). Linear and quadratic y-axis gradient controls were varied (two-dimensional simplex) with dramatic results.

Although originally developed for experimental optimization, the simplex method and its modifications have been used extensively for mathematical optimizations such as the nonlinear least-squares fitting of data. Figure 8 shows the movement of a simplex on a least-squares response surface. Parameters A_∞ and k in the equation

$$A = A_\infty(1 - e^{-kt}) \tag{7}$$

were adjusted by the simplex to give the best fit for the data in Table II. Numerical results are given in Table III. The initial simplex, indicated in Figure 8 by dots at its vertices, was intentionally made small and remote from the point of convergence (represented by the circled point) to illustrate the movement of the simplex. Expansions in directions that are favorable and contractions when they become unfavorable are clearly in evidence for the 23 moves shown in Figure 8.

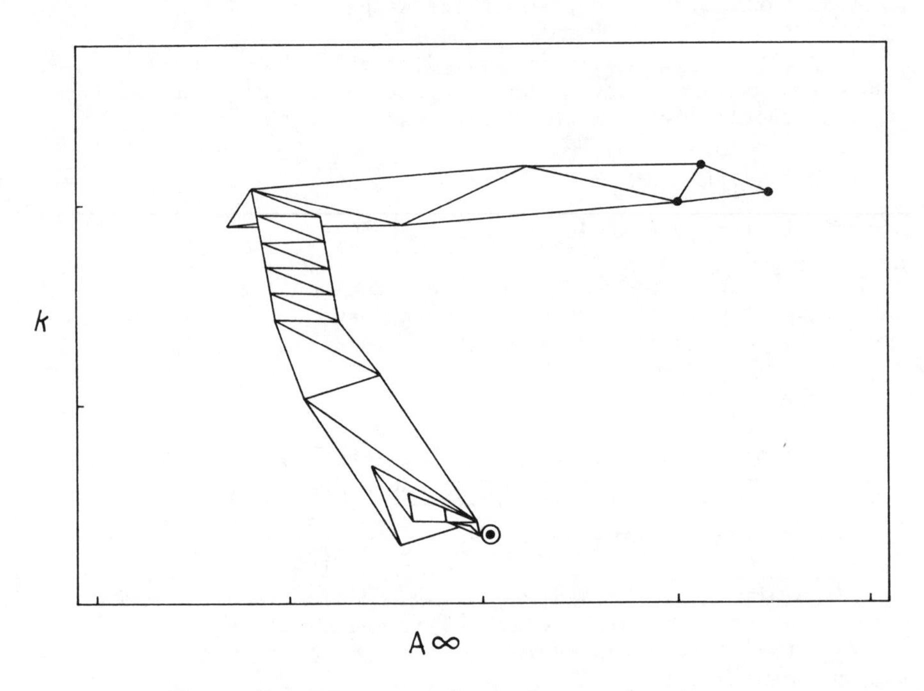

Figure 8. Progress of simplex on least-squares response surface.

CONCLUSIONS

Although there are many methods available for experimental optimization (1), none is as conceptually simple as the sequential simplex method or its modifications. The calculations involved are trivial and are easily programmed on small digital computers for use in automated instrumentation. The coordinates and responses of only $k+1$ points need be retained, thus obviating large storage requirements. If computers are not involved, the necessary calculations can be done quickly and easily by hand.

Potential applications of the simplex method in analytical chemistry are perhaps unlimited. The inherent simplicity of the method makes otherwise difficult experimental optimizations more attractive.

REFERENCES

(1) G.S.G. Beveridge and R.S. Schechter, "Optimization: Theory and Practice," McGraw-Hill, New York, N.Y., 1970.

(2) H. Hotelling, Ann. Math. Statist., 12, 20 (1941).

(3) M. Friedman and L.J. Savage, "Techniques of Statistical Analysis," C. Eisenhart, M.W. Hastay, and W.A. Wallis, Eds., Chap. 13, McGraw-Hill, New York, N. Y., 1947.

(4) G.E.P. Box and K.B. Wilson, J. Roy. Statist. Soc., Series B, 13, 1 (1951).

(5) G.E.P. Box, Appl. Statist., 6, 81 (1957).

(6) W. Spendley, G.R. Hext, and F.R. Himsworth, Technometrics, 4, 441 (1962).

(7) D.E. Long, Anal. Chim. Acta, 46, 193 (1969).

(8) S.L. Morgan and S.N. Deming, unpublished data.

(9) T.C. Huang, C.P. Chen, V. Wefler, and A. Raftery, Anal. Chem., 33, 1405 (1961).

(10) J.A. Nelder and R. Mead, Computer J., 7, 308 (1965).

(11) R.R. Ernst, Rev. Sci. Instrum., 39, 998, (1968).

Support is acknowledged from National Science Foundation Grant No. GP-32911.

Table I. Movement of Three-Dimensional Simplex

Vertex no.	Vertices retained from previous simplex	Variables: 1 Vol H_2SO_4, %	2 Na_2SO_4 mg	3 Obsv time, min	4 Response, absorbance
1		0.00	0.00	2.00	0.000
2		2.36	11.79	5.02	0.041
3		9.43	11.79	2.75	0.093
4		2.36	47.15	2.75	0.018
5	2,3,4	9.43	47.15	5.02	0.235
6	2,3,5	11.79	0.00	5.77	0.367
7	3,5,6	18.11	27.55	4.01	0.414
8	5,6,7	16.73	38.00	7.11	0.452
9	6,7,8	21.70	-3.44[a]	6.25	-0.100[a]
10	7,8,9[b]	25.88	41.41	5.81	0.527
11	7,8,10	18.82	74.71	5.04	0.456
12	8,10,11	22.85	75.20	7.96	0.483
13	10,11,12	28.25	89.55	5.43	0.503
14	10,12,13	32.49	62.72	7.76	0.450
15	10,13,14[b]	34.90	53.92	4.70	0.445
16	10,13,15[b]	26.89	60.53	2.87	0.485
17	10,13,16	19.29	73.73	4.70	0.474

[a]Boundary violation. [b]Newest vertex in the previous simplex was the worst. Next-t0-the-worst vertex was rejected to form present simplex.

Table II. Data for Least-Squares Fitting

t	A	t	A
1.5	0.110	9.0	0.325
1.5	0.109	12.0	0.326
3.0	0.169	12.0	0.330
3.0	0.172	15.0	0.362
4.5	0.210	15.0	0.383
4.5	0.210	18.0	0.381
6.0	0.251	18.0	0.372
6.0	0.255	24.0	0.422
9.0	0.331	24.0	0.411

Vertex Number	Vertices Retained from Previous Simplex	Variables		Response
		A_∞	k	Sum of Squares
1		0.500	1.000	0.8189
2		0.548	1.026	1.1988
3		0.513	1.097	0.9398
4	1,3	0.423	1.093	0.4093
5	1,4	0.358	0.947	0.1826
6	4,5	0.281	1.040	0.1360
7	5,6	0.268	0.943	0.1409
8	6,7	0.317	0.969	0.1315
9	6,8	0.284	0.974	0.1307
10	8,9	0.319	0.903	0.1262
11	9,10	0.286	0.908	0.1249
12	10,11	0.321	0.837	0.1201
13	11,12	0.288	0.842	0.1185
14	12,13	0.324	0.771	0.1129
15	13,14	0.291	0.776	0.1115
16	14,15	0.326	0.705	0.1045
17	15,16	0.293	0.710	0.1039
18	16,17	0.347	0.571	0.0945
19	17,18	0.308	0.511	0.0717
20	18,19	0.397	0.203	0.0072
21	19,20	0.358	0.142	0.0477
22	20,21	0.343	0.342	0.0332
23	20,22	0.364	0.207	0.0099
24		0.362	0.273	0.0197[a]
23'		0.381	0.205	0.0059
24'	20,23'	0.380	0.238	0.0118
25	20,23'	0.399	0.170	0.0039
26	23',25	0.394	0.195	0.0050
27	25,26	0.411	0.160	0.0038
33		0.403	0.172	0.0036
48		0.404	0.170	0.0036
53		0.404	0.170	0.0036

Table 3. A new vertex is the worst in the present simplex. Simplex contracted toward best vertex.

LINEAR MODELS AND MATRIX LEAST SQUARES IN CLINICAL CHEMISTRY

Stanley N. Deming
Department of Chemistry, Emory University
Atlanta, Georgia 30322

Clinical chemists are often required to fit mathematical models to experimental data. In some studies, the individual model parameters and their uncertainties are of primary importance; for example, the initial reaction rate (slope with respect to time) of a kinetic method of analysis (*1*). In other studies, the graph of the whole model (and the associated uncertainty) is of interest—e.g., a calibration curve relating measured values of response to a property of a material (*2*). In still other studies, statistical measures of how well the model fits the data are desired; the correlation coefficient obtained in methods comparison studies is an example (*3*).

In each of these examples, least-squares methods can be used to fit mathematical models to experimental data. Unfortunately, many clinical chemists have been introduced to least-squares methods through elusive normal equations, "simplifying" algebraic manipulations, and tedious summations of products; it is not surprising that most workers are reluctant to use least-squares methods for relationships more complicated than the simple straight line. Fortunately, there is an alternative approach to least squares that avoids normal equations, algebraic manipulations, and (given the availability of modern calculators) manual summations.

The intent of this paper is to present a unified approach to the use of linear models and matrix least squares in clinical chemistry, and to provide a better understanding of the statistics that arise from these techniques.

Linear Models

A common misconception involves the meaning of the words "linear model." Many individuals understand the term "linear model" to mean (and to be limited to) straight-line relationships of the form

$$y = a + bx \tag{1}$$

where y is generally considered to be the dependent variable, x is the independent variable, and a and b are the parameters of the model (intercept and slope, respectively). Although it is true that equation 1 is a linear model, the reason for this is *not* that its graph is a straight line. Instead, it is a linear model because it is constructed of additive terms, each of which contains one and only one multiplicative parameter (*4*). That is, the model is first-order or linear *in the parameters*. This definition of "linear model" includes models that are not first-order in the independent variables. The model

$$y = a + bx + c(10^x) + d\log(x) + ex^2 \tag{2}$$

is a linear model by the above definition. The model

B. R. Kowalski (ed.), Chemometrics. Mathematics and Statistics in Chemistry, 267–304.

$$y = a[\exp(-bx)] \quad (3)$$

however, is a nonlinear model because it contains more than one parameter in a single term. For some nonlinear models it is possible to make transformations on the dependent variable, on the independent variable, or on both, to "linearize" the model. Taking the natural logarithm of both sides of equation 3, for example, gives a model that is linear in the parameters a' and b: $\log_e(y) = a' - bx$, where $a' = \log_e(a)$.

This paper is limited to models that are linear in the parameters.

Terminology and Symbology

The model represented by equation 1 can be rewritten as

$$y_{1i} = \beta_0 + \beta_1 x_{1i} + r_{1i} \quad (4)$$

Here, the symbols a and b in equation 1 have been replaced by the equivalent symbols β_0 and β_1. The two parameters, β_0 and β_1, are said to be parameters of the model. We will use the symbol p to indicate the number of parameters of the model. For the model of equation 4, $p = 2$.

The subscript i is used as an index to refer to a particular experiment. We will use the symbol n to refer to the total number of experiments in a set of experiments. Thus, i ranges from 1 to n.

The symbol y_{1i} is used to represent measured values of *response* (the "dependent variable"); y_{11} would represent the response obtained in the first experiment, y_{12} would be the response obtained in the second experiment, and so on. The subscript "1" on y_{1i} might seem unnecessary, but in some experimental situations (methods comparisons, for example) more than one response might be obtained from each experiment; the symbols for the additional responses would be y_{2i}, y_{3i}, etc. We will limit our discussions to situations in which only one response is of interest, and will therefore be concerned with y_{1i}'s only.

In an analogous manner, the symbol x_{1i} is used to represent specified values of an "independent variable." We will use the term *factor* when referring to an independent variable. Thus, x_{11} would represent the value of the factor x_1 in the first experiment, x_{12} would be the value in the second experiment, and so on. The first subscript on x is used to designate a particular factor. For example, x_1 might be the factor pH, x_2 might be the factor temperature, x_3 might be substrate concentration, and so on.

The subscripts on the β's are now seen to be significant. The subscript "0" on β_0 in equation 4 indicates that the parameter is not associated with any of the factors. The subscript "1" on β_1 indicates that it is associated with the first factor, x_1. If a term of the form $\beta_2 x_{2i}$ were included, the subscript "2" on β_2 would indicate that β_2 is associated with the second factor, x_2. The subscript "12" on β_{12} in the term $\beta_{12} x_{1i} x_{2i}$ is similarly meaningful, as is the subscript "11" on β_{11} in the term $\beta_{11} x_{1i}^2$ $(= \beta_{11} x_{1i} x_{1i})$.

The form of equation 4 differs from the form of equation 1 by the additional term r_{1i}. This term allows for uncertainty in the corresponding measured response y_{1i}. Models that take into account the possibility of uncertainty are said to be probabilistic or stochastic models, as opposed to deterministic models, which do not allow for uncertainty. We will consider only probabilistic models in this paper. The r stands for a "residual" or a "deviation"; the meaning of this term will be discussed later. The r_{1i}'s are *not* parameters of the model.

In the statistical literature, the word "value" is often replaced by the word "level." Thus, for example, one speaks of a "specified factor level," or "measured levels of response." The conditions for a given experiment are described by the set of factor levels associated with that experiment; e.g., if Experiment Three were carried out at $x_1 = 2$ and $x_2 = 37$, the conditions of the experiment could be described by the set $\{x_{13} = 2, x_{23} = 37\}$. The traditional statistical literature calls this set of factor levels a "treatment" or a "treatment combination," primarily because many of the pioneering statistical studies involved factors that were real agricultural treatments, such as fertilizer, sunlight, water, etc. In this paper, the term *factor combination* will be used to refer to the set of factor levels associated with an experiment.

Finally, *replication* means carrying out more than one experiment at a given factor combination. If Experiment Four were carried out at the factor combination $\{x_{14} = 2, x_{24} = 37\}$, then this experiment and the previous experiment would be said to be replicates. There are two replicates, each a replicate of the other; neither is considered as the "original," with the other its "duplicate." We will use the symbol f to indicate the number of distinctly different factor combinations in a set of experiments. If there is no replication, $f = n$; if there is replication, $f < n$.

It is important to distinguish between replication and what has been called "duplicity" (*5*). In a set of experiments designed to evaluate the effect of diet on uric acid excretion, for example, replication might involve feeding the same diet to each of two persons and measuring the uric acid excretion of each; "duplicity," or false replication, might be involved if the uric acid excretion of one person were measured twice and the results treated as if they were two independent experiments.

Design Matrix

As indicated above, the conditions for each experiment can be described by a set of factor levels; that is, by a particular factor combination. This set of factor levels can be written conveniently as a row vector, in which each column corresponds to a particular factor. The row vector for Experiment Three above might be represented symbolically as

$$\boldsymbol{D}_3 = (x_{13} \quad x_{23}) \tag{5}$$

or numerically as

$$\begin{matrix} & x_1 & x_2 \\ \boldsymbol{D}_3 = & (2 & 37) \end{matrix} \tag{6}$$

The designations x_1 and x_2 above the numerical values in equation 6 are not part of the row vector; they simply indicate which factors are associated with which columns.

The row vector that represents the factor combination of Experiment Four would be $\boldsymbol{D}_4 = (2 \;\; 37)$, which is identical to $\boldsymbol{D}_3$ because they are replicates.

Suppose the factor combinations for Experiments One and Two are $\boldsymbol{D}_1 = (1 \;\; 37)$ and $\boldsymbol{D}_2 = (3 \;\; 37)$. If these are the only experiments in the set, we could stack the row vectors in order of experiment number and arrange them in an array, or matrix, in which *each row corresponds to an experiment and each column corresponds to a factor:*[3]

$$\boldsymbol{D} = \begin{pmatrix} \boldsymbol{D}_1 \\ \boldsymbol{D}_2 \\ \boldsymbol{D}_3 \\ \boldsymbol{D}_4 \end{pmatrix} = \begin{pmatrix} x_{11} & x_{21} \\ x_{12} & x_{22} \\ x_{13} & x_{23} \\ x_{14} & x_{24} \end{pmatrix} = \begin{pmatrix} 1 & 37 \\ 3 & 37 \\ 2 & 37 \\ 2 & 37 \end{pmatrix} \quad (7)$$

(columns of the last matrix headed x_1, x_2)

Such an array is said to be an *experimental design matrix* because it contains the factor combinations of each experiment in the set.

In the example of equation 7, the total number of experiments, n, is equal to four. The number of distinctly different factor combinations, f, is three (Experiments Three and Four are replicates). Because two factors are involved in the design, there are two columns in the design matrix.

Response Matrix

In a similar way, it is possible to construct a *matrix of observed responses*, $\boldsymbol{Y}$, in which each row again corresponds to an experiment, but now each column corresponds to a response. Because we will be considering only one response from each experiment, our response matrixes will have only one column. If the four experiments in equation 7 have responses of 2, 10, 6, and 8, respectively, the response matrix is

$$\boldsymbol{Y} = \begin{pmatrix} y_{11} \\ y_{12} \\ y_{13} \\ y_{14} \end{pmatrix} = \begin{pmatrix} 2 \\ 10 \\ 6 \\ 8 \end{pmatrix} \quad (8)$$

Choice of Model

Two types of probabilistic models are often used in clinical chemical studies.

Mechanistic models are based upon some known or presumed transformation between factors and responses. Examples of mechanistic models are the Michaelis–Menton equation, sometimes used in enzyme–substrate studies, and the exponential equation (equation 3) that describes first-order kinetics.

[3] Because of the number and size of matrixes presented in this paper, the largest matrixes have been reproduced from typewritten copy supplied by the authors. Underlined letters correspond to italic letters set in type, and letters with both an underline and a "squiggle" correspond to boldface italic letters in type.

Empirical models are usually used when the exact form of the transformation between factors and responses is not known; they are generally used in the hope that they will sufficiently approximate the real transformation, whatever it might be. Examples of empirical models are the "much desired" straight line of equation 4, the logit-log plot in radioimmunoassay, and second-order (or parabolic) relationships such as

$$y_{1i} = \beta_0 + \beta_1 x_{1i} + \beta_{11} x_{1i}^2 + r_{1i} \tag{9}$$

Generalized second-order polynomials (similar to equation 9) are popular empirical models.

Let us assume that we do not know what the transformation is between the factor combinations of $\boldsymbol{D}$ (equation 7) and the responses of $\boldsymbol{Y}$ (equation 8). We will therefore use an empirical model.

Examination of the design matrix $\boldsymbol{D}$ suggests that it would be unwise to try to relate the response to the factor x_2. Although such an effect might exist, the design shown in equation 7 would not allow us to estimate the magnitude of the effect. In general, although not always, the experimental design must incorporate variation in a factor if the effect of that factor is to be estimated. Our model, then, will be based upon the single factor x_1. For simplicity (and to illustrate the use of matrix least squares for a familiar model), we will use equation 4 to approximate the relationship between the factor x_1 and the response y_1.

Matrix of Parameter Coefficients

We now begin the task of estimating the parameters β_0 and β_1 of equation 4. It is important to realize that although the factor x_1 is an "independent variable," the factor levels that appear in a given design matrix are not variable but instead are constant for the particular study. *The factor levels are fixed:* we can no longer adjust them. What we can adjust, however, are the numerical values of the parameters β_0 and β_1 in our model to obtain a "best fit" of the model to the data. It is interesting that some researchers still "fit the data to a model." Taken literally, this suggests a lack of scientific integrity. What is meant, of course, is that they "fit a model to the data."

If we write equation 4 for each of our four experiments, we get

$$\begin{aligned}
\underline{y}_{11} &= 1\times\beta_0 + \underline{x}_{11}\beta_1 + \underline{r}_{11} \\
\underline{y}_{12} &= 1\times\beta_0 + \underline{x}_{12}\beta_1 + \underline{r}_{12} \\
\underline{y}_{13} &= 1\times\beta_0 + \underline{x}_{13}\beta_1 + \underline{r}_{13} \\
\underline{y}_{14} &= 1\times\beta_0 + \underline{x}_{14}\beta_1 + \underline{r}_{14}
\end{aligned} \tag{10}$$

In each case, the coefficient of β_0 is implicitly unity and is explicitly written as such in equation 10. The coefficient of β_1 is the specific factor level, x_{1i}, for each experiment. If numerical values are substituted for algebraic symbols, we have

$$
\begin{aligned}
2 &= 1\times\beta_0 + 1\times\beta_1 + r_{11}\\
10 &= 1\times\beta_0 + 3\times\beta_1 + r_{12}\\
6 &= 1\times\beta_0 + 2\times\beta_1 + r_{13}\\
8 &= 1\times\beta_0 + 2\times\beta_1 + r_{14}
\end{aligned}
\tag{11}
$$

Equation 11 is a set of simultaneous equations; it would appear to be four equations in two unknowns (β_0 and β_1) and seems to be overdetermined. However, not only do we not know β_0 and β_1, but also we do not know r_{11}, r_{12}, r_{13}, and r_{14}; the system now appears to be underdetermined, with four equations and six unknowns. How can these equations be solved for β_0 and β_1?

The usual procedure is to impose a constraint on the system such that the sum of squares of the residuals be minimum; hence, the term least squares. When dealing with models that are linear in the parameters (as we are here), the term *linear least squares* is used; when dealing with models that are not linear in the parameters, the term *nonlinear least squares* is used to describe methods for obtaining the minimum sum of squares of residuals. Because least-squares estimates of parameters in linear models are easily obtained with use of matrix techniques, linear least squares is sometimes called *matrix least squares*. (The direct matrix solution is not applicable to nonlinear models.)

Let us define a *matrix of residuals*, $\boldsymbol{R}$:

$$
\boldsymbol{R} = \begin{pmatrix} r_{11} \\ r_{12} \\ r_{13} \\ r_{14} \end{pmatrix}
\tag{12}
$$

and a *matrix of parameters*, $\boldsymbol{B}$:

$$
\boldsymbol{B} = \begin{pmatrix} \beta_0 \\ \beta_1 \end{pmatrix}
\tag{13}
$$

Note that the left-hand side of equation 11 is $\boldsymbol{Y}$, and that the column of residuals in the same equation is $\boldsymbol{R}$. We can thus write

$$
\boldsymbol{Y} = \begin{bmatrix} 1\times\beta_0 + 1\times\beta_1 \\ 1\times\beta_0 + 3\times\beta_1 \\ 1\times\beta_0 + 2\times\beta_1 \\ 1\times\beta_0 + 2\times\beta_1 \end{bmatrix} + \boldsymbol{R}
\tag{14}
$$

as an abbreviated form of equation 11.

In matrix multiplication, the elements of the product matrix are formed by multiplying a *row* of the left-hand matrix by a *column* of the right-hand matrix, element by element, and adding together the intermediate products. It is then evident that the matrix containing the β's in equation 14 is the product of two other matrixes:

$$\begin{pmatrix} 1 & 1 \\ 1 & 3 \\ 1 & 2 \\ 1 & 2 \end{pmatrix} \begin{pmatrix} \beta_0 \\ \beta_1 \end{pmatrix} = \begin{pmatrix} 1\times\beta_0 + 1\times\beta_1 \\ 1\times\beta_0 + 3\times\beta_1 \\ 1\times\beta_0 + 2\times\beta_1 \\ 1\times\beta_0 + 2\times\beta_1 \end{pmatrix}$$

The matrix containing the two parameters is ***B***, defined in equation 13. The left-most matrix is called the *matrix of parameter coefficients*, ***X***, and for the present model is of the form

$$\underset{\sim}{X} = \overset{\beta_0 \quad \beta_1}{\begin{pmatrix} 1 & x_{11} \\ 1 & x_{12} \\ 1 & x_{13} \\ 1 & x_{14} \end{pmatrix}} = \overset{\beta_0 \quad \beta_1}{\begin{pmatrix} 1 & 1 \\ 1 & 3 \\ 1 & 2 \\ 1 & 2 \end{pmatrix}} \qquad (15)$$

Each row of the ***X*** matrix corresponds to an experiment and each column corresponds to a parameter coefficient (the parameters associated with each column are indicated above the matrixes in equation 15). Equation 11 may now be written in matrix form as

$$\boldsymbol{Y} = \boldsymbol{XB} + \boldsymbol{R} \qquad (16)$$

This is the general matrix notation that is used to represent all probabilistic linear models. *What follows is entirely general and applies to all linear models.* We will use the data and model of equation 11 to illustrate certain points.

Matrix Least-Squares Solution

Statisticians usually use Greek letters for the true values of parameters. True values are usually never known with absolute certainty; instead, estimates of the true values are obtained. Italic letters are used to represent parameter estimates. The caret (ˆ) is also used to designate estimated quantities. For example, the *matrix of parameter estimates*, $\hat{\boldsymbol{B}}$, is often written

$$\hat{\boldsymbol{B}} = \begin{pmatrix} b_0 \\ b_1 \end{pmatrix} \qquad (17)$$

It can be shown (*4*) that the set of parameter estimates that gives the minimum sum of squares of residuals is obtained as

$$\hat{\boldsymbol{B}} = (\boldsymbol{X}'\boldsymbol{X})^{-1}(\boldsymbol{X}'\boldsymbol{Y}) \qquad (18)$$

The prime (′) indicates the *transpose* of a matrix; the superscript "reciprocal" ($^{-1}$) indicates the *inverse* of a matrix.

The transpose of an n-row by p-column matrix is a p-row by n-column matrix with row and column elements interchanged:

$$\boldsymbol{X}' = \begin{pmatrix} 1 & 1 & 1 & 1 \\ x_{11} & x_{12} & x_{13} & x_{14} \end{pmatrix} = \begin{pmatrix} 1\,1\,1\,1 \\ 1\,3\,2\,2 \end{pmatrix} \quad (19)$$

Thus,

$$(\underset{\sim}{X}'\underset{\sim}{Y}) = \begin{bmatrix} 1 & 1 & 1 & 1 \\ 1 & 3 & 2 & 2 \end{bmatrix} \begin{bmatrix} 2 \\ 10 \\ 6 \\ 8 \end{bmatrix}$$

$$= \begin{bmatrix} (1\times2) + (1\times10) + (1\times6) + (1\times8) \\ (1\times2) + (3\times10) + (2\times6) + (2\times8) \end{bmatrix}$$

$$= \begin{bmatrix} 26 \\ 60 \end{bmatrix}.$$

Similarly,

$$(\underset{\sim}{X}'\underset{\sim}{X}) = \begin{bmatrix} 1 & 1 & 1 & 1 \\ 1 & 3 & 2 & 2 \end{bmatrix} \begin{bmatrix} 1 & 1 \\ 1 & 3 \\ 1 & 2 \\ 1 & 2 \end{bmatrix}$$

$$= \begin{bmatrix} \{(1\times1)+(1\times1)+(1\times1)+(1\times1)\} & \{(1\times1)+(1\times3)+(1\times2)+(1\times2)\} \\ \{(1\times1)+(3\times1)+(2\times1)+(2\times1)\} & \{(1\times1)+(3\times3)+(2\times2)+(2\times2)\} \end{bmatrix}$$

$$= \begin{bmatrix} 4 & 8 \\ 8 & 18 \end{bmatrix}$$

The $(\boldsymbol{X}'\boldsymbol{X})$ matrix is always symmetric about the diagonal from the upper left to the lower right of the matrix.

The inverse of a 2 by 2 matrix such as $(\boldsymbol{X}'\boldsymbol{X})$ is obtained as follows:

$$\text{if } (\boldsymbol{X}'\boldsymbol{X}) = \begin{pmatrix} a & b \\ c & d \end{pmatrix} \quad (20)$$

$$\text{then } (\boldsymbol{X}'\boldsymbol{X})^{-1} = \begin{pmatrix} d/D & -b/D \\ -c/D & a/D \end{pmatrix} \quad (21)$$

$$\text{where } D = (a \times d) - (c \times b) \quad (22)$$

is the determinant of the $(\boldsymbol{X}'\boldsymbol{X})$ matrix (the determinant D is not the same as the design matrix $\boldsymbol{D}$). It is evident from equation 21 that the determinant must be nonzero if a solution is to be obtained.

The inverse of a larger matrix is not so easily calculated, and computer programs are helpful.

Numerically, for the present example, $D = (4 \times 18) - (8 \times 8) = 72 - 64 = 8$, and

$$(X'X)^{-1} = \begin{pmatrix} 18/8 & -8/8 \\ -8/8 & 4/8 \end{pmatrix} = \begin{pmatrix} 2.25 & -1 \\ -1 & 0.5 \end{pmatrix} \tag{23}$$

Finally,

$$\hat{B} = \begin{pmatrix} b_0 \\ b_1 \end{pmatrix} = (X'X)^{-1}(X'Y)$$

$$= \begin{pmatrix} 2.25 & -1 \\ -1 & 0.5 \end{pmatrix}\begin{pmatrix} 26 \\ 60 \end{pmatrix} = \begin{pmatrix} -1.5 \\ 4 \end{pmatrix} \tag{24}$$

The estimate of β_0 (the intercept at $x_1 = 0$) is $b_0 = -1.5$; the estimate of β_1 (the slope with respect to the factor x_1, i.e., the factor effect) is $b_1 = 4$. The fitted model is thus of the form

$$y_{1i} = -1.5 + 4 \times x_{1i} \tag{25}$$

Figure 1 is a plot of the experimental points and the best least-squares estimate of the relationship between the response y_1 and the factor x_1.

Estimated Responses

The *estimated response*, $\hat{y}_1$, for any value of x_1 can now be estimated by substituting the appropriate parameter coefficients in the model and solving the resulting equation. For the

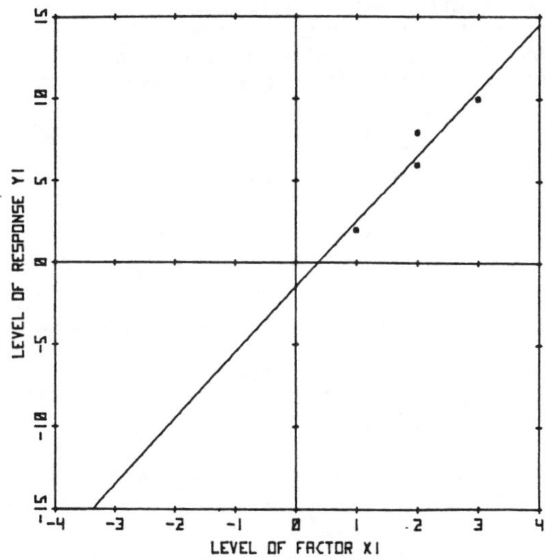

Fig. 1. Experimental points and least-squares straight line through them

Intercept $b_0 = -1.5$; slope $b_1 = 4$

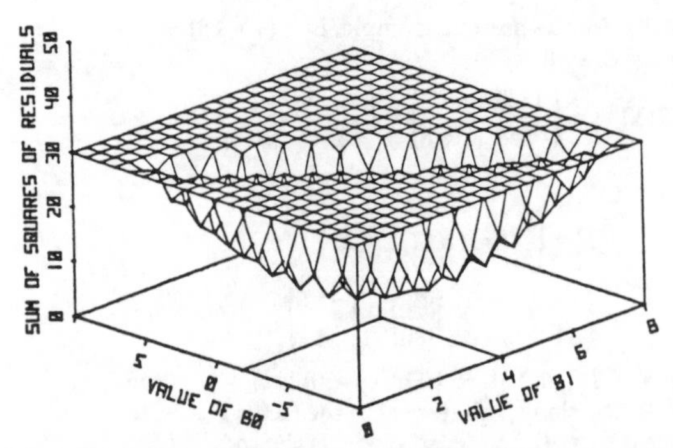

Fig. 2. Sum of squares of residuals as a function of different values of b_0 and b_1

Sums of squares values greater than 30 set equal to 30

model of equation 25, this simply involves inserting the desired level of x_1. If we want to estimate the response at $x_1 = 4$, for example, then $\hat{y}_{10} = -1.5 + 4 \times x_{10} = -1.5 + 4 \times 4 = 14.5$. The second subscript "0" on $\hat{y}_1$ and x_1 is used to indicate that the factor combination of interest does not necessarily correspond to one of the experiments that was previously carried out.

An equivalent means of estimating a response involves matrix techniques. Let the matrix X_0 contain only one row; let X_0 have columns that correspond to the columns of the X matrix (the matrix of parameter coefficients); and let the elements of X_0 correspond to the factor combination of interest. Then

$$\hat{y}_{10} = X_0\hat{B} \tag{26}$$

For the above example at $x_{10} = 4$, the matrix X_0 is of the form

$$\begin{matrix} & \beta_0 & \beta_1 & \\ X_0 = (& 1 & 4 &) \end{matrix} \tag{27}$$

$$\text{and } \hat{y}_{10} = (1\ 4)\begin{pmatrix} -1.5 \\ 4 \end{pmatrix} = (-1.5 + 16) = 14.5$$

An extension of this approach allows estimation of the responses at the original experimental factor combinations:

$$\hat{Y} = X\hat{B} \tag{28}$$

where $\hat{Y}$ is the *matrix of estimated responses*. For the data we have been using,

$$\hat{Y} = \begin{pmatrix} \hat{y}_{11} \\ \hat{y}_{12} \\ \hat{y}_{13} \\ \hat{y}_{14} \end{pmatrix} = X\hat{B}$$

$$= \begin{pmatrix} 1 & 1 \\ 1 & 3 \\ 1 & 2 \\ 1 & 2 \end{pmatrix} \begin{pmatrix} -1.5 \\ 4 \end{pmatrix} = \begin{pmatrix} 2.5 \\ 10.5 \\ 6.5 \\ 6.5 \end{pmatrix} \tag{29}$$

Residuals

As indicated previously, the models considered in this paper are probabilistic or stochastic, in the sense that they allow for uncertainty in the measured response. This uncertainty is expressed as the r_{1i} term in the equation that represents the model. These "residuals" or "deviations" are the differences between what was actually observed (the y_{1i}'s) and what is predicted by the fitted model (the $\hat{y}_{1i}$'s):

$$r_{1i} = y_{1i} - \hat{y}_{1i} \tag{30}$$

or, using matrixes to calculate the residuals for the complete set of experiments,

$$\boldsymbol{R} = \boldsymbol{Y} - \hat{\boldsymbol{Y}} \tag{31}$$

Because $\hat{\boldsymbol{Y}} = \boldsymbol{X}\hat{\boldsymbol{B}}$, this is sometimes written as $\boldsymbol{R} = \boldsymbol{Y} - \boldsymbol{X}\hat{\boldsymbol{B}}$. For the data we have been using,

$$R = Y - \hat{Y} = \begin{pmatrix} 2 \\ 10 \\ 6 \\ 8 \end{pmatrix} - \begin{pmatrix} 2.5 \\ 10.5 \\ 6.5 \\ 6.5 \end{pmatrix} = \begin{pmatrix} -0.5 \\ -0.5 \\ -0.5 \\ +1.5 \end{pmatrix}$$

Note that the residuals add up to zero. This is always true if the linear model contains a β_0 term.

Recall that the constraint imposed to allow estimation of the parameters in equation 11 was that the sum of squares of residuals be minimum. If we let SS_r represent the sum of squares of residuals (often called the sum of squares about regression), then

$$SS_r = \boldsymbol{R}'\boldsymbol{R}$$

$$= (-0.5 \quad -0.5 \quad -0.5 \quad +1.5) \begin{pmatrix} -0.5 \\ -0.5 \\ -0.5 \\ +1.5 \end{pmatrix} \tag{32}$$

$$= (0.25 + 0.25 + 0.25 + 2.25) = 3$$

The result represents the minimum sum of squares of residuals obtainable for our data set using the chosen model: *all combinations of* b_0 *and* b_1 *other than* $b_0 = -1.5$ *and* $b_1 = 4$ *will give a larger sum of squares of residuals.* This is illustrated in Figure 2, which shows the sum of squares of residuals plotted against different values of the parameter estimates b_0 and b_1. Note that the minimum lies at $b_0 = -1.5$ and $b_1 = 4$. Note also that the resulting surface is paraboloid in shape, flattened so that a slice through it is elliptical, and rotated so that it lies diagonally with respect to the b_0- and b_1-axes.

Before leaving this section, we would point out that the residuals are often referred to as "errors." This is somewhat presumptuous because it assumes that the model is the correct model and all values predicted by the model are true values—any measured values of response not in agreement with the model must therefore be in error. It is important to realize that the uncertainties between what is observed and what is predicted might exist (*a*) because the model is not correct, or (*b*) because of imprecision in the measurement process. The term "residual" is preferred and refers to uncertainties, not necessarily to errors (*6*).

Pure Experimental Uncertainty

Replication allows the estimation of the magnitude of imprecision in the measurement process. Because replication is by definition the carrying out of two or more experiments at the same factor combination (i.e., at the same set of experimental conditions), any variation in response among the replicates cannot be caused by variation in any of the controlled factors but instead must be attributed to purely experimental uncertainties. We will refer to this variation as pure experimental uncertainty (statisticians usually call it pure experimental error, or pure error).

If the pure experimental uncertainty is the same for all factor combinations, it is said to be *homoscedastic*. If it varies as a function of the factors, the pure experimental uncertainty is said to be *heteroscedastic*. The techniques that are discussed in this paper assume that the pure experimental uncertainty is approximately homoscedastic, and in fact many systems are sufficiently homoscedastic (and the data treatment techniques are sufficiently rugged) that this approximation is valid. However, for some systems the assumption of homoscedasticity is not appropriate, and "weighted least squares" must be used (*2, 4*).

One measure of pure experimental uncertainty is the sum of squares due to pure experimental uncertainty, SS_{pe}, which is defined as the sum of squares of the deviations between the individual replicate responses and the corresponding mean (average) response. It is useful to define a *matrix of mean replicate response,* $\boldsymbol{J}$, which is structured the same as the $\boldsymbol{Y}$ matrix, but contains mean values of responses from replicates. For those factor combinations that were not replicated, the "mean" response is simply the single value of response. In the data set we have been using, Experiments Three and Four are replicates; thus,

$$\boldsymbol{J} = \begin{pmatrix} 2 \\ 10 \\ 7 \\ 7 \end{pmatrix}$$

where the lower two elements are the average response (7) of the corresponding replicates (6 and 8). The *matrix of pure experimental deviations*, $\boldsymbol{P}$, is obtained by subtracting $\boldsymbol{J}$ from $\boldsymbol{Y}$:

$$\underset{\sim}{P} = \underset{\sim}{Y} - \underset{\sim}{J}$$

$$= \begin{bmatrix} 2 \\ 10 \\ 6 \\ 8 \end{bmatrix} - \begin{bmatrix} 2 \\ 10 \\ 7 \\ 7 \end{bmatrix} = \begin{bmatrix} 0 \\ 0 \\ -1 \\ +1 \end{bmatrix} \tag{33}$$

Notice that zeros appear in the $\boldsymbol{P}$ matrix for those experiments that were not replicated. It is now a simple matter to calculate SS_{pe} by means of matrix techniques:

$$SS_{pe} = \underset{\sim}{P}'\underset{\sim}{P}$$

$$= \begin{pmatrix} 0 & 0 & -1 & +1 \end{pmatrix} \begin{bmatrix} 0 \\ 0 \\ -1 \\ +1 \end{bmatrix} = 2 \tag{34}$$

Lack of Fit

In a sense, calculating the mean replicate responses removes the effect of pure experimental uncertainty from the data. It is not unreasonable, thus, to expect that the deviation of these mean replicate responses from the estimated responses is due to a *lack of fit* of the model to the data. A measure of this inadequacy of the model is the sum of squares due to lack of fit, SS_{lof}. The *matrix of lack-of-fit deviations*, $\boldsymbol{L}$, is obtained by subtracting $\hat{\boldsymbol{Y}}$ from $\boldsymbol{J}$:

$$\underset{\sim}{L} = \underset{\sim}{J} - \underset{\sim}{\hat{Y}}$$

$$= \begin{bmatrix} 2 \\ 10 \\ 7 \\ 7 \end{bmatrix} - \begin{bmatrix} 2.5 \\ 10.5 \\ 6.5 \\ 6.5 \end{bmatrix} = \begin{bmatrix} -0.5 \\ -0.5 \\ +0.5 \\ +0.5 \end{bmatrix} \tag{35}$$

SS_{lof} is now easily calculated:

$$SS_{lof} = L'L$$

$$= \begin{pmatrix} -0.5 & -0.5 & +0.5 & +0.5 \end{pmatrix} \begin{pmatrix} -0.5 \\ -0.5 \\ +0.5 \\ +0.5 \end{pmatrix} = 1 \quad (36)$$

Other Sums of Squares

Anticipating a later discussion, we define four other sums of squares: the total sum of squares, the sum of squares due to the mean, the sum of squares corrected for the mean, and the sum of squares due to the factors.

The clinical chemist is usually most interested in how variations in the factor levels result in variations in the response; that is, the absolute values of the response are usually not as important as the variations in the response. When this is the case, a β_0 term is usually provided in the model so that the model is not forced to go through the origin but instead can be offset up or down the response axis by some amount. It is possible to offset the raw data in a similar way by subtracting the mean value of response from each of the individual responses. When this is done, the data is said to be "corrected for the mean." Two sums of squares are associated with this process, the sum of squares due to the mean, SS_{mean}, and the sum of squares corrected for the mean, SS_{corr} (also called the sum of squares about the mean). It is convenient to define a *matrix of mean response*, $\bar{Y}$, of the same form as the response matrix Y, but containing for each element the mean response. For the present example, the mean response is (2 + 10 + 6 + 8)/4 = 6.5 and

$$\bar{Y} = \begin{pmatrix} 6.5 \\ 6.5 \\ 6.5 \\ 6.5 \end{pmatrix}$$

The sum of squares due to the mean is simply

$$SS_{mean} = \bar{Y}'\bar{Y}$$

$$= \begin{pmatrix} 6.5 & 6.5 & 6.5 & 6.5 \end{pmatrix} \begin{pmatrix} 6.5 \\ 6.5 \\ 6.5 \\ 6.5 \end{pmatrix} = 169 \quad (37)$$

It is also convenient to define a *matrix of responses corrected for the mean*, **C**:

$$\underset{\sim}{C} = \underset{\sim}{Y} - \underset{\sim}{\bar{Y}}$$

$$= \begin{pmatrix} 2 \\ 10 \\ 6 \\ 8 \end{pmatrix} - \begin{pmatrix} 6.5 \\ 6.5 \\ 6.5 \\ 6.5 \end{pmatrix} = \begin{pmatrix} -4.5 \\ +3.5 \\ -0.5 \\ +1.5 \end{pmatrix} \tag{38}$$

The sum of squares corrected for the mean is

$$SS_{corr} = \underset{\sim}{C}'\underset{\sim}{C}$$

$$= \begin{pmatrix} -4.5 & +3.5 & -0.5 & +1.5 \end{pmatrix} \begin{pmatrix} -4.5 \\ +3.5 \\ -0.5 \\ +1.5 \end{pmatrix} = 35 \tag{39}$$

The total sum of squares, SS_T, is simply the sum of squares of responses and is obtained as

$$SS_T = \underset{\sim}{Y}'\underset{\sim}{Y}$$

$$= \begin{pmatrix} 2 & 10 & 6 & 8 \end{pmatrix} \begin{pmatrix} 2 \\ 10 \\ 6 \\ 8 \end{pmatrix} = 204 \tag{40}$$

It was suggested earlier that some of the variation of the responses about their mean is caused by variation of the factors. A measure of the variation caused by the factors as they appear in the model is the sum of squares due to the factors, SS_{fact} (sometimes called the sum of squares due to regression). We define a *matrix of factor contributions*, **F**:

$$\underset{\sim}{F} = \underset{\sim}{\hat{Y}} - \underset{\sim}{\bar{Y}}$$

$$= \begin{pmatrix} 2.5 \\ 10.5 \\ 6.5 \\ 6.5 \end{pmatrix} - \begin{pmatrix} 6.5 \\ 6.5 \\ 6.5 \\ 6.5 \end{pmatrix} = \begin{pmatrix} -4 \\ +4 \\ 0 \\ 0 \end{pmatrix} \tag{41}$$

The sum of squares due to the factors is

$$SS_{\underline{fact}} = \underset{\sim}{F}'\underset{\sim}{F}$$

$$= \begin{pmatrix} -4 & +4 & 0 & 0 \end{pmatrix} \begin{bmatrix} -4 \\ +4 \\ 0 \\ 0 \end{bmatrix} = 32 \tag{42}$$

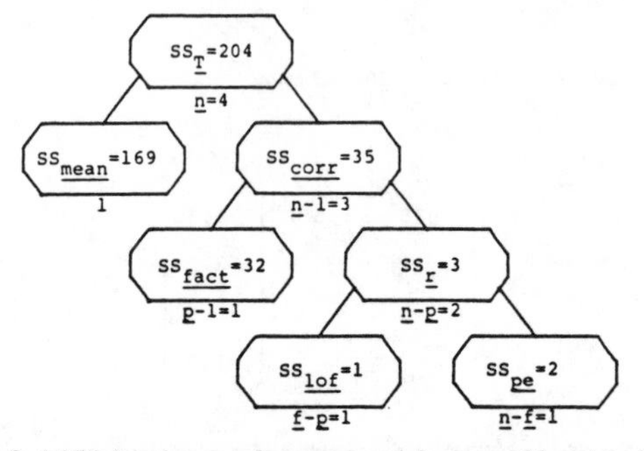

Fig. 3. Additivity of sums of squares and degrees of freedom for linear models containing a β_0 term

Numerical values are for the model and data represented by Fig. 1: $n = 4$, $p = 2$, $f = 3$

It is a characteristic of linear least squares that the sums of squares are additive, as shown in Figure 3, for linear models that contain a β_0 term. Numerical values are for the example we have been using. The matrix operations involved in the calculation of these sums of squares are summarized in Table 1. The number of degrees of freedom associated with each sum of squares is discussed in the following section.

Degrees of Freedom

Suppose someone asks you to pick two integers. You could choose 6 and 8, or 28 and 365, or any other set of two integer numbers. You would have complete freedom in your choice of values.

Suppose you are now asked to pick two integers such that their mean is equal to 7. You could choose 6 and 8, or 3 and 11, or other sets of two integer numbers. However, you do not have as much freedom in your choices as you did before. Once one integer is specified, the other integer is uniquely defined by the requirement that the mean of the two numbers be equal to 7.

Suppose you are asked to pick two integers such that their mean is equal to 7 and their product is equal to 48. You could choose 6 and 8. There are no other numbers that satisfy these requirements. You have no more freedom in your choice of values.

Of the two original *degrees of freedom*, one was taken away by the requirement that the mean be equal to 7 and another was taken away by the requirement that the product be equal to 48. This is a general phenomenon associated with data sets: when parameters that describe the data set are estimated, some degrees of freedom are taken away from the data set and are assigned to those estimates.

The formulas for the degrees of freedom are especially important when designing experiments (see Figure 3). For example, if a measure of lack of fit of the model to the data is desired, some degrees of freedom must be provided for this estimate (SS_{lof}). Thus, f (the number of distinctly different factor combinations) must be greater than p (the number of parameters in the model). Similarly, if a measure of pure experimental uncertainty is desired, f must be less than n (the total number of experiments in the set of experiments); this, of course, implies replication of experiments.

Analysis of Variance (ANOVA) Tables for Linear Models

The information contained in Figure 3 is commonly presented in an analysis of variance (ANOVA) table similar to that shown in Table 2. The column farthest to the right contains the values that are obtained by dividing a sum of squares by its associated degrees of freedom. These values are called "mean squares." Statistically, the mean squares are estimates of variances. Although ANOVA tables are often used to present matrix least-squares results for linear models, the diagrammatic tree-structure presentation in Figure 3 better illustrates several important statistical concepts.

Correlation Coefficient

Let us examine Figure 3 in detail. For the moment, focus on the sum of squares corrected for the mean (SS_{corr}), the sum of squares due to the factors (SS_{fact}), and the sum of squares of residuals (SS_r). How would these sums of squares be partitioned if the factors had no effect at all? In the example we have been using, "having no effect" is equivalent to $\beta_1 = 0$; that is, it doesn't matter what the level of x_1 is, it does not affect the response. If the factors have very little effect on the response, we would expect that the sum of squares removed from SS_{corr} by SS_{fact} would be small, and therefore SS_r would be large—about the same size as SS_{corr}. Conversely, if the factors do a very good job of explaining the responses, we would expect the residuals to be very small and SS_{fact} to be relatively large—about the same size as SS_{corr}.

The *coefficient of multiple determination, R^2*, is a measure of how much of SS_{corr} is accounted for by the factor effects (7):

$$R^2 = SS_{fact}/SS_{corr} \qquad (43)$$

The coefficient of multiple determination ranges from 0 (indicating that the factors, as they appear in the model, have no effect upon the response) to 1 (indicating that the model "explains" the data "perfectly"). The square root of the coefficient of multiple determination is the *coefficient of multiple correlation, R*.

Table 1. Summary of Matrix Operations Used To Calculate Sums of Squares

Sum of squares	Matrix operation	Degrees of freedom
SS_T	$\mathbf{Y'Y}$	n
SS_{mean}	$\mathbf{\overline{Y}'\overline{Y}}$	1
SS_{corr}	$\mathbf{C'C} = (\mathbf{Y}-\mathbf{\overline{Y}})'(\mathbf{Y}-\mathbf{\overline{Y}})$	n-1
SS_{fact}	$\mathbf{F'F} = (\mathbf{\hat{Y}}-\mathbf{\overline{Y}})'(\mathbf{\hat{Y}}-\mathbf{\overline{Y}}) = (\mathbf{X\hat{B}}-\mathbf{\overline{Y}})'(\mathbf{X\hat{B}}-\mathbf{\overline{Y}})$	p-1
SS_r	$\mathbf{R'R} = (\mathbf{Y}-\mathbf{\hat{Y}})'(\mathbf{Y}-\mathbf{\hat{Y}}) = (\mathbf{Y}-\mathbf{X\hat{B}})'(\mathbf{Y}-\mathbf{X\hat{B}})$	n-p
SS_{lof}	$\mathbf{L'L} = (\mathbf{J}-\mathbf{\hat{Y}})'(\mathbf{J}-\mathbf{\hat{Y}}) = (\mathbf{J}-\mathbf{X\hat{B}})'(\mathbf{J}-\mathbf{X\hat{B}})$	f-p
SS_{pe}	$\mathbf{P'P} = (\mathbf{Y}-\mathbf{J})'(\mathbf{Y}-\mathbf{J})$	n-f

If the model is the two-parameter (β_0 and β_1) single-factor straight-line relationship we are using in our present example, R^2 is given the symbol r^2 and is called the *coefficient of determination*. It is defined the same way R^2 is defined:

$$r^2 = SS_{fact}/SS_{corr} = 32/35 = 0.914 \quad (44)$$

Again, r^2 ranges from 0 to 1. The square root of the coefficient of determination is the *coefficient of correlation, r,* often called the *correlation coefficient:*

$$r = \text{SGN}(b_1) \times \sqrt{r^2} = +0.956 \quad (45)$$

where SGN(b_1) is the sign of the slope of the straight-line relationship. Although r ranges from -1 to $+1$, only the absolute value is indicative of how much the factors explain the data; the sign of r simply indicates the sign of b_1.

It is important to realize that r might give a false sense of how well the factors explain the data. For example, the r value of 0.956 arises because the factors explain 91.4% of the sum of squares corrected for the mean. An r value of 0.60 indicates that only 36% of SS_{corr} has been explained by the factors.

Although the coefficient of determination and the correlation coefficient are conceptually simple and attractive and are frequently used as a measure of how well a model fits a set of data, they are not, by themselves, a true measure of goodness of fit of the model, primarily because they do not take into account the degrees of freedom.

F-Test for Goodness of Fit

A statistically valid measure of the goodness of fit of a model to the data is given by the variance ratio

$$F_{(p-1,n-p)} = [SS_{fact}/(p-1)]/[SS_r/(n-p)] \quad (46)$$

for linear models containing a β_0 term. Although it is beyond the scope of this presentation, it can be shown that SS_{corr}, SS_{fact}, and SS_r would all be expected to have the same value if the factors had no effect upon the response (*8*). However, if the factors do have an effect upon the response, then $SS_{fact}/(p-1)$ will become larger than $SS_r/(n-p)$. The more significant the factor effects, the larger will be the F-ratio given in equation 46. In the present example, $F_{(1,2)} = (32/1)/(3/2) = 21.33$, which is significant at the 95.62% level of confidence (*4*). In general, this means the risk (α) is only (100% − 95.62%)/100% = 0.0438 that all of the factor effects (β's) are equal to zero. In the present model, there is only one factor effect (β_1), and thus we can be 95.62% confident that the effect of x_1 is real. The F-test for the significance of the factor effects is often called the "test for the significance of the regression" (*9*).

F-Test for Lack of Fit

In a similar way, it can be shown that

$$F_{(f-p,n-f)} = [SS_{lof}/(f-p)]/[SS_{pe}/(n-f)] \quad (47)$$

is a statistically valid measure of the significance of the lack of fit of the model to the data (see Figure 3). In the present example, $F_{(1,1)} = (1/1)/(2/1) = 0.5$, which is not highly significant.

We emphasize that if the lack of fit of a model is to be tested,

Table 2. Analysis of Variance (ANOVA) Table for Linear Models

Source	Sum of squares	Degrees of freedom	Mean square
Total	$SS_T = 204$	$n = 4$	51
Due to the mean	$SS_{mean} = 169$	1	169
Corrected for the mean	$SS_{corr} = 35$	$n-1 = 3$	11.67
Due to the factors	$SS_{fact} = 32$	$p-1 = 1$	32
Residuals	$SS_r = 3$	$n-p = 2$	1.5
Lack of fit	$SS_{lof} = 1$	$f-p = 1$	1
Pure experimental uncertainty	$SS_{pe} = 2$	$n-f = 1$	2

$f - p$ (the degrees of freedom associated with SS_{lof}) and $n - f$ (the degrees of freedom associated with SS_{pe}) must be greater than zero; that is, the number of factor combinations must be greater than the number of parameters in the model, and there should be replication to provide an estimate of the variance due to pure experimental uncertainty. Some types of experiments do not lend themselves to replication (e.g., the measurement of absorbance as a function of time in a reaction rate system), and the variance due to pure experimental uncertainty must be estimated independently.

Statistical Significance and Practical Significance

The F-tests for goodness of fit and lack of fit sometimes give seemingly conflicting results: with some sets of data, it will happen that each of the F-tests will be highly significant. The question then arises, "How can a model exhibit both a highly significant goodness of fit and a highly significant lack of fit?"

Such a situation will often arise if the model does indeed fit the data well, and if the measurement process is highly precise. Recall that the F-test for lack of fit compares the variance due to lack of fit with the variance due to pure experimental uncertainty. The reference point of this comparison is the precision with which measurements can be made. Thus, although the lack of fit might be so small as to be of no practical importance, the F-test for lack of fit will show that it is statistically significant if the estimated variance due to pure experimental uncertainty is relatively very small.

It is important in this case to keep in mind the distinction between "statistical significance" and "practical significance." If, *in a practical sense,* the residuals are small enough to be considered acceptable for the particular application, it is not necessary to test for lack of fit.

Variance–Covariance Matrix

Matrix least-squares methods provide a rapid means of obtaining estimates of the variances and covariances associated with the parameter estimates. We will use V to designate the *variance–covariance matrix:*

$$V = s_{pe}^2 \, (X'X)^{-1} \tag{48}$$

where

$$s_{pe}^2 = SS_{pe}/(n - f) \tag{49}$$

is the estimate of the pure experimental variance, σ_{pe}^2, based upon true replicates. If the model does not show a serious lack of fit (in a statistical or practical sense), then

$$s_r^2 = SS_r/(n - p) \tag{50}$$

is also a valid estimate of σ_{pe}^2, and the equation

$$V = s_r^2 \, (X'X)^{-1} \tag{51}$$

is often used. *It is important to realize that this is a valid estimate only if the model is adequate!* In the present example, using s_r^2 with $n - p$ degrees of freedom gives

$$V = (3/2)\begin{pmatrix} 2.25 & -1 \\ -1 & 0.5 \end{pmatrix} = \begin{pmatrix} 3.375 & -1.5 \\ -1.5 & 0.75 \end{pmatrix}$$

The diagonal elements are the variances of the parameter estimates, from upper left to lower right in the same order as they appear from top to bottom in the B matrix; the off-diagonal elements are the corresponding covariances. For the present model,

$$V = \begin{pmatrix} s_{b_0}^2 & s_{b_0b_1}^2 \\ s_{b_1b_0}^2 & s_{b_1}^2 \end{pmatrix} \tag{52}$$

where $s_{b_0}^2$ is the estimated variance associated with b_0, $s_{b_1}^2$ is the estimated variance associated with b_1, and $s_{b_0b_1}^2 = s_{b_1b_0}^2$ is the estimated covariance between the two parameters.

The variances of the parameter estimates can be used to set confidence intervals that would include the true value of the parameter a certain percentage of the time. In general, the confidence interval for β, based on s_r^2, is given by

$$b \pm \sqrt{F_{(1,n-p)} \times s_b^2} \tag{53}$$

where β is the true value of the parameter, b is its estimated value, s_b^2 is the corresponding variance from the diagonal of the variance–covariance matrix, and $F_{(1,n-p)}$ is the tabular value of F at the desired level of confidence. For example, the 95% confidence interval for β_0 is

$$-1.5 \pm \sqrt{18.51 \times 3.375} = -1.5 \pm 7.9 \tag{54}$$

or

$$-9.4 \le \beta_0 \le 6.4 \tag{55}$$

Similarly, the confidence interval for β_1 is

$$4 \pm \sqrt{18.51 \times 0.75} = 4 \pm 3.7,$$

or

$$0.3 \le \beta_1 \le 7.7. \tag{56}$$

Figure 2 suggests one meaning of the negative covariance between b_0 and b_1. If b_0 were increased, the sum of squares can be made to remain relatively small if b_1 is decreased; similarly, decreasing b_0 compensates for increases in b_1. The "good" estimates of b_0 and b_1 (those estimates that give small sums of squares of residuals) lie in an elliptical area of the b_0–b_1 plane, an elliptical area that lies diagonally with respect to the b_0- and b_1-axes. This is further suggested in Figure 1, where it is seen that increasing the y_1-intercept (that is, increasing b_0) and decreasing the slope (b_1) will cause the line to fit the data reasonably well, though not as well as the best estimates of the parameters do. Decreasing the y_1-intercept and increasing the slope will also give a rather good fit (but again, not the best fit). Any uncertainty in estimating one of the parameters has an effect upon the best estimate of the other parameter—hence the covariance.

Coding Transformations

There are some situations in which the intercept and the uncertainty of its estimate in the original coordinate system are of little real value. An example is in methods development, in which the origin corresponds to conditions that are seldom, if ever, found in practical situations—that is, conditions corresponding to a complete absence of the substance being determined (*3*). What is often of greater interest is the best estimate of response and its associated uncertainty near the midpoint of factor values that are likely to be encountered. There are two ways of treating this situation: coding transformations, and confidence bands. We will discuss coding first.

Coding transformations are usually of the form

$$x^*_{1i} = (x_{1i} - c_1)/d_1 \qquad (57)$$

where x^*_{1i} is the coded factor level, c_1 corresponds to the value of x_1 about which the values of x_1 are to be referenced, and d_1 is a normalizing factor that scales the units. The autoscaling function (*10*) is of this form, where c_1 is the mean of the x_1 values in the data set, and d_1 is their standard deviation.

Let us code our x_1 data about its mean ($c_1 = \bar{x}_1 = 2$) and set $d_1 = 1$. Then, $x^*_{11} = (1 - 2)/1 = -1$, for example, and

$$X^* = \begin{pmatrix} 1 & -1 \\ 1 & +1 \\ 1 & 0 \\ 1 & 0 \end{pmatrix}$$

where X^* is the *matrix of coded parameter coefficients.*

$$(X^{*\prime}X^*) = \begin{bmatrix} 1 & 1 & 1 & 1 \\ -1 & +1 & 0 & 0 \end{bmatrix} \begin{bmatrix} 1 & -1 \\ 1 & +1 \\ 1 & 0 \\ 1 & 0 \end{bmatrix} = \begin{bmatrix} 4 & 0 \\ 0 & 2 \end{bmatrix}$$

$$(\underset{\sim}{X}^{*\prime}\underset{\sim}{Y}) = \begin{pmatrix} 1 & 1 & 1 & 1 \\ -1 & +1 & 0 & 0 \end{pmatrix} \begin{pmatrix} 2 \\ 10 \\ 6 \\ 8 \end{pmatrix} = \begin{pmatrix} 26 \\ 8 \end{pmatrix}$$

$$\underline{D} = 4 \times 2 - 0 \times 0 = 8$$

$$(\underset{\sim}{X}^{*\prime}\underset{\sim}{X}^{*})^{-1} = \begin{pmatrix} 2/8 & -0/8 \\ -0/8 & 4/8 \end{pmatrix} = \begin{pmatrix} 0.25 & 0 \\ 0 & 0.5 \end{pmatrix}$$

$$\underset{\sim}{B}^{*} = (\underset{\sim}{X}^{*\prime}\underset{\sim}{X})^{-1}(\underset{\sim}{X}^{*\prime}\underset{\sim}{Y}) = \begin{pmatrix} 0.25 & 0 \\ 0 & 0.5 \end{pmatrix} \begin{pmatrix} 26 \\ 8 \end{pmatrix} = \begin{pmatrix} 6.5 \\ 4 \end{pmatrix}$$

Thus, $b_0^* = 6.5$ corresponds to the y_1-intercept at the new origin, $x_1^* = 0$. The value is the same as that predicted at $x_1 = 2$ in the uncoded data system (see equation 29). The slope $b_1^* = 4$ remains unchanged, as expected.

The variance–covariance matrix is now

$$V^{*} = (3/2)\begin{pmatrix} 0.25 & 0 \\ 0 & 0.5 \end{pmatrix} = \begin{pmatrix} 0.375 & 0 \\ 0 & 0.75 \end{pmatrix} \tag{58}$$

We see that b_0^* is now estimated with greater precision than b_0; this is partly because $x_1^* = 0$ now lies in the middle of the data set. (Before, $x_1 = 0$ lay to the left, outside the region of experimentation, and extrapolation added to the uncertainty of estimating b_0.)

The confidence interval for β_0^* is

$$6.5 \pm \sqrt{18.51 \times 0.375} = 6.5 \pm 2.6 \tag{59}$$

or

$$3.9 \leq \beta_0^* \leq 9.1 \tag{60}$$

Similarly, the confidence interval for β_1^* is

$$4 \pm \sqrt{18.51 \times 0.75} = 4 \pm 3.7$$

or

$$0.3 \leq \beta_1^* \leq 7.7 \tag{61}$$

as before (coding has not changed the uncertainty of the slope).

The covariance between b_0^* and b_1^* is equal to zero in equation 58, which suggests that uncertainties in estimating b_0^* do not have an effect on the best estimate of b_1^* and vice versa. The effect of changes in the slope upon the sum of squares of residuals cannot be compensated by changing the intercept at $x_1^* = 0$. Similarly, changes in the intercept cannot be compensated by changes in the slope.

Figure 4, a plot of the sum of squares of residuals as a function of different values of b_0^* and b_1^* for this coded data system, further confirms this. Although the shape of the least-squares surface is still a paraboloid, the major and minor axes of the elliptical cross sections now lie parallel to the parameter axes.

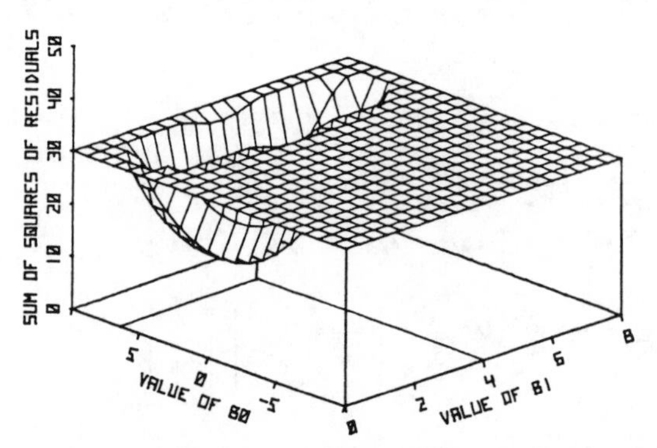

Fig. 4. Sum of squares of residuals as a function of different values of b_0^* and b_1^* (coded data system).
Sums of squares values greater than 30 set equal to 30

There are two points to be made about coding. First, the intercept can be made to occur at any level of x_1 simply by setting c_1 in equation 57 equal to the desired level and coding the data accordingly; b_0^* then corresponds to the predicted response at this level of the factor x_1. This is relevant to methods comparisons, where an offset at the midpoint of the data set is a more meaningful indicator of bias than any offset near the origin (*3*).

Second, covariance between parameter estimates is not necessarily undesirable or harmful. The numerical values and the precisions of estimating β_1 and β_1^* are the same in equations 56 and 61; the covariance between β_1 and β_0 arises because of the location of the data set with respect to the origin. This is relevant in the area of pseudo-zero-order kinetic methods of analysis where $\delta A/\delta t$ (the slope) contains all of the analytical information and A_0 (the absorbance at time zero, the intercept) is unimportant (*1*).

Confidence Bands

It can be shown (*7*) that, for any factor combination, the interval within which the true mean response will be found a given percentage of the time is

$$\hat{y}_{10} = X_0\hat{B} \pm \sqrt{F_{(1,n-p)} \times s_r^2 \times [X_0(X'X)^{-1}X_0']} \qquad (62)$$

where X_0 has been defined previously (see equation 27) and

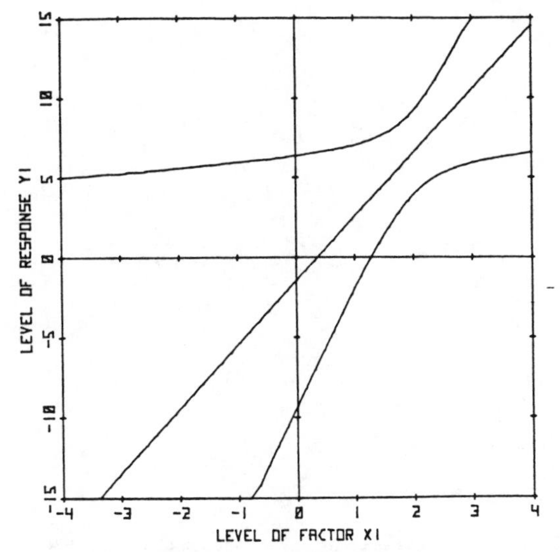

Fig. 5. 95% confidence bands for data of Fig. 1

$F_{(1,n-p)}$ is the critical Fisher ratio for the desired confidence level. In the uncoded data system, when $x_{10} = 0$, then $\hat{y}_{10}$ corresponds to b_0. The confidence interval for $\hat{y}_{10}$ (and therefore for b_0) is obtained as follows:

$$\underset{\sim}{X}_0(\underset{\sim}{X}'\underset{\sim}{X})^{-1} = (1 \quad 0)\begin{pmatrix} 2.25 & -1 \\ -1 & 0.5 \end{pmatrix}$$

$$= (2.25 \quad -1)$$

$$\{\underset{\sim}{X}_0(\underset{\sim}{X}'\underset{\sim}{X})^{-1}\}\underset{\sim}{X}_0' = (2.25 \quad -1)\begin{pmatrix} 1 \\ 0 \end{pmatrix} = 2.25$$

$$\underset{\sim}{X}_0\hat{\underset{\sim}{B}} = (1 \quad 0)\begin{pmatrix} -1.5 \\ 4 \end{pmatrix} = -1.5$$

$$\hat{y}_{10} = -1.5 \pm \sqrt{18.51\times(3/2)\times2.25} = -1.5 \pm 7.9$$

or $-9.4 \leq \hat{y}_{10} \leq 6.4$, at the origin of the uncoded data system (see equations 54 and 55). Likewise, when $x_{10} = 2$ (which corresponds in the uncoded data system to the origin of the coded data system),

$$\underset{\sim}{X}_0(\underset{\sim}{X}'\underset{\sim}{X})^{-1} = (1 \quad 2)\begin{pmatrix} 2.25 & -1 \\ -1 & 0.5 \end{pmatrix}$$

$$= (0.25 \quad 0)$$

$$\{\underset{\sim}{X}_0(\underset{\sim}{X}'\underset{\sim}{X})^{-1}\}\underset{\sim}{X}_0' = (0.25 \quad 0)\begin{pmatrix} 1 \\ 2 \end{pmatrix} = 0.25$$

$$\underset{\sim}{X}_0\hat{\underset{\sim}{B}} = (1 \quad 2)\begin{pmatrix} -1.5 \\ 4 \end{pmatrix} = 6.5$$

$$\hat{y}_{10} = 6.5 \pm \sqrt{18.51 \times (3/2) \times 0.25} = 6.5 \pm 2.6$$

or $3.9 \leq \hat{y}_{10} \leq 9.1$ at the origin of the coded data system (see equations 59 and 60).

If X_0 is made to move across all factor combinations, confidence bands can be generated. For the present example, the 95% confidence band is plotted in Figure 5. Identical confidence bands would be obtained if the coded data system were used.

Design of Experiments

The precision of estimating parameters (equation 48) and the precision of estimating confidence bands (equation 62) are dependent upon both s^2_{pe} (or s^2_r; see equation 51) and $(X'X)^{-1}$. This suggests two ways of obtaining improved precision (i.e., decreased uncertainty) in parameter estimates or confidence bands.

The first is related to the measurement process used to obtain the response values, y_{1i}. If the estimate of variance due to pure experimental uncertainty, s^2_{pe}, is large, it will be propagated up the sum-of-squares tree (Figure 3) through SS_r to SS_{corr}. This will tend to make R^2 (and R, r^2, and r) smaller than they might otherwise be; the F-ratio for goodness of fit will be smaller than it would be if SS_{pe} were smaller; and the F-ratio for lack of fit will also be smaller. In addition, the estimated variances and covariances of the parameter estimates will be large, and the confidence bands will be wide. Improving s^2_{pe} will make the lack-of-fit test more sensitive, the F-ratio for goodness of fit more representative of what it is intended to measure, and the R^2 value more indicative of factor effects; the uncertainties in the parameter estimates will be smaller; and the confidence bands will be narrower. Greatly improved precision in measurement (i.e., improved s^2_{pe}) is often possible by making relatively simple modifications (by using scale expansion in photometry or potentiometry, for example, if the readout device is the limiting module).

The second method of obtaining improved precision involves the $(X'X)^{-1}$ matrix, which depends upon the matrix of parameter coefficients, X, which in turn is obtained from the design matrix, D. In general, the broader the domain of the factor levels and the greater the number of experiments, the smaller will be the elements of the $(X'X)^{-1}$ matrix, the better will be the precision of the parameter estimates, and the tighter will be the confidence bands.

If the original four-experiment design were broadened to cover twice the factor domain (but still centered at $x_1 = 2$), then

$$\underset{\sim}{X} = \begin{bmatrix} 1 & 0 \\ 1 & 4 \\ 1 & 2 \\ 1 & 2 \end{bmatrix}$$

$$(\underset{\sim}{X}'\underset{\sim}{X}) = \begin{bmatrix} 1 & 1 & 1 & 1 \\ 0 & 4 & 2 & 2 \end{bmatrix} \begin{bmatrix} 1 & 0 \\ 1 & 4 \\ 1 & 2 \\ 1 & 2 \end{bmatrix} = \begin{bmatrix} 4 & 8 \\ 8 & 24 \end{bmatrix}$$

$$\underline{D} = 4 \times 24 - 8 \times 8 = 96 - 64 = 32$$

$$(\underset{\sim}{X}'\underset{\sim}{X})^{-1} = \begin{bmatrix} 24/32 & -8/32 \\ -8/32 & 4/32 \end{bmatrix} = \begin{bmatrix} 0.75 & -0.25 \\ -0.25 & 0.125 \end{bmatrix}$$

which leads to more precise estimates of b_0 and b_1 than the narrower design (see equation 23). The resulting confidence bands are shown in Figure 6, assuming the same slope, intercept, and s_r^2 as in Figure 5.

If each of the four experiments in the original example is replicated (and identical responses are obtained), the 8 by 2 matrix of parameter coefficients becomes

$$\underset{\sim}{X} = \begin{bmatrix} 1 & 1 \\ 1 & 1 \\ 1 & 3 \\ 1 & 3 \\ 1 & 2 \\ 1 & 2 \\ 1 & 2 \\ 1 & 2 \end{bmatrix}$$

$$(\underset{\sim}{X}'\underset{\sim}{X}) = \begin{bmatrix} 1 & 1 & 1 & 1 & 1 & 1 & 1 & 1 \\ 1 & 1 & 3 & 3 & 2 & 2 & 2 & 2 \end{bmatrix} \begin{bmatrix} 1 & 1 \\ 1 & 1 \\ 1 & 3 \\ 1 & 3 \\ 1 & 2 \\ 1 & 2 \\ 1 & 2 \\ 1 & 2 \end{bmatrix} = \begin{bmatrix} 8 & 16 \\ 16 & 36 \end{bmatrix}$$

$$\underline{D} = 8\times 36 - 16\times 16 = 288 - 256 = 32$$

$$(\underset{\sim}{X}'\underset{\sim}{X})^{-1} = \begin{bmatrix} 36/32 & -16/32 \\ -16/32 & 8/32 \end{bmatrix} = \begin{bmatrix} 1.125 & -0.5 \\ -0.5 & 0.25 \end{bmatrix}$$

which is exactly one-half the original four-experiment $(X'X)^{-1}$ matrix (equation 23). Note that in this case, $n - p = 8 - 2 = 6$, $F_{(1,6)} = 5.99$, and (because in this example identical responses were assumed) $s_r^2 = 6/6 = 1$. The resulting confidence bands are shown in Figure 7.

Considerations such as these lead naturally into the area of "optimum design." In this regard, the cautions of Box et al. (*4*) are worth heeding: in general, it is advantageous to work with a *good* experimental design, in the sense of being *adequate* for a given purpose; it is probably not worth the effort to try to find the truly optimum design.

Second-Order Example

Up to this point, we have dealt only with first-order (straight-line) relationships. In this section and the next, the application of matrix least squares to higher-order linear models is briefly illustrated.

The data in Table 3 show a non-first-order relationship between a factor x_1 and a response y_1; although the factor levels increase in equal intervals, the response increases by generally larger and larger intervals, suggesting that the response might be a geometric function of the factor. We will choose the empirical second-order probabilistic polynomial model of equation 9 as an approximation of the true relationship between the factor x_1 and the response y_1. The matrix of parameter coefficients X contains columns for the three coefficients β_0, β_1, and β_{11}:

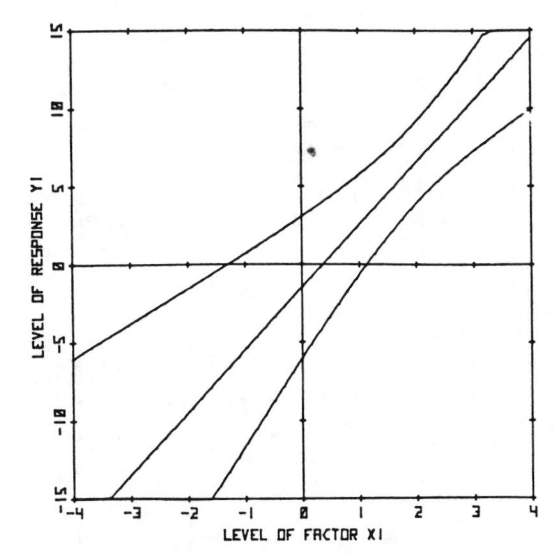

Fig. 6. 95% confidence bands for broader data set

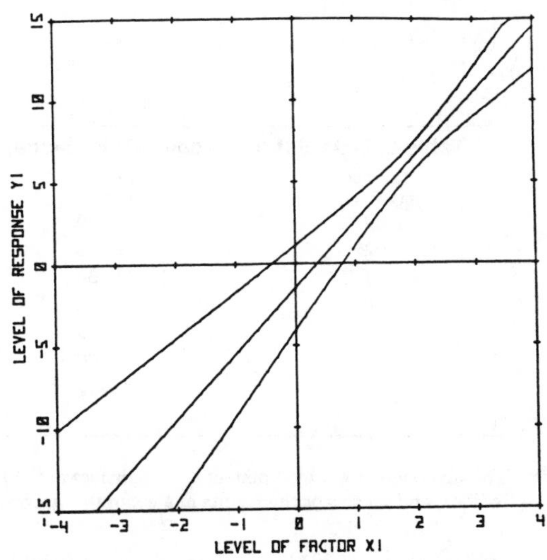

Fig. 7. 95% confidence bands for eight-experiment data set

$$\underset{\sim}{X} = \begin{bmatrix} 1 & -3 & 9 \\ 1 & -2 & 4 \\ 1 & -1 & 1 \\ 1 & 0 & 0 \\ 1 & +1 & 1 \\ 1 & +2 & 4 \\ 1 & +3 & 9 \end{bmatrix}$$

$$(\underset{\sim}{X}'\underset{\sim}{X}) = \begin{bmatrix} 1 & 1 & 1 & 1 & 1 & 1 & 1 \\ -3 & -2 & -1 & 0 & +1 & +2 & +3 \\ 9 & 4 & 1 & 0 & 1 & 4 & 9 \end{bmatrix} \begin{bmatrix} 1 & -3 & 9 \\ 1 & -2 & 4 \\ 1 & -1 & 1 \\ 1 & 0 & 0 \\ 1 & +1 & 1 \\ 1 & +2 & 4 \\ 1 & +3 & 9 \end{bmatrix}$$

$$= \begin{bmatrix} 7 & 0 & 28 \\ 0 & 28 & 0 \\ 28 & 0 & 196 \end{bmatrix}$$

Table 3. Data Set for Second-Order Model

Level of factor x_1	Level of response y_1
−3	30
−2	48
−1	68
0	98
+1	120
+2	160
+3	232

The inversion of a 3 by 3 matrix is straightforward (*9*) but tedious, and we accept the results of a computer program:

$$(\underset{\sim}{X}'\underset{\sim}{X})^{-1} = \begin{bmatrix} 0.333333 & 0 & 0.0476190 \\ 0 & 0.0357143 & 0 \\ -0.0475190 & 0 & 0.0119048 \end{bmatrix}$$

We retain six digits to avoid rounding errors propagated through the later addition or subtraction of large numbers. After calculation of the (***X′Y***) matrix, we can calculate

$$\hat{B} = \begin{bmatrix} b_0 \\ b_1 \\ b_{11} \end{bmatrix} = (X'X)^{-1}(X'Y) = \begin{bmatrix} 91.1429 \\ 31.5000 \\ 4.21429 \end{bmatrix}$$

Thus, the fitted model is

$$\hat{y}_{1i} = 91.1429 + 31.5000 \times x_{1i} + 4.21429 \times x_{1i}^2 \qquad (63)$$

A sum-of-squares and degrees-of-freedom tree is shown in Figure 8. In this example, $n = 7$, $p = 3$, and the degrees of freedom for the variance of residuals is $n - p = 7 - 3 = 4$; thus, the 95% confidence level F-statistic for equation 62 is $F_{(1,4)} = 7.71$. The graph of the experimental points, the fitted model, and the 95% confidence bands are shown in Figure 9. The parameter $b_0 = 91.1429$ is the intercept at $x_1 = 0$. The parameter b_1 is the tangential slope at $x_1 = 0$, and b_{11} is a measure of the curvature.

The following quantities, $R^2 = 0.9887$ and $F_{(2,4)} = 175.8$, can be calculated for the goodness of fit, which is significant at the 99.99% level of confidence. The lack of fit cannot be tested because there is no replication.

Two-Factor Full Second-Order Polynominal Example

Table 4 contains the concentrations of calcium and magnesium (two factors) and the corresponding absorbance responses for a set of experiments in a recent study investigating

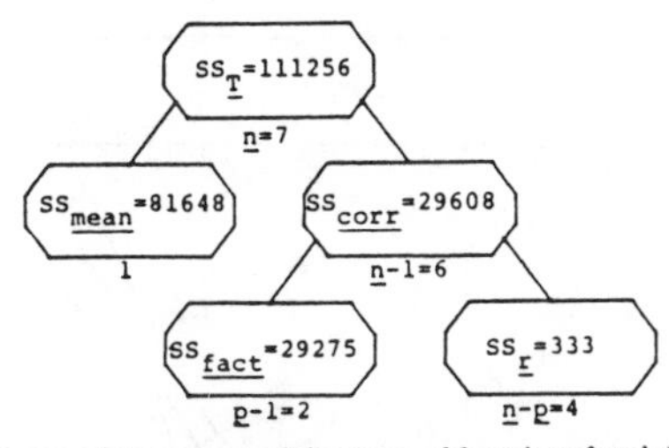

Fig. 8. Sums of squares and degrees of freedom for data of Table 3 and model of equation 9

the cresolphthalein complexone method for the determination of serum calcium (*11*). The experimental design is shown in Figure 10. We will illustrate here the fitting of a two-factor full second-order polynominal model to the data for the purpose of predicting absorbance response as a function of calcium and magnesium concentrations. Thus, we will not be concerned with the precision of the estimated parameters, or with the possibility that the model might contain more parameters than are really necessary to adequately describe the relationship between absorbance response and calcium and magnesium concentrations.

Table 4. Factor Levels and Responses for Investigating the Effects of Calcium and Magnesium Concentrations on the Absorbance of the Cresolphthalein Complex

[Ca], mg/L	[Mg], mg/L	Absorbance
157.5	30	0.943
157.5	30	0.959
137.5	10	0.825
137.5	10	0.839
137.5	50	0.825
137.5	50	0.829
117.5	30	0.740
117.5	30	0.718
117.5	30	0.720
117.5	30	0.726
117.5	70	0.726
117.5	90	0.708
97.5	10	0.607
97.5	10	0.593
97.5	50	0.607
97.5	50	0.645
77.5	30	0.489
77.5	30	0.493
57.5	30	0.366
57.5	30	0.364

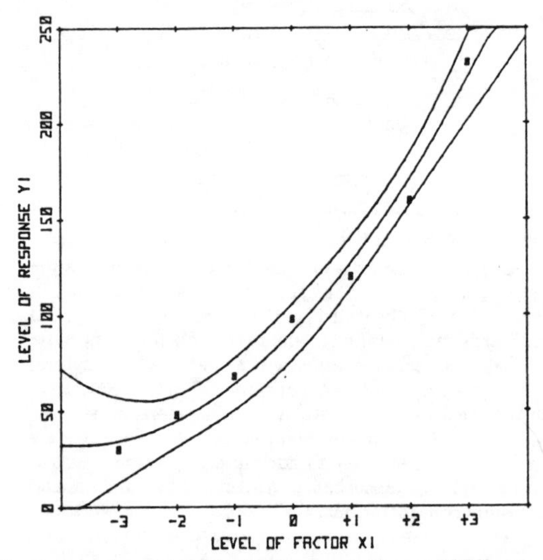

Fig. 9. Experimental points, least-squares line, and 95% confidence bands for data of Table 3 and model of equation 9

Letting x_1 designate the concentration of calcium, x_2 the concentration of magnesium, and y_1 the absorbance response, we describe the model to be fit as

$$y_{1i} = \beta_0 + \beta_1 x_{1i} + \beta_2 x_{2i} + \beta_{11} x_{1i}^2 + \beta_{22} x_{2i}^2 + \beta_{12} x_{1i} x_{2i} + r_{1i} \quad (64)$$

The model includes an offset term (β_0), linear terms (β_1 and β_2) and quadratic terms (β_{11} and β_{22}) in the factors, and an interaction term (β_{12}) between the two factors. The $\boldsymbol{X}$ matrix thus contains 20 rows and six columns:

$$\underset{\sim}{X} = \begin{pmatrix} 1 & 157.5 & 30 & 24806.25 & 900 & 4725 \\ 1 & 157.5 & 30 & 24806.25 & 900 & 4725 \\ 1 & 137.5 & 10 & 18906.25 & 100 & 1375 \\ 1 & 137.5 & 10 & 18906.25 & 100 & 1375 \\ 1 & 137.5 & 50 & 18906.25 & 2500 & 6875 \\ 1 & 137.5 & 50 & 18906.25 & 2500 & 6875 \\ 1 & 117.5 & 30 & 13806.25 & 900 & 3525 \\ 1 & 117.5 & 30 & 13806.25 & 900 & 3525 \\ 1 & 117.5 & 30 & 13806.25 & 900 & 3525 \\ 1 & 117.5 & 30 & 13806.25 & 900 & 3525 \\ 1 & 117.5 & 70 & 13806.25 & 4900 & 8225 \\ 1 & 117.5 & 90 & 13806.25 & 8100 & 10575 \\ 1 & 97.5 & 10 & 9506.25 & 100 & 975 \\ 1 & 97.5 & 10 & 9506.25 & 100 & 975 \\ 1 & 97.5 & 50 & 9506.25 & 2500 & 4875 \\ 1 & 97.5 & 50 & 9506.25 & 2500 & 4875 \\ 1 & 77.5 & 30 & 6006.25 & 900 & 2325 \\ 1 & 77.5 & 30 & 6006.25 & 900 & 2325 \\ 1 & 57.5 & 30 & 3306.25 & 900 & 1725 \\ 1 & 57.5 & 30 & 3306.25 & 900 & 1725 \end{pmatrix}$$

The parameter estimates are

$$\hat{\underset{\sim}{B}} = \begin{pmatrix} b_0 \\ b_1 \\ b_2 \\ b_{11} \\ b_{22} \\ b_{12} \end{pmatrix} = \begin{pmatrix} -0.102034 \\ 0.00758425 \\ 0.00313789 \\ -0.00000569437 \\ -0.00000973555 \\ -0.0000193750 \end{pmatrix}$$

A sum-of-squares tree is given in Figure 11. From it, R^2 = 0.9969 and $F_{(5,14)}$ = 899.4 for the goodness of fit, which is significant at the 100.00% level of confidence. In this example, there is replication and we can test the lack of fit: $F_{(4,10)}$ = 0.5627, which is not highly significant.

Figure 12 is a plot of the predicted absorbance response as a function of two factors, calcium concentration and magnesium concentration. The second-order effects (b_{11} and b_{22}) are both negative; thus, the response surface of Figure 12 curves downward in both directions. The interaction effect (b_{12}) is evident in Figure 12—the effect of each factor depends upon the level of the other factor.

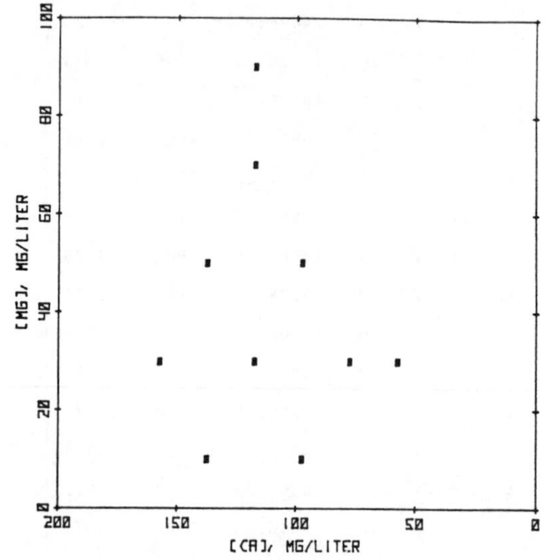

Fig. 10. Experimental design for cresolphthalein complexone study

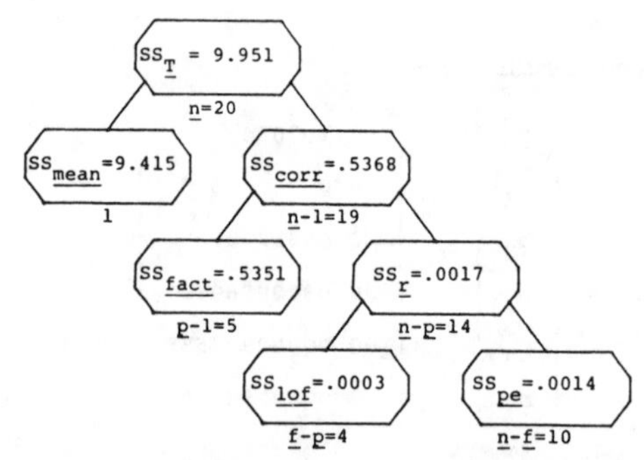

Fig. 11. Sums of squares and degrees of freedom for cresolphthalein complexone study

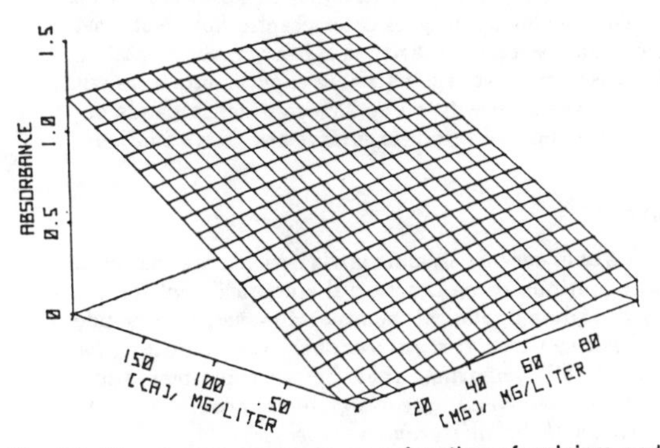

Fig. 12. Absorbance response as a function of calcium and magnesium concentrations for the cresolphthalein complexone study

Absorbance values less than zero set equal to zero

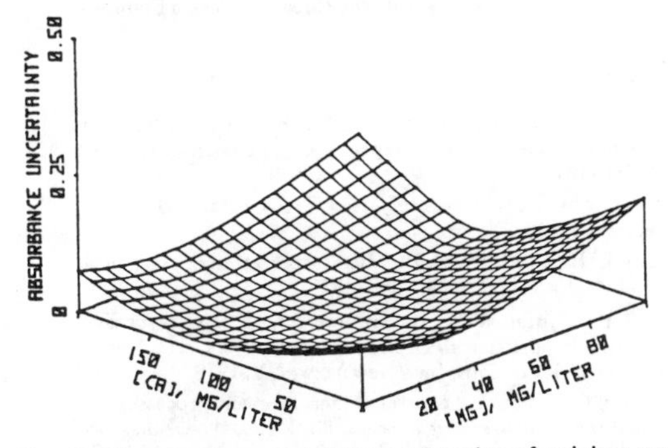

Fig. 13. Absorbance uncertainty as a function of calcium and magnesium concentrations

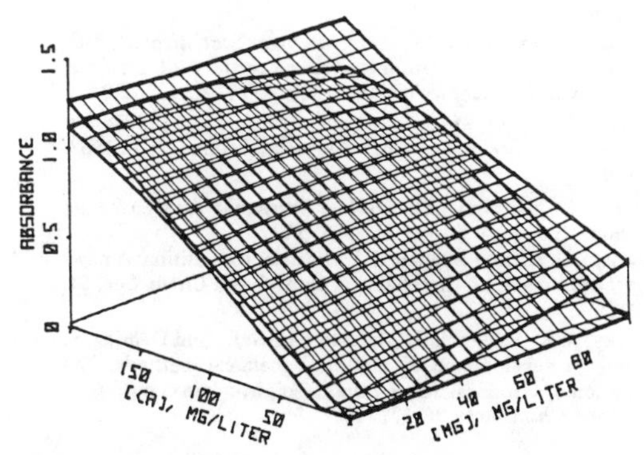

Fig. 14. 95% confidence bands for cresolphthalein complexone study

Figure 13 plots the 95% uncertainty in absorbance as a function of calcium and magnesium concentrations. Note that the uncertainty is least in the region of heavy experimentation and increases in those regions where experiments were not performed (see Figure 10). The 95% confidence bands enclosing the estimated response surface are shown in Figure 14.

Summary

The use of linear models and matrix least squares has been and will continue to be a powerful statistical tool for the treatment of data by clinical chemists. However, we conclude by emphasizing that *statistical treatment is no substitute for good data:* it is important that, before experimentation, consideration be given to the *design of experiments* and to the *precision of the measurement process.*

The authors thank D. Leggett, R. Deming, J. Jansen, and B. Sachok for helpful discussions and suggestions. S.N.D. gratefully acknowledges support from Grant E-644 from the Robert A. Welch Foundation.

References

1. Davis, R. B., Thompson, J. E., and Pardue, H. L., Characteristics of statistical parameters used to interpret least-squares results. *Clin. Chem.* **24**, 611 (1978).

2. Mandel, J., *The Statistical Analysis of Experimental Data*, Interscience, New York, NY, 1964.

3. Eppstein, L. B., and Levy, G. B., Misinterpretation of statistical intercept values. *Clin. Chem.* **24**, 1286 (1978).

4. Box, G. E. P., Hunter, W. G., and Hunter, J. S., *Statistics for Experiments. An Introduction to Design, Data Analysis, and Model Building*, John Wiley & Sons, Inc., New York, NY, 1978.

5. Wernimont, G., Statistical control of measurement processes. In *Validation of the Measurement Process*, **63**, J. R. Devoe, Ed., ACS Symposium Series, American Chemical Society, Washington, DC, 1977, pp 1–29.

6. Natrella, M. G., *Experimental Statistics*, National Bureau of Standards Handbook 91, U.S. Govt. Printing Office, Washington, DC, 1963, chap. 6, pp 6–10.

7. Neter, J., and Wasserman, W., *Applied Linear Statistical Models. Regression, Analysis of Variance, and Experimental Designs*, Richard D. Irwin, Inc., Homewood, IL, 1974, pp 89–92, 228–229.

8. Mendenhall, W., *Introduction to Linear Models and the Design and Analysis of Experiments*, Duxbury Press, Belmont, CA, 1968, pp 176–179.

9. Draper, N. R., and Smith, H., *Applied Regression Analysis*, John Wiley & Sons, Inc., New York, NY, 1966, p 24.

10. Kowalski, B. R., and Bender, C. F., Pattern recognition: A powerful approach to interpreting chemical data. *J. Am. Chem. Soc.* **94**, 5632 (1972).

11. Olansky, A. S., Parker, L. R., Jr., Morgan, S. L., and Deming, S. N., Automated development of analytical chemical methods. The determination of serum calcium by the cresolphthalein complexone method. *Anal. Chim. Acta* **95**, 107 (1977).

Appendix

α = risk of stating that a hypothesis is false when it is true
β = a parameter of a model
β_0 = intercept parameter
β_1 = linear parameter for factor no. one
β_{11} = quadratic parameter for factor no. one
β_{12} = interaction parameter between factor no. one and factor no. two
b = an estimated parameter value
$\boldsymbol{B}$ = matrix of parameters
$\hat{\boldsymbol{B}}$ = matrix of parameter estimates
c = central value about which a factor is coded
$\boldsymbol{C}$ = matrix of responses corrected for the mean
d = normalizing value for coding
D = determinant of $(\boldsymbol{X}'\boldsymbol{X})$ matrix
$\boldsymbol{D}$ = experimental design matrix
$\boldsymbol{D}_i$ = row vector giving design of experiment i
f = number of different factor combinations in a set of experiments
F = Fisher variance ratio
$\boldsymbol{F}$ = matrix of factor contributions
i = subscript index indicating experiment no.
$\boldsymbol{J}$ = matrix of mean replicate responses
$\boldsymbol{L}$ = matrix of lack-of-fit deviations
n = total no. of experiments in a set
p = no. of parameters in model
$\boldsymbol{P}$ = matrix of pure experimental deviations
r = a residual or deviation $(y - \hat{y})$; also, coefficient of correlation, or correlation coefficient
r_1 = residual of response no. one
r_{1i} = residual of response no. one in ith experiment
r^2 = coefficient of determination
R = coefficient of multiple correlation
R^2 = coefficient of multiple determination
$\boldsymbol{R}$ = matrix of residuals
σ^2_{pe} = variance due to pure experimental uncertainty
s^2_{pe} = estimated variance due to pure experimental uncertainty
s^2_r = estimated variance of residuals
SS_{corr} = sum of squares corrected for mean
SS_{fact} = sum of squares due to factors
SS_{lof} = sum of squares due to lack of fit
SS_{mean} = sum of squares due to mean
SS_{pe} = sum of squares due to pure experimental uncertainty
SS_r = sum of squares of residuals
SS_T = total sum of squares
$\boldsymbol{V}$ = variance-covariance matrix
x = level of an independent variable (factor)
x_1 = level of factor no. one
x_{1i} = level of factor no. one in ith experiment
x_{2i} = level of factor no. two in ith experiment
$\boldsymbol{X}$ = matrix of parameter coefficients

$\boldsymbol{X}_0$ = row vector of parameter coefficients for a factor combination of interest
y = level of a dependent variable (response)
y_1 = level of response no. one
y_{1i} = level of response no. one in ith experiment
$\bar{y}$ = the average (mean) response
$\hat{y}$ = an estimated response
$\boldsymbol{Y}$ = matrix of observed responses
$\overline{\boldsymbol{Y}}$ = matrix of mean response
$\hat{\boldsymbol{Y}}$ = matrix of estimated responses
* = symbols with an asterisk refer to coded data sets

Credit

Adapted from Clin. Chem., 25, 840-855 (1979) with permission of the American Association for Clinical Chemistry.

Adapted from Anal. Chem., 45, 278A-283A (1973) with permission of the American Chemical Society.

DATA ANALYSIS IN FOOD CHEMISTRY

Michele Forina and Silvia Lanteri

Istituto di Analisi e Teconologie Farmaceutiche ed
Alimentari V. le Bendetto XV, 3, 16132 GENOVA

Every day, numerous analyses are carried out on food products to answer a host of problems, in modern food science. These analyses are ultimately aimed at the improvement of food quality, reduction in price, increase in production yield, the elimination of undesirable effects during preservation, and determining the correlation between the food and the frequency of some diseases. In addition, food scientists are requested to assess some kind of guarantee of food quality, with regard to its content, geographic origin and age.

Many parameters determine the composition of food: class, order, family, species, variety of breed of the food producers (plant or animal), age, soil composition or nourishment, climate, preservation and treatments of the food, etc.; these parameters constitute the "cause space".

On the other hand, food composition determines its quality, preservation characteristics, price, food value, appearance, taste, and odor; these parameters constitute the "space of effects".

The chemical composition of samples we collect is the "chemical space" which represents the intermediate between cause and effect space, and which is necessary for the interpretation of cause-effect relationships.

The chemical space is described qualitatively and quantitatively and its description can be rough or very detailed. We can obtain the determination of all components, at the trace level as well. Real descriptions in practice, however, are

B. R. Kowalski (ed.), Chemometrics. Mathematics and Statistics in Chemistry, 305–349.

made by measuring a few components, in some cases belonging to a particular chemical class, or by measuring physical quantities related to composition, such as spectral variables.

Cause space, as well as chemical space, can be described more or less in detail: e.g., climate description can range from a brief description (warm, cold) to daily temperature, moisture, lighting intensity at every incident frequency, etc., for all samples. Currently, the cause space is described roughly (mean temperature, etc.) or with a little more detail (seasonal or monthly temperature, precipitation, thermal swing, etc.).

Effect space is often the hardest to describe. Only recently were nutrients systematically connected with longevity or with the frequency of some diseases. Food quality is rarely broken down into its components (taste, smell, appearance, etc.), which are very difficult to measure. However, also in this space, feature description is tending to become quantitative (panel scores, quality index).

Improved descriptions of these three spaces are necessary to determine cause-effect relationships. Today, we can obtain very efficient chemical descriptions and possibly also good descriptions of some features of the cause and effect spaces. Because of the consequent high number of descriptors and of samples, powerful means are necessary to find useful relationships in this great quantity of data.

These tools are chemometric methods by which one may find relationships between cause space and chemical features, and between these and effect space, explaining them in the light of chemical composition.

However, one cannot think of obtaining more and more detailed descriptions for a greater and greater number of samples, because of economics and time considerations. So, chemometricians must study food problems to evaluate the performance of reduced sets of chemical components.

Moreover, in spite of the great improvement in analytical instrumentation, and the spreading of the principles of multivariate analysis in chemistry, a great many food analyses are carried out by univariate criteria, that is, seeking the parameters that are correlated to particular effects on the basis of experience. But this experience is old. For example, in the study of many foods such as fats and beverages, the ratio between absorption at two wavelengths is used; the whole information contained by remaining part of the spectrum is ignored, because the old experience could only recognize the most evident aspect of the information.

Therefore, the aim of chemometric studies in the field of food science is not only the extraction of the greatest amount of useful information from analytical data, but, also:

- to program the measurement of a more rational chemical information;
- to show that multivariate techniques also allow the extraction of more information poor data;
- to characterize bad information sources that often irreparably mask the information contained in the chemical composition.

In this regard one must observe that analytical errors, the lack of rational sampling, errors during sample storage, and the use of non-standardized methods all obscure useful information, and one can often come to the conclusion that a lot of analytical work has been simply wasted.

Computerized multivariate data analysis in food chemistry has been widely used after 1973, as shown in the bibliography of Martens and Harries (1).

The main topics are classification and relationships between chemical parameters and the results of tasting. Beer, wine, spirits, juices, milk, soy sauces, essential oils, fats were studied. Under the wide term "classification" we meet many kinds of relationships between cause variables and chemical composition; when the cause variable is discontinuous it can be used to assign a category to a food sample. For example, the following are some typical problems in food chemistry that show the evolution of these chemometric techniques. Castino (2) used linear statistical discriminant analysis (LSDA) to classificate two Italian wines, that come from the same region and are made up with the same vinification technique so that the only difference is the grape-cultivar. The results are shown in Fig. 1. We have two major classification problems in wines: - to identify the ampelographic origin of wine, i.e., the cultivar of the grapes used; - to identify the geographical origin of wines of the same cultivar.

Smeyers and co-workers (3) used LSDA in the classification of cow and sheep milks by their fatty acid composition. The very good separation obtained allows the detection of mixtures of milks. The loadings of variables in the directions of the maximum useful information (discriminant scores) were used as a feature extraction method, so that the parameters to be measured were reduced to five medium- and long-chain fatty acids. GLC analysis time was consequently shortened.

Saxberg et al. (4) used eigenvector projection for the classification of whiskies. Here an "effect variable", the price, was used as classification criterion. The loadings on the principal components were used to eliminate the unimportant chemical features to simplify the analysis method.

Forina and co-workers (5) applied some methods to the geographical classification of virgin olive oils. This classification uses the term "geographic" for many cause variables that are not measured or identified, mainly variety, climate and soil composition. Eigenvector projection used as clustering methods sometimes allowed the detection of true, well-separated, oil-producing regions. Derde and colleagues (6) applied SIMCA to the study of olive oils. The outlier detection in the second pattern recognition level increases greatly the usefulness of multidimensional analysis in food chemistry, where we usually have many possible categories, but we cannot have samples from all categories. Practical problems are often of the type rejection (outliers) - acceptance (classification).

These few examples and a more detailed overview of the bibliography show a generalized evolution from LSDA to Bayesian and SIMCA classification, from simple classification to classification plus outlier detection, from canonical correlation analysis to PLS, always with the aim to increase simplicity and reliability of the analytical procedure.

Some results of data analysis on three data sets will be now reported to show some problems in food chemistry, some techniques we have used or changed and some plots we use to follow data analysis.

Data analysis was performed with PARVUS, a package developed at Genova University on a desk computer for low-dimensional data sets and with ARTHUR (7).

The data set FISH (8) contains thirty datavectors of ten fatty acid percentages in the lipids extracted from the meats of thirty different water animals (Table 1). Each sample was obtained from several animals of the same species.

The ten fatty acids are those with the greatest mean percentages and were selected among thirty fatty acids detected by GC after lipid extraction, saponification and esterification. The separation of the samples into the three classes, sea water fishes, fresh water fishes and cephalopoda, was not the main objective in this study. In fact, we looked for relationships between variables and type of animals in order to correlate these variables with human nutrition. So, discarded acids have a discriminating power higher than that of retained variables, but the selected major components are more representative of nutritional characteristics. However, the projection on the two first eigenvectors of the generalized covariance matrix (Fig. 2) shows good separation along the first component.

The main variables in the separation of the classes are palmitoleic acid (C20:5), oleic acid (C18:1), stearic (C18:0) and docosaenoic acid (C22:6). An increased content of eicosanpentaenoic acid (C20:5) shifts the samples towards the low left corner of the plot.

Eicosanpentaenoic acid is believed to have a very important effect on the interaction between thrombocytes (platelets) and vessel wall. In fact, eicosanpentaenoic acid should be a precursor of a thrombossan and of a prostacyclin with a very favourable action on the reduction of arteriosclerosis and ischemia.

Let us now consider the plot (Fig. 3) of the discriminant scores of linear statistical discriminant analysis and the loadings of the variables on the eigenvectors of the generalized covariance matrix (Fig. 4), on the eigenvectors of the covariance matrix of class barycentres (Fig. 5) and on the discriminant scores (Fig. 6). The last two plots are very similar, so that the discriminant directions are about the same directions of class barycentres. The first discriminant direction holds only about thirteen percent of the total information, and the loadings of acids C16:0 and C22:5 (that are very important compounds of the single classes) are very low. So, we can imagine the space of datavectors as constituted by three hyperdiscoids with the minor axes in discriminating direction and with a higher spread in the other directions. C20:5 is an important constituent of the discriminating directions and we can say that a diet rich in sea fish, or better in cephalopoda, is suitable to reduce circulation diseases. Moreover, the loadings on the variables on the discriminant scores give a measure of how the environment acts on the chemical composition of the fat and could be useful for the study of the relationship between environment and metabolism.

In our studies, we sometimes use a simplified method of nonlinear mapping (NLM). NLM was introduced in chemistry by Kowalski and Bender (9) and is a very useful display method. However, it requires the computation of $N(N-1)/2$ (N = number of datavectors) interpoint distances in each step of the iteration required to find the optimum position of the points in the map plane. The simplified method (SNLM) was designed to reduce computation time (especially with desk computers) and memory storage requirements with a minimum effect on method performance. In many simulated cases, we have seen (10) that the method gives a good representation of the original space regarding category separation. In the case of real data sets we have seen a good correspondence between the visually evaluated separation and that measured by classification methods such as KNN.

The method is simple. Few base-points (no more than ten class barycenters, random points or arbitrarily selected points) are mapped by usual NLM. The iteration time is very short because of the low number of points. Then, the datavector points are placed one at a time in the map plane, so that the distances between the datavector and the basepoints in the map plane preserve at best the corresponding distances in the hyperspace of the variables.

The optimization can be made with the simplex method or with the conjugate gradient method, from a starting point selected by a geometrical procedure (11).

The principal fault in SNLM is that frequently a class splits into two apparent sub-classes, especially when the number of base-points is very low (3-4). This fault can be reduced by increasing the base-point number or by using a correction coefficient for the distances in the hyperspace. In the present case the correction factor is SQR (number of variables/number of base-points). Figures 7-8 show SNLM without and with the correction coefficient. The map in Figure 8 gives a good visual representation of the data set in that the class distances, the "outliers" in the class and the class variances agree well with the quantitative results of classification methods (especially KNN, based on euclidean distances).

Portuguese Olive Oils

The second dataset contains four hundred eighty five datavectors of ten variables, the percentages of fatty acids and sterols in Portuguese olive oils.

Samples were of four harvest (1977/78, 1978/79, 1979/80, 1980/81) and were collected both at the beginning and at the end of the harvest time. The variables (palmitic, palmitoleic, margaric, stearic, oleic, linoleic and eicosanoic acid, campesterol and stigmasterol) were obtained from the original ones by deleting some acids at trace level and betasitosterol, the major component of the sterolic fraction.

The Portuguese Institute of olive oil (12) collected samples over all of Portugal by a well-planned random sampling. Thus, we have a good representation of Portuguese olive oil and, since the analyses were made in the same laboratory, we do not suspect any problem arising from inter-laboratory variance.

The principal cause variables are the harvest year, the harvest period, and the geographical position. The latter is an apparent factor; the underlying factors are climate and variety.

The eigenvector projection of Figure 9 shows the datavectors categorized on the basis of harvest year. Few samples lie in the high left corner of the plot and these are the samples of Tavira, characterized by a very low content of oleic acid. Most samples lie in a cluster on the right side of the plot, where, at the left, we see a second more scattered cluster separated from the first cluster by a region with a low point density in the middle of the first eigenvector. The loadings of variables on the first components give other information.

The eighty percent of the variance contained in the first eigenvector is stored in almost equal parts among four acids: an increase of stearic and linoleic acid shifts points towards the left in Figure 9, an increase of palmitic and palmitoleic acid towards the right. The remaining variance is shared between eicosanoic acid (left) and margaric acid (right). The loadings are almost unchanged as a function of harvest year and time of collection. Thus, this first component is very characteristic, and we can name it "chain-length component". The second eigenvector is characterized by a very high contribution of oleic acid, linoleic acid and sterols. The loading of oleic acid is almost independent of the year and the harvest time; in contrast, the loadings of the other variables depend on these two factors. This is the "oleic acid component".

The third eigenvector is characterized by a high contribution of linoleic acid and stigmasterol. The other loadings change significantly with the year and the harvest time. The shift of '77 samples on the second and the third eigenvector (Fig. 10) is due mainly to a high percentage of stigmasterol.

When samples are divided in four categories according to the harvest year and the first components of each category are computed, we can see (Table 2) that the scatter of the first two-three components is very similar. The greatest difference is among '77 and '78. To visualize the difference between the four years, we represent each year as a rectangle having as side lengths the standard deviations on the two first components. A common vertex of rectangles is the barycenter of the hyperspace of the variables. The other vertices of rectangles are points in the hyperspace of the variable. The representation of Figure 11 was obtained by minimizing the error in intervertex distances (as in non linear mapping) with the constraints of the right angle and side lengths for rectangles. Two years appear to be very similar ('79/'80 and '80/'81), while the others show deviation in the opposite direction.

The second and the third eigenvectors bring about the same quantity of information (percent of the total variance). They can easily reverse their order of importance, as shown in Figures 12 and 13. Here, samples are categorized on the basis of the geographic origin according to the Portugal plan in Figure 14. Tavira samples were excluded from the computation of the means and the variances used in autoscaling and the eigenvector loadings. Because Tavira samples are characterized by a very low oleic acid percentage, the importance of the oleic acid component lowers and becomes the third component.

In Figure 12 we see that the two main clusters detected in Figure 9 are well related to the geographical origin: the chain-length component separates the oils of Douro Valley from those of the remainder of Portugal (afterward improperly named South-Portugal oils). The separation of the two categories comes into better view when a single year is studied as shown in Figure 15, where the second eigenvector is the oleic acid component.

We used many classification methods of the package ARTHUR to measure the recognition and predictive abilities of our data concerning the geographical origin of the Portuguese oils subdivided in Douro-Valley oils and South-Portugal oils. The results shown in Tables 3 and 4 were obtained in the following way. One at a time, each harvest year was designated the training set, and the other three years were three separate evaluation sets. Bayesian analysis performed by the program BACLAS of ARTHUR behaves very well; we do not observe the problem of having a very great recognition ability at the expense of a relatively low prediction ability, frequently noticed in Bayesian analysis.

Bayesian analysis in BACLAS assumes the variables are independent, and the asymmetry of their distribution is taken into account. The results in Tables 3 and 4 were confirmed within about one percent by using a different model for Bayesian analysis (program BAYES of PARVUS) in which the distributions are supposed to be multivariate normal, and the correlation between variables is taken into account by computing the probability densities with the covariance matrix of the category. To avoid problems from collinearity (the sum of the percentages of retained acids is very close to 100% and almost constant), before Bayesian analysis the datavectors were projected from the space of the ten variables to that of the nine first eigenvectors of the generalized covariance matrix.

Both the results of eigenvector projection and of Bayesian analysis can be interpreted on the basis of Portugal's climate, inferred from latitude and topography. Oleic acid component seems to be the most evident climate-affected parameter (the increase in the percentage of oleic acid in olive oil with the decreasing in mean temperature has been well-recognized for many years). A direction (A in Figure 15) shows the maximum correlation with latitude. However, at higher latitudes (oil production areas n. 61 and 62) oleic acid percentage decreases.

In recent years a study was made (13) on the effect of latitude on the composition of olive oil of the Italian variety "Coratina". The increase in the percentage of oleic acid with latitude was confirmed with the exception of the sample at the maximum latitude of Italian olive growing territory. This "reversal" of latitude effect on oleic acid percentage may be a general rule. A second direction (B in Figure 15) could be correlated with the climate. This direction is the same as the first component of Douro-Valley oils. Frequently, Middle and South-Portugal oils move in this direction when sampled at the end of the harvest time; so direction B could be dependent on the climate during the last period of fruit ripening. To test this hypothesis a detailed description of the climate must be obtained.

Also, classification errors with Bayesian analysis agree well with climatic interpretation: e.g., some errors occur in the prediction of oils from area 59. This area is close to Douro river and in the North of the province of Viseu, at the end of a valley open to the warm South winds. Prediction and classification errors, however, strongly depend on the year. Data in Table 3 show that 78 oils are well recognized by the other years, while the model of 78 oil has the smallest predictive ability. So, for South-Portugal the '78 samples form a compact class, whose model is consistent with the model of the other years. Thus, the '78 crop is the more "typical".

However, detailed discussion of classification analysis in terms of "typical" years is very difficult because of the interactions between the class models: e.g. the low predictive ability of Douro-Valley 80 towards Douro-Valley 77 depends also on South-Portugal 80. The use of class modeling techniques such as SIMCA (14) removes these kinds of interactions.

SIMCA approximates each class by a single principal component model, whose dimensionality is estimated by cross validation. Let us consider the case of one component model. The limits of the model are obtained in the following way (Fig. 16). A "base model" is obtained by the range $t_1 - t_2$ of the scores t_i (i: index of a datavector in the training set) of the datavectors on the component, and the standard deviation s_0 of the residuals s_i is computed. The standard deviation s_t of the scores t_i is then added to both sides of the base model to obtain the classic SIMCA augmented model, with limits t_{lim}:$t_{min}=t_1-s_t$ and $t_{max} = t_2 + s_t$. When a new datavector is fitted to the class model the score t_p

on the class component and the residual s_p are evaluated. Then the augmented residual d_p is computed as:

$$d_p^2 = s_p^2 + (t_p - t_{lim})^2 \, s_o^2/s_t^2 \qquad t_p > t_{max} \quad t_p < t_{min} \qquad (1a)$$

$$d_p^2 = s_p^2 \qquad t_{min} \leq t_p \leq t_{max} \qquad (1b)$$

The fit of the object p to the class is obtained using the F value

$$F = d_p^2 \; / \; s_o^2 \qquad (2)$$

at a selected confidence level. This corresponds to the construction of a confidence box (SIMCA box) around the class model. By substituting equation (1a) in equation (2) we obtain:

$$F = s_p^2 \; / \; s_0^2 + (t_p - t_{lim})^2 \; / \; s_t^2 \qquad (3)$$

which is the equation of an ellipse. The shape of the SIMCA box in the case of many variables and one component is that of a hypercylinder around the component with half-hyperellipsoids on each end.

In the case of some problems in food chemistry we used a modified SIMCA model, the "reduced model" in which the limits are obtained by a contraction of the base model:

$$t_{min} = t_1 + s_0$$

$$t_{max} = t_2 - s_0$$

This means that we believe we have a high number of objects in the training set with an almost regular distribution along the model components. Thus errors drive the objects away from the true model, in much the same way as measurement error causes an experimental result to deviate from the true value (zero dimensional model).

Moreover, some food identity problems may require impenetrable box borders, e.g. to separate typical products of high quality (or great prestige), provided that the training set is really representative of the class.

The F test on the reduced model is performed by the use of the augmented residual:

$$d_p^2 = s_p^2 + (t_p - t_{lim})^2$$

where the vertical and horizontal distances from the model are added without multiplying by the ratio between the standard deviation of the class and that of the scores. Thus,

$$F = s_p^2 \,/\, s_0^2 + (t_p - t_{lim})^2 \,/\, s_0^2$$

is the equation of an hypersphere.

SIMCA was applied to evaluate the year effect on the Portuguese olive oil. Some results are reported here for Douro-Valley samples. In Figure 17 the projection of two-dimensional SIMCA base model of the four years is shown. The coordinates are the eigenvectors of the generalized covariance matrix of all Douro-Valley samples. A Coomans diagram (15) in Figure 18 emphasizes the anomaly of '77, the resemblance of 79 and 80, the closeness of 78.

Quantitative data in Tables 5 and 6 confirm the results deducted from the diagram.

Italian Wines

The use of modeling techniques will now be illustrated with a problem of the authentification of wines (Table 7).

Bayesian modeling was performed on the first four eigenvectors of the generalized covariance matrix, to obtain a graphical display of the confidence hyperellipsoids as shown in Figure 19. The objects in the training set, which show a poor fit to their class model, were detected by the histogram of their significance level, as shown in Figure 20. Here eight samples (1, 4,......53) have a significance level below 10%, all other samples have a significance level higher than 20%. So, these eight samples were considered as outliers to obtain the model of typical Barolo wine (obviously, the significance level of these samples is not low enough to reject the hypothesis they are authentic Barolo wines).

After outlier deletion (Fig. 21) we obtain the misclassification matrix of Table 8.

Table 8.

Computed class	Barolo	Grignolino	Barbera
True class			
Barolo	59	0	0
Grignolino	0	68	3
Barbera	0	1	47

Two classification errors out of four are due to non-typical samples discarded as outliers. Three classification errors are associated with a relatively low classification probability, i.e. only one sample is misclassified without doubt. To follow SIMCA treatment we use graphic display of the type shown in Figures 22 and 23. The projection of the 2-dimensional base model around the projection of the one-dimensional reduced model gives a rough idea of SIMCA box around the one-component model. Figure 23 notes an improper weighting of variables. In the case of multiclass problems (as frequently happens in food chemistry) the use of mean discriminating power (or mean Fisher weights, etc.) can make the classification worse. In fact, the weighting gives a great importance to eigenvector 1. So, the F ratios d_p^2/s_0^2 of the samples of class 1 from the model of class 3 are increased (but they were very high also before weighting). On the contrary, the reduced importance of eigenvector 2 makes the separation worse between classes 1 and 2, and between classes 2 and 3. So, in multiclass problems, single class-to-class discriminating powers must be considered, and the importance of the variables that have high discriminating power for class pairs with low SIMCA interclass distance can be increased by a suitable weight. In this case, however, the classification becomes better only with the use of a 2-component (unweighted) SIMCA model, as shown in Figures 24-26. In Figure 26 the half-hemispheric ends of the box are cut out, so that some objects outside the parallelepiped can be in the true box.

After outlier detection (only 9 with SIMCA against 18 with Bayesian analysis) we obtained the misclassification matrix in Table 9.

Table 9.

Computed class	Barolo	Grignolino	Barbera
True class			
Barolo	59	0	0
Grignolino	2	68	1
Barbera	0	1	47

This matrix has the same classification success rate of Bayesian analysis. The SIMCA model, however, is slightly more impenetrable to foreign samples of their own class. Because all samples were authentic wines, the number of outliers is considered an important parameter to evaluate the performance of the technique.

The examples shown here of Pattern Recognition in food chemistry, together with the few hundred works that appear in literature, are just a few drops from the barrel of food chemistry problems requiring chemometric methods.

Today most data in food science show an univariate cultural bias and few chemists have had access to chemometric techniques up to now. As D. Duewer (16) recently pointed out, for a multivariate statistical technique to be really accepted by bench chemists, it has to be made invisible inside the analytical instrument. In the mean time chemometricians must encourage, with good results from analytical data now available, the improvement of sample collection design, the measurement of analytical parameters more suitable to multivariate analysis, and the improvement of the description of non chemical variables on the cause and effect spaces.

REFERENCES

1) H. Martens and J.M. Harries, in H. Martens and H. Russwurm Jr Ed., "Food research and data analysis", Applied Science Publ., Barking (U.K.) (1983).

2) M. Castino, Att Acc. It. Vite e Vino, 27 (1975) pp. 3.

3) I. Smeyers-Verbeke, D.L. Massart and D. Coomans, J. Ass. Off. Anal. Chem., 60 (1977) pp. 1382-85.

4) B.E.H. Saxberg, D.O. Duewer, J.L. Booker and B.R. Kowalski, Anal. Chim. Acta, 103 (1978) pp. 1382-85.

5) M. Forina and E. Tiscornia, Ann. Chim., 72 (1982) pp. 143.

6) M.P. Derde, D. Coomans and D.L. Massart, J. Agricul. Food Chem., in press.

7) B.R. Kowalski, "Chemometrics: theory and application", A.C.S. Symposium Series 52, Am. Chem. Soc., Washington, D.C., (1977) pp. 14.

8) G. Forneris, L. Morisio Guidetti and C. Sarra, Riv. Soc. Ital. Sc. Alimentazione, 10 (1981) pp. 155.

9) B.R. Kowalski and C.F. Bender, J. Amer. Chem. Soc., 95 (1973) pp. 686.

10) M. Forina, C. Armanino, S. Lanteri and C. Calcagno, Ann. Chim., (Rome), in press.

11) M. Forina and C. Armanino, Ann. Chim., 72 (1982) pp. 127.

12) Instituto do Azeite e Produtos Oleaginosos, XII, XIII, XIV, XV Cadastro Oleicola (1979-1982).

13) M. Forina, C. Armanino, S. Lanteri, C. Calcagno and E. Tiscornia, Riv. Ital. Sostanze Grasse, 60 (1983) pp. 607.

14) S. Wold, First International Symposium on Data Analysis and informatics, 7-9 September 1977, Versailles.

15) D. Coomans, in "Patroonherkenning in de Medische Diagnose aan de hand van klinische laboratorium ondrzoeken", Pharm. Thesis, Vrije Universiteit Brussel (1982).

16) S.A. Borman, Anal. Chem., 54 (1982) pp. 1379A.

Table 1: Dataset Fish

Samples	Variables
1) Gobius cobitis	C14:0
2) Cypsilurus rondeleti	C16:0
3) Mugil cephalus	C16:1
4) Platichtys flesus flesus	C18:0
5) Pagrus pagrus	C18:1
6) Merluccius merluccius	C18:2
7) Scorpaena scrofa	C22:1
8) Boops boops	C20:5
9) Aspitriglia cuculus	C22:5
10) Sparus auratus	C22:6
11) Trachinus draco	
12) Mullus surmuletus	
13) Trachurus trachurus	
14) Sepia officinalis	
15) Loligo vulgaris	
16) Octopus vulgaris	
17) Chondrostoma soetta	
18) Salvellinus alpinus	
19) Anguilla anguilla	
20) Leuciscus cephalus	
21) Ictalurus melas	
22) Alburnus alburnus	
23) Exos lucius	

Table 1: Dataset Fish (Continued)

Samples	Variables
24) Cobitis taenia	
25) Ciprinus carpio	
26) Perca fluvialis	
27) Chondrostoma toxostoma	
28) Pagodobius martensi	
29) Salmo gairdneri	
30) Scardinius arytrhophtalmus	

Table 2: Percent Variance on the First Eigenvectors (Portugal)

Eig. number	1	2	3	Sum
YEAR				
1977/78	44.1	15.0	13.0	72.0
1978/79	52.0	15.0	12.2	79.1
1979/80	45.7	15.9	12.2	73.8
1980/81	46.9	14.5	12.9	74.2

Table 3: Results of Classification Analysis with BACLAS of ARTHUR (South Portugal)

Evaluation Set / Training Set	1977	1978	1979	1980	Row means
1977	97.8&	100	98.3	94.6	97.3
1978	93.5	100&	93.3	89.3	92.0
1979	98.9	100	93.3&	94.6	97.8
1980	98.0	100	98.3	99.1&	99.1
Column means	97.1	100	96.6	92.8	

& : Recognition ability, not used for means

Table 4: Results of Classification Analysis with BACLAS of ARTHUR (Douro Valley)

Evaluation Set / Training Set	1977	1978	1979	1980	Row means
1977	100&	92.3	86.2	89.7	89.4
1978	92.6	100&	89.7	93.1	91.8
1979	92.6	84.6	89.7&	89.7	90.7
1980	66.7	84.6	82.8	93.1&	78.0
Column means	84.0	87.2	86.2	90.8	

& : Recognition ability, not used for means

Table 5: SIMCA Distances (Douro Valley samples)

Between Class : and Class	1978	1979	1980
1977	.82	1.84	1.46
1978		.86	.47
1979			.18

Table 6 : SIMCA Classification of Douro Valley samples

% Datavector of class : that fit class	1977	1978	1979	1980
1977	92	55	27	37
1978	72	89	50	75
1979	36	45	92	93
1980	60	82	85	96

Table 7 : Three Well Known Italian Wines

Class		
1)	Barolo	59 samples
2)	Grignolino	71 samples
3)	Barbera	48 samples

Variables	
1)	Total alcohol
2)	Total poliphenols
3)	Flavonoids
4)	Colour intensity
5)	Tonality
6)	Optical density ratio (280/315 nm) diluted wine
7)	Optical density ratio (280/315 nm) flavonoids
8)	Proline

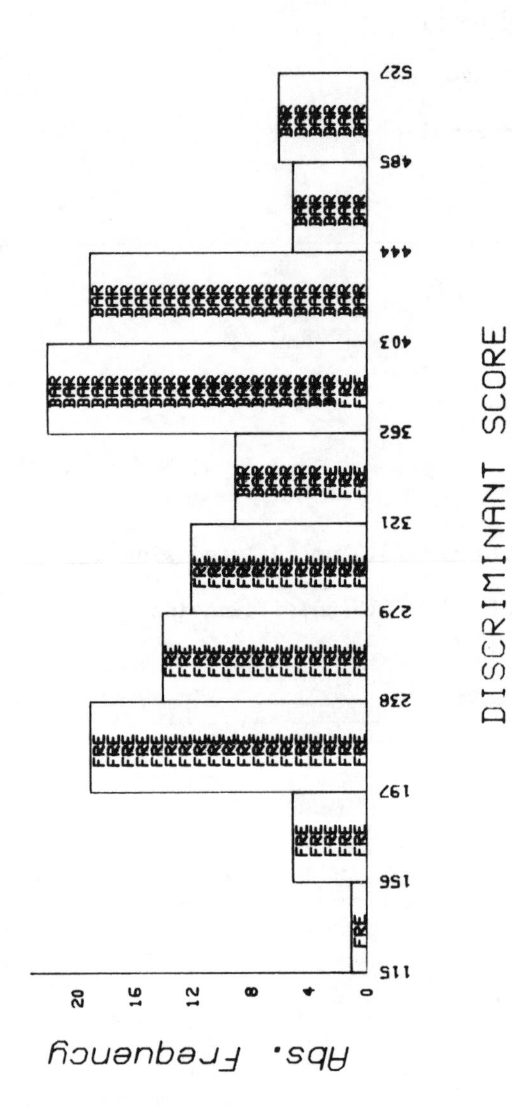

FRE: Freisa BAR: Barbera

Figure 1.

Separation of two red wines by linear discriminant analysis (2).

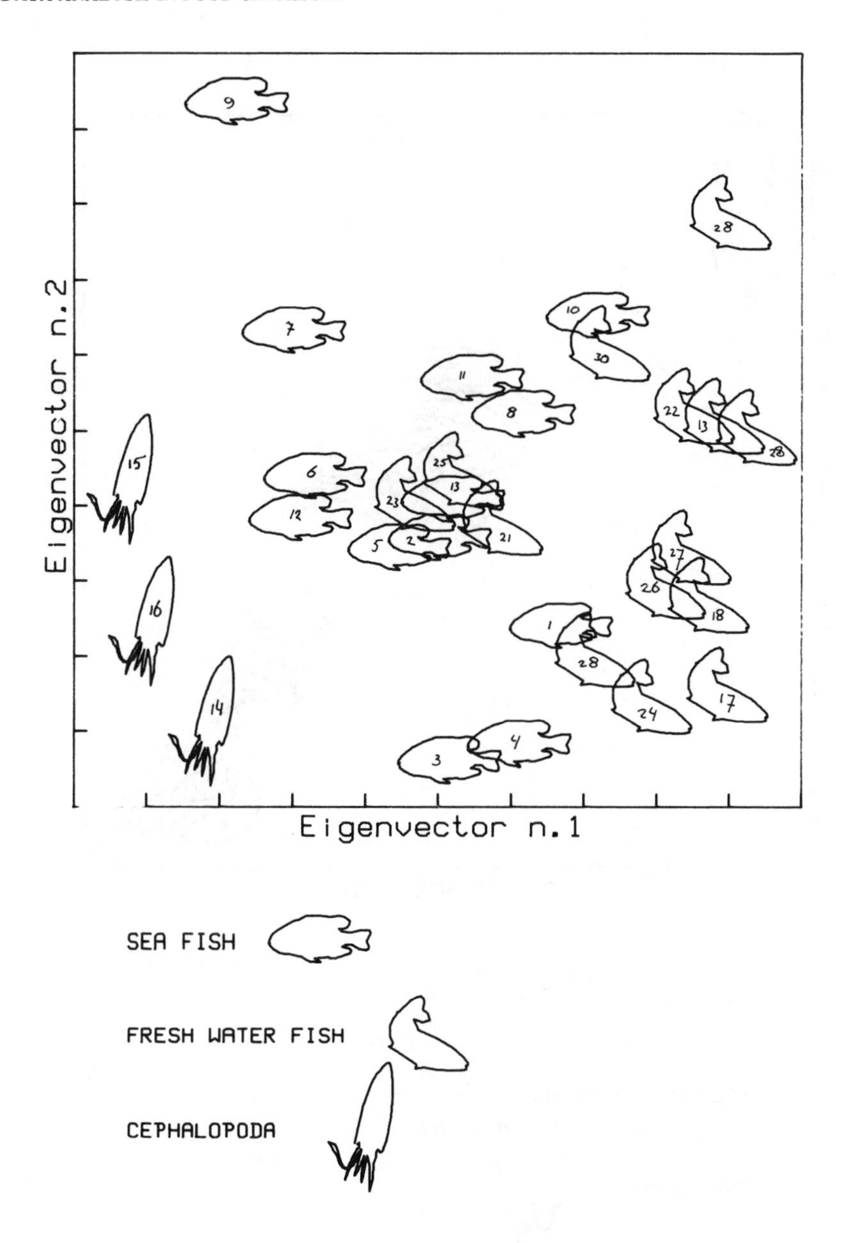

Figure 2.

Projection on the first eigenvectors of the generalized covariance matrix. Autoscaled data.

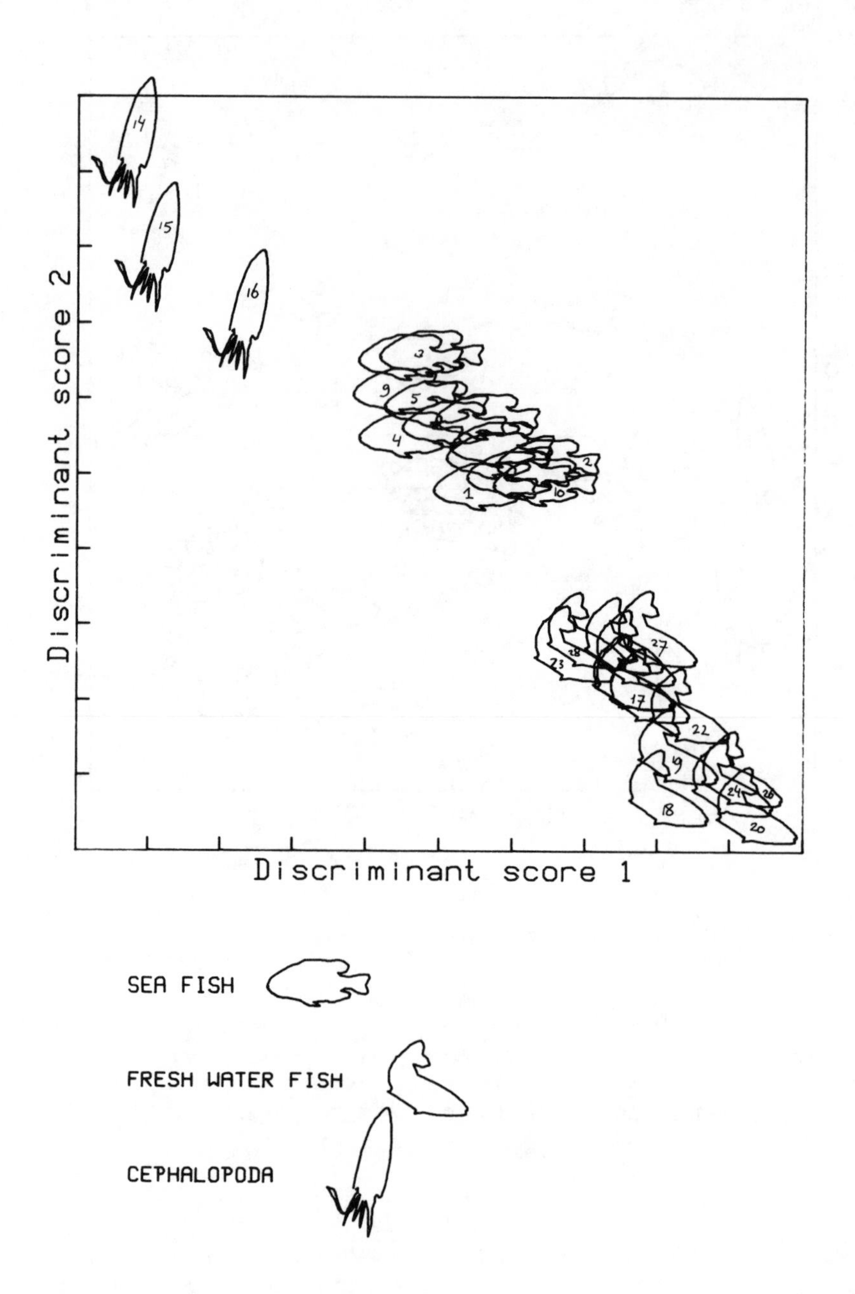

Figure 3.

Shoals reordered by linear discriminant analysis.

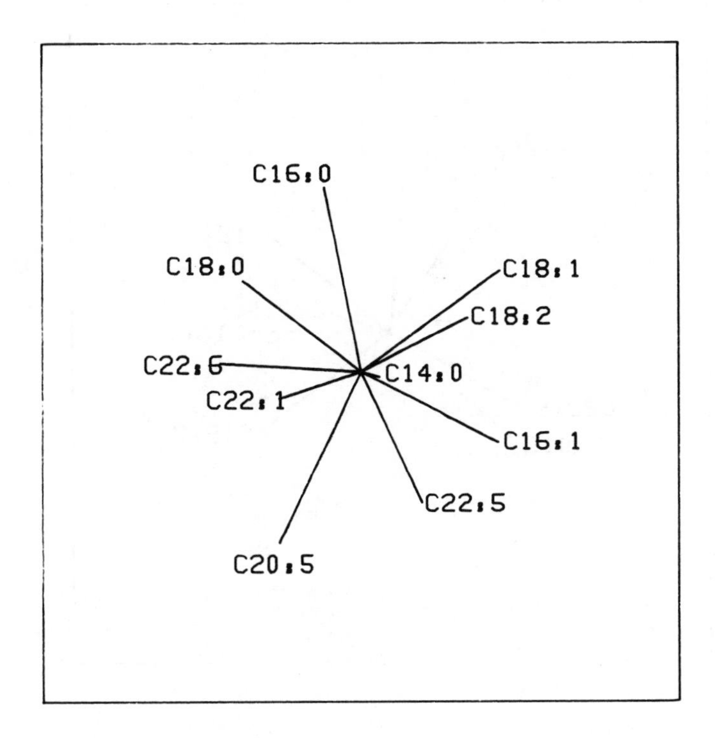

Figure 4.

FAT composition of sea animals. Loadings of the variables on the two first eigenvectors of the generalized covariance matrix. Autoscaled data.

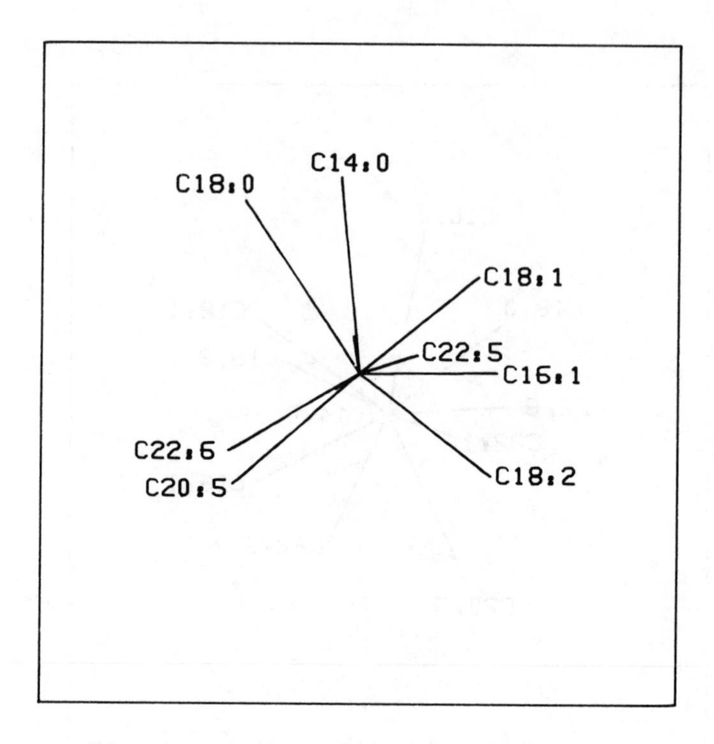

Figure 5.

Plot of the variables loadings on the two eigenvectors of the variance matrix of class barycenters.

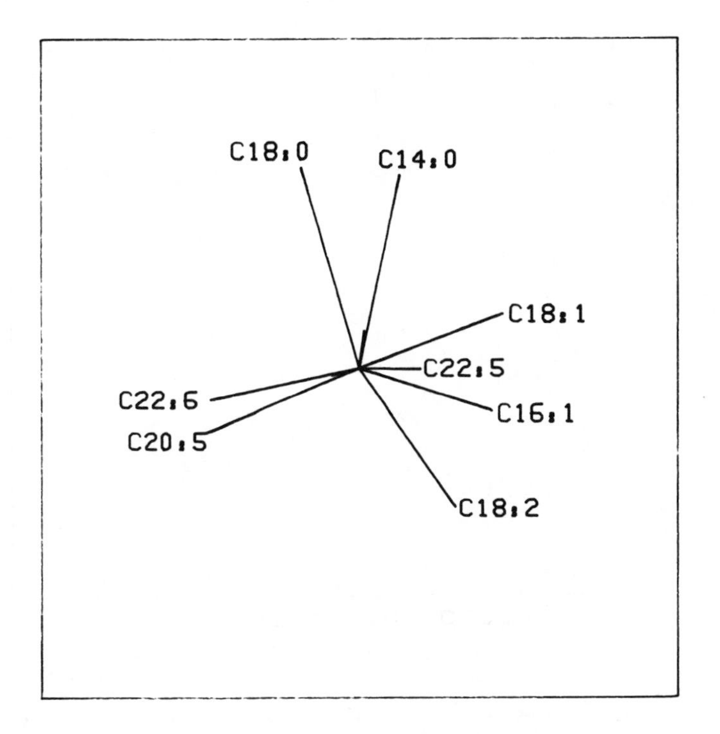

Figure 6.

Plot of the loadings of the variables on the two discriminant directions.

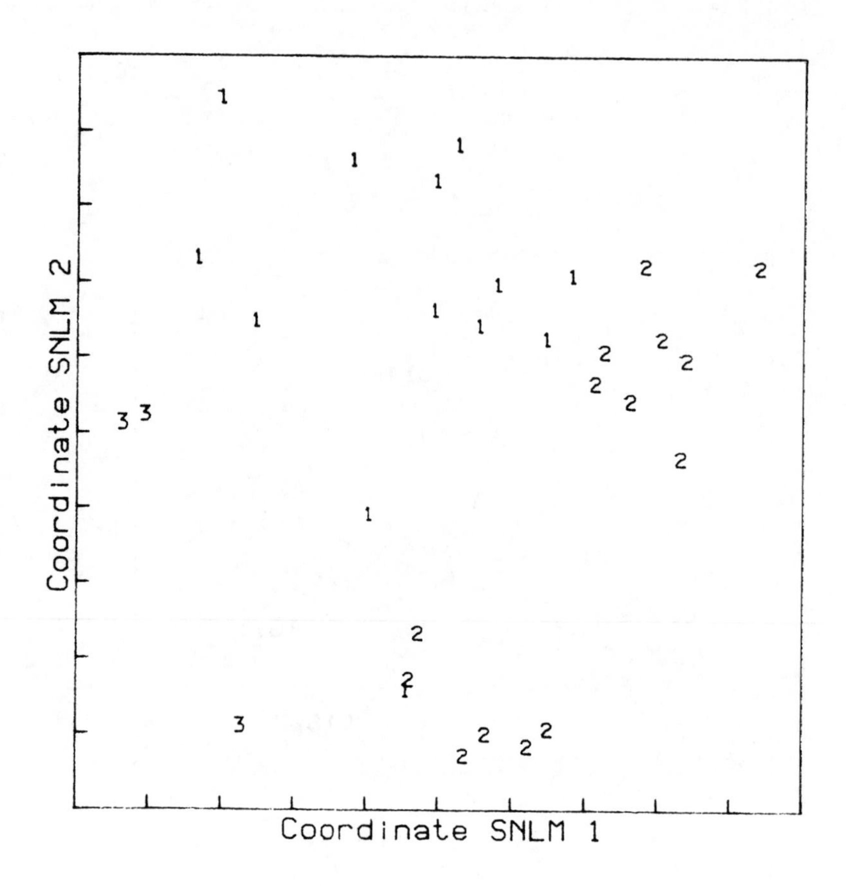

Figure 7.

SNLM: the base-points are the three barycenters.

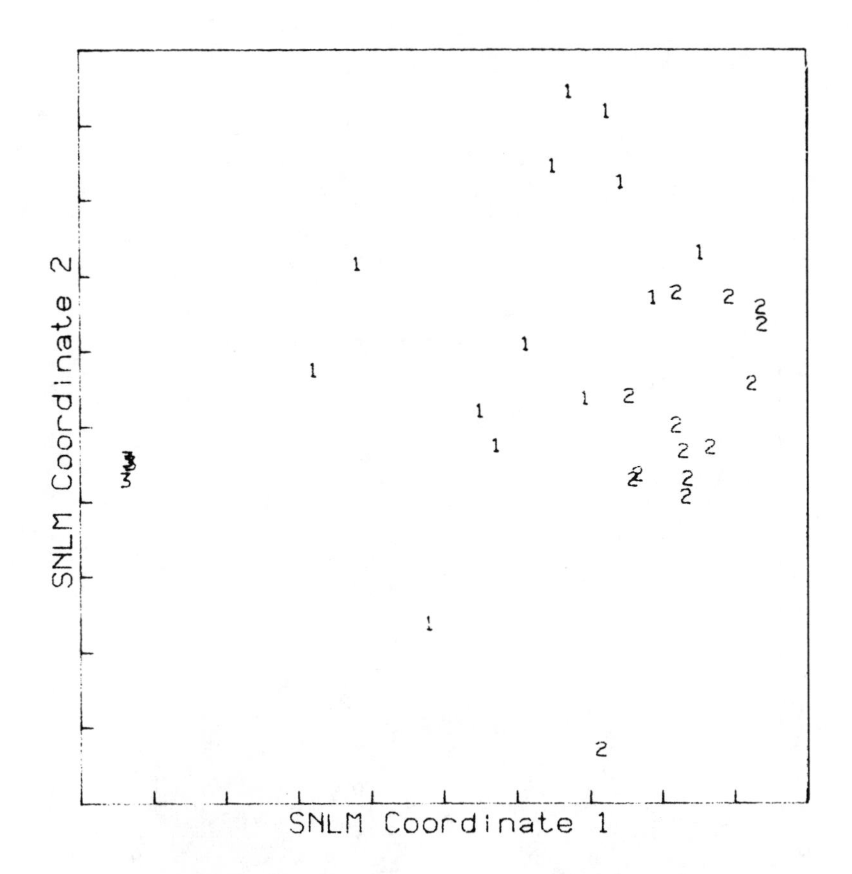

Figure 8.

SNLM plot with barycenters as base-points. The interpoint distances in the hyperspace are multiplied by a correction factor SQR(number of base-points / number of variables).

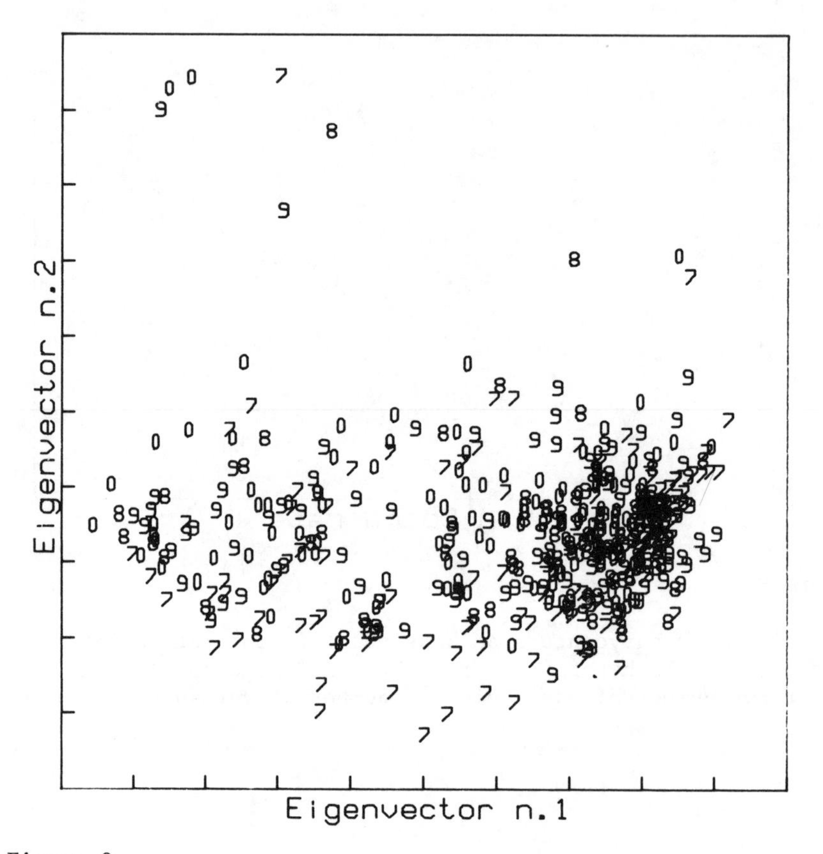

Figure 9.

Eigenvector plot of Portugal oils.
7: year 1977/78; 8: year 1978/79; 9: 7 year 1979/80;
0: year 1980/81.

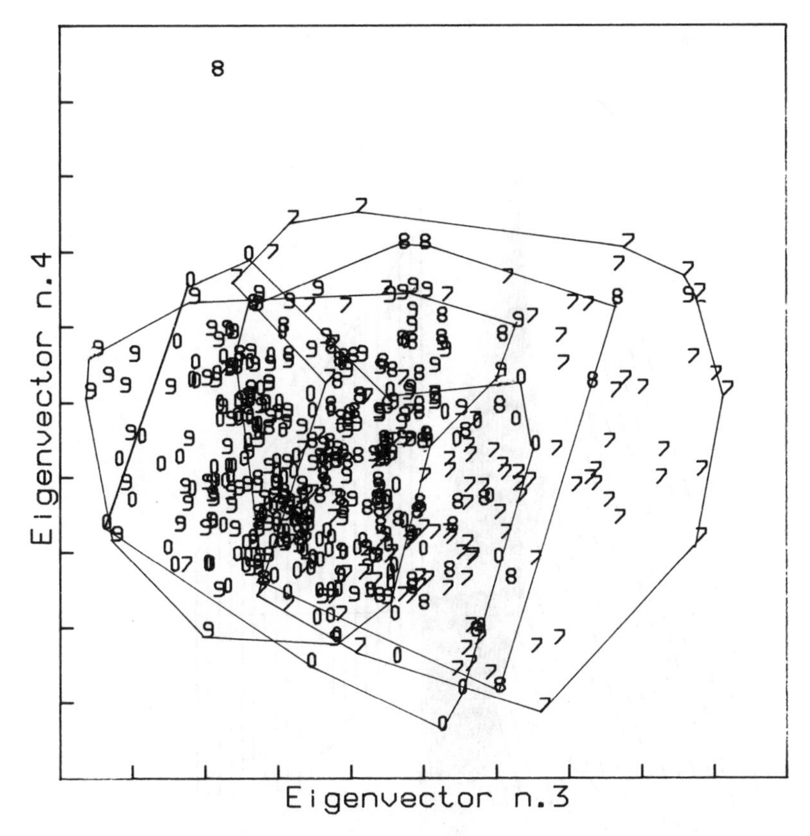

Figure 10.

Eigenvector plot of Portugal oils.
7: year 1977/78; 8: year 1978/79; 9: year 1979/80;
0: year 1980/81

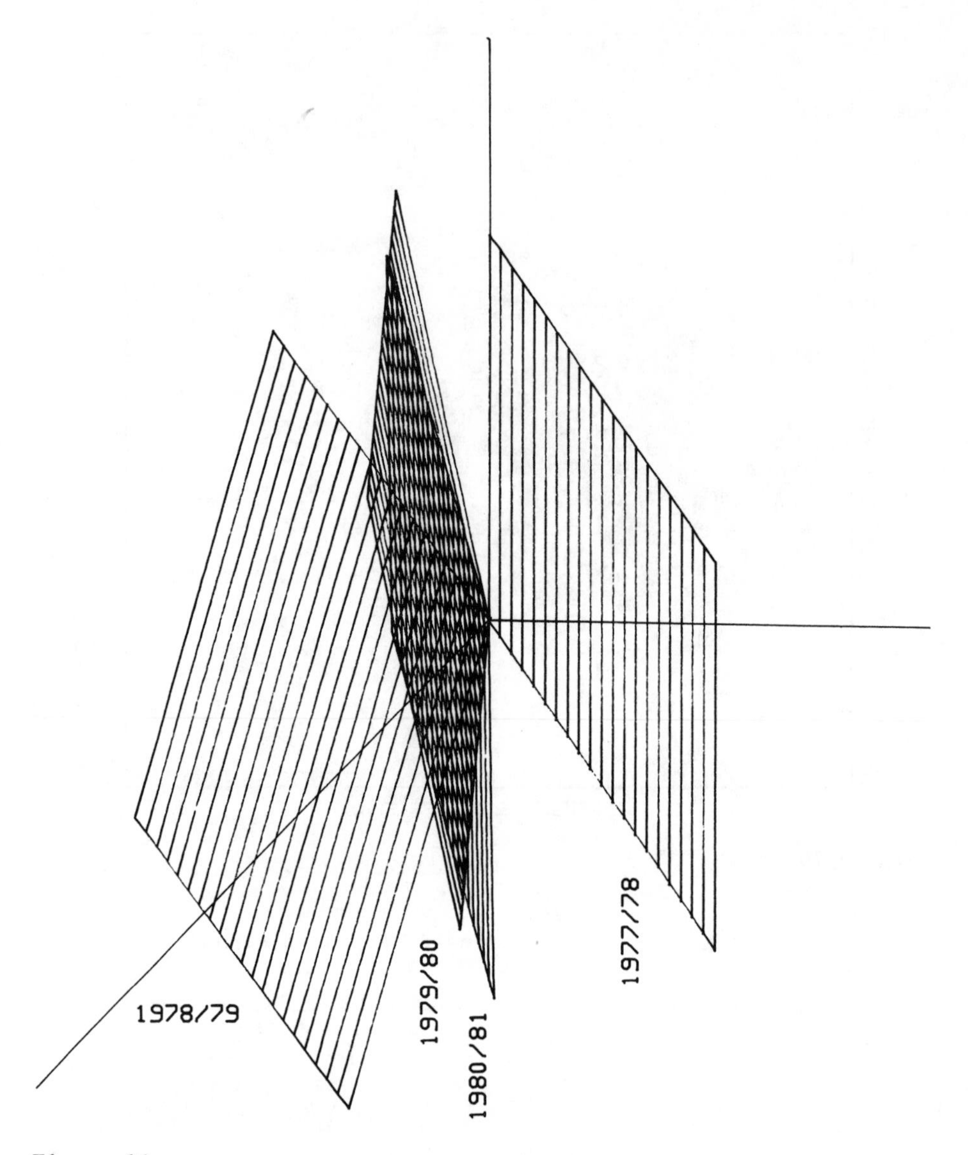

Figure 11.

Models of the four years of Portugal oils. From left (lower corner) 1978/79, 1979/80, 1980/81, 1977/78.

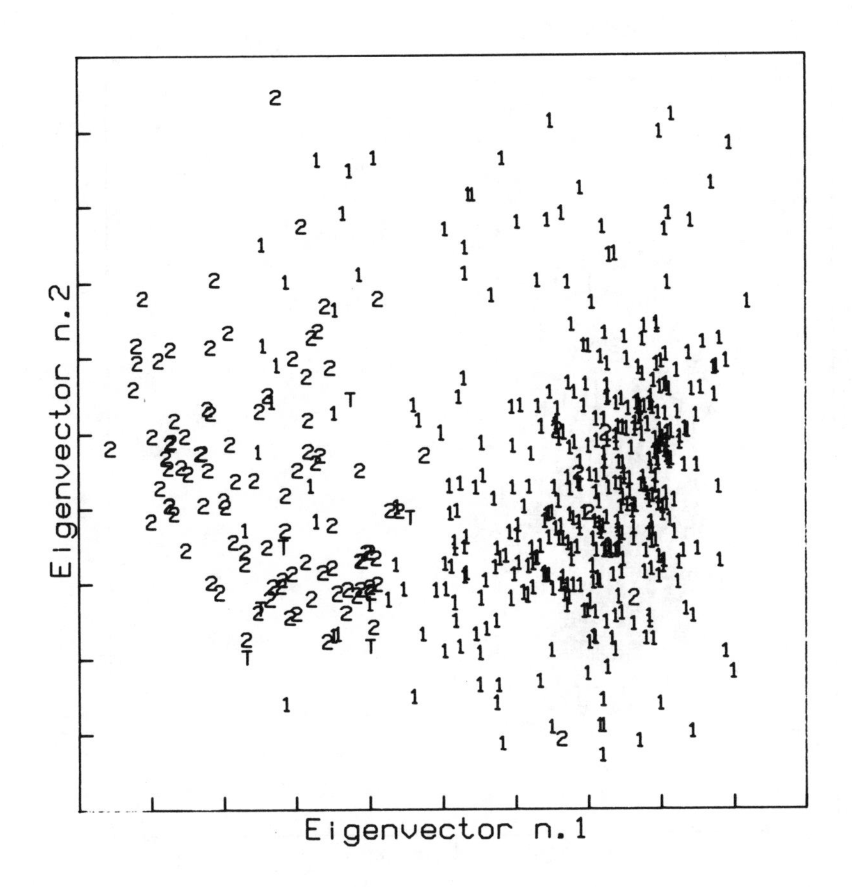

Figure 12.

Eigenvector plot of Portugal oils. 1: Coast and South Portugal; 2: Douro Valley; T: Tavira.

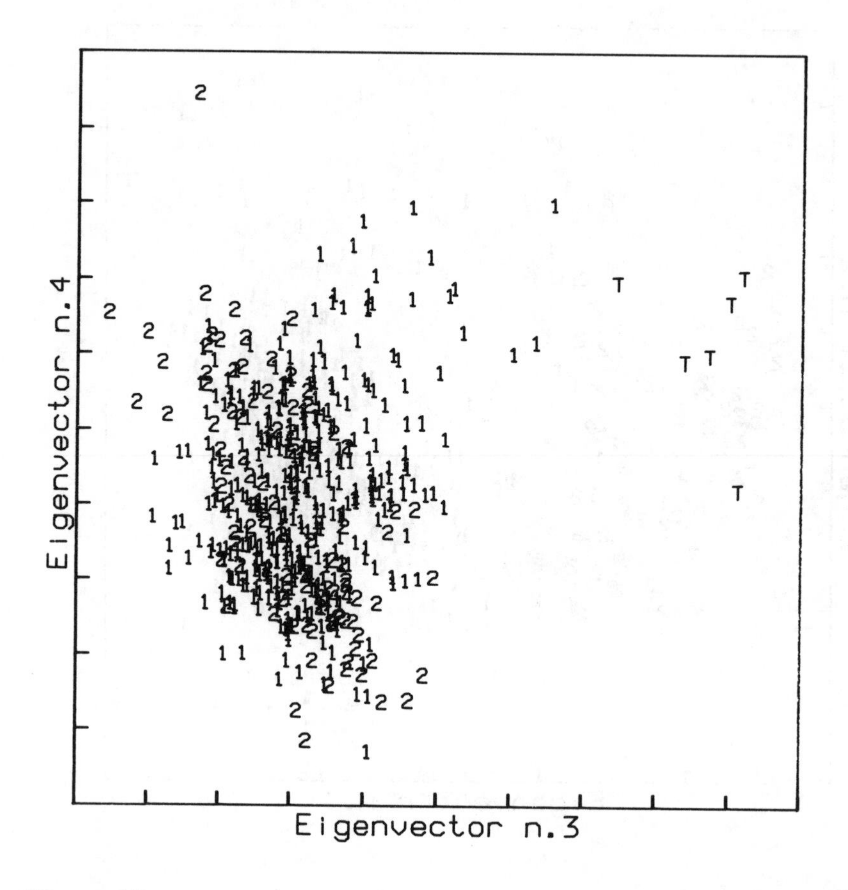

Figure 13.

Eigenvector plot of Portugal oils. 1: Coast and South Portugal; 2: Douro Valley; T: Tavira.

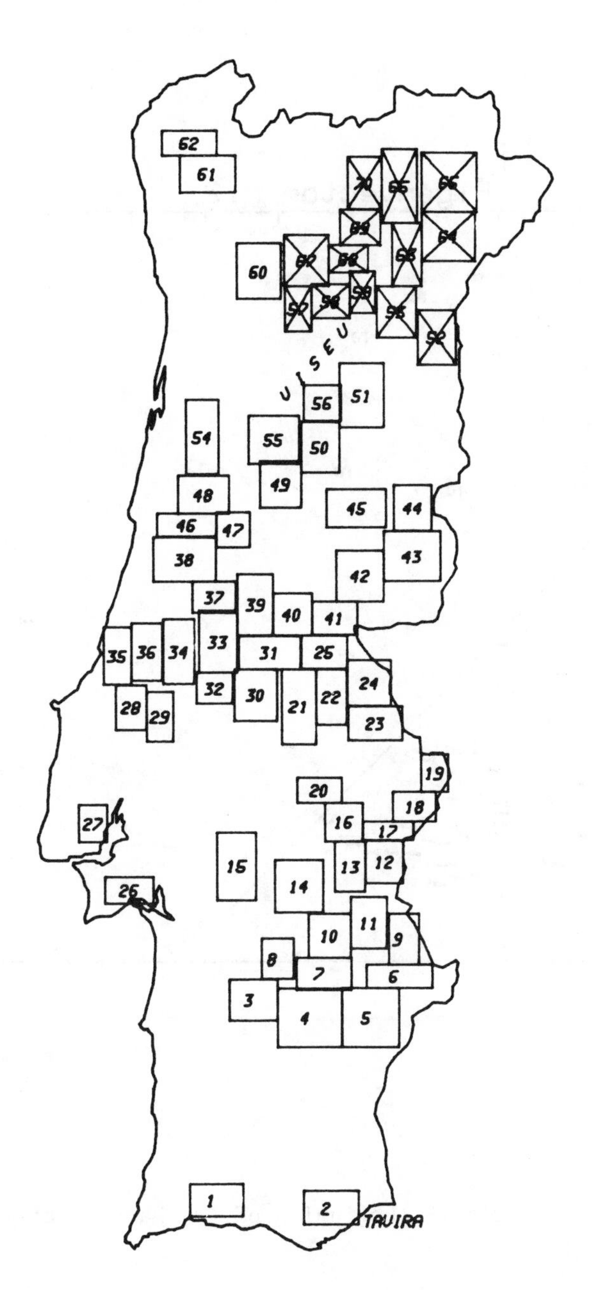

Figure 14.

Portugal plan with oil production areas. Crossed areas: Douro Valley.

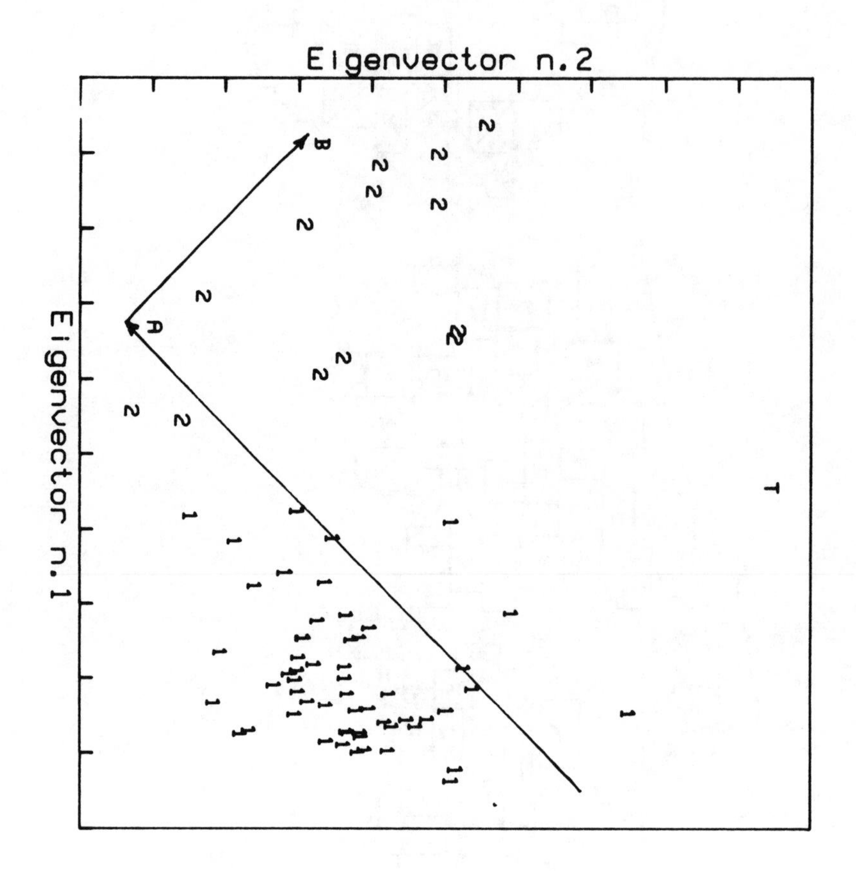

Figure 15.

Eigenvector plot of 1978/79 (beginning of the harvest time) oil samples.

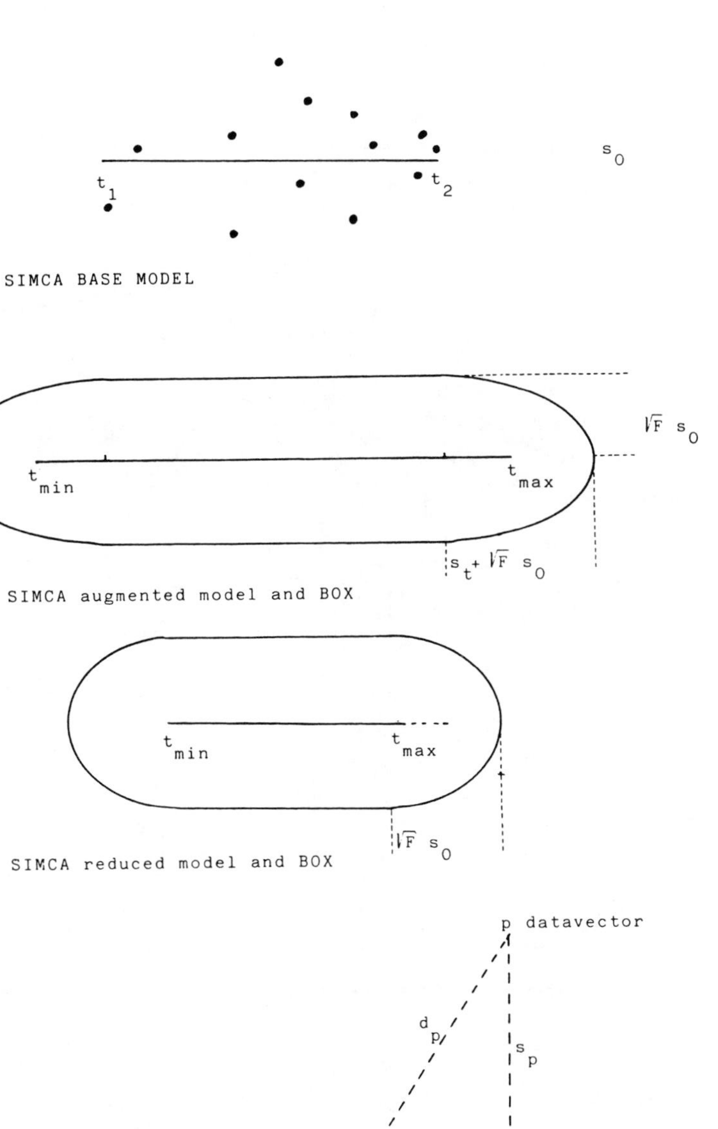

Figure 16.

SIMCA models and boxes.

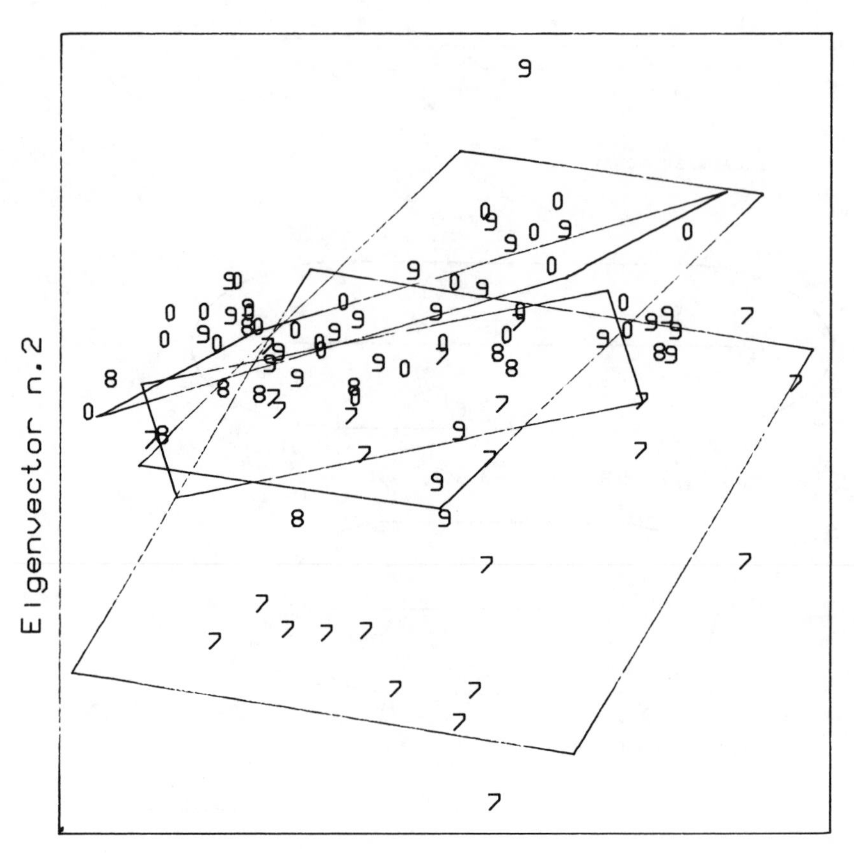

Figure 17.

Projection of 2-component SIMCA base-models after outlier deletion. Douro Valley samples.

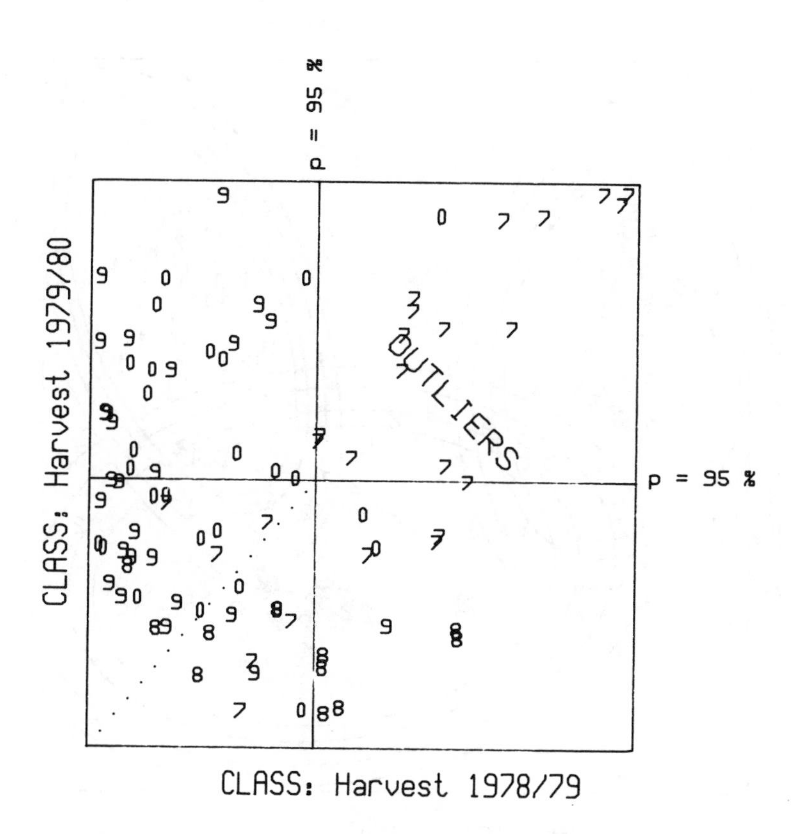

Figure 18.

Coomans diagram of Douro Valley oils. (3-component SIMCA reduced-models).

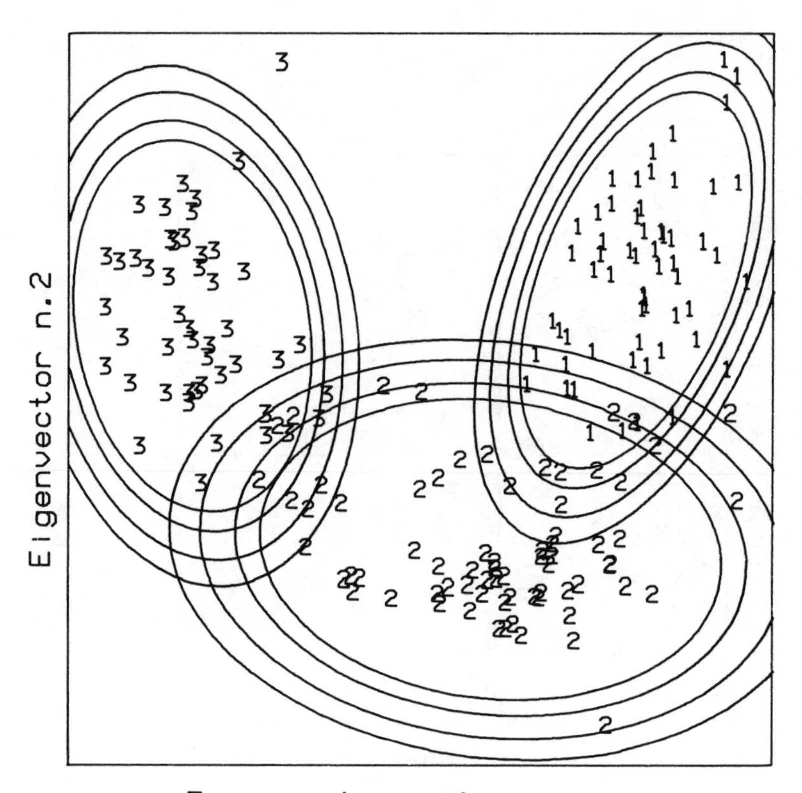

Figure 19.

Projections of confidence hyperellipsoids (70-80-90-95% confidence level). 1: Barolo; 2: Grignolino; 3: Barbera.

Figure 20.

Significance level of BAROLO samples on their bayesian class model; datavector index reported.

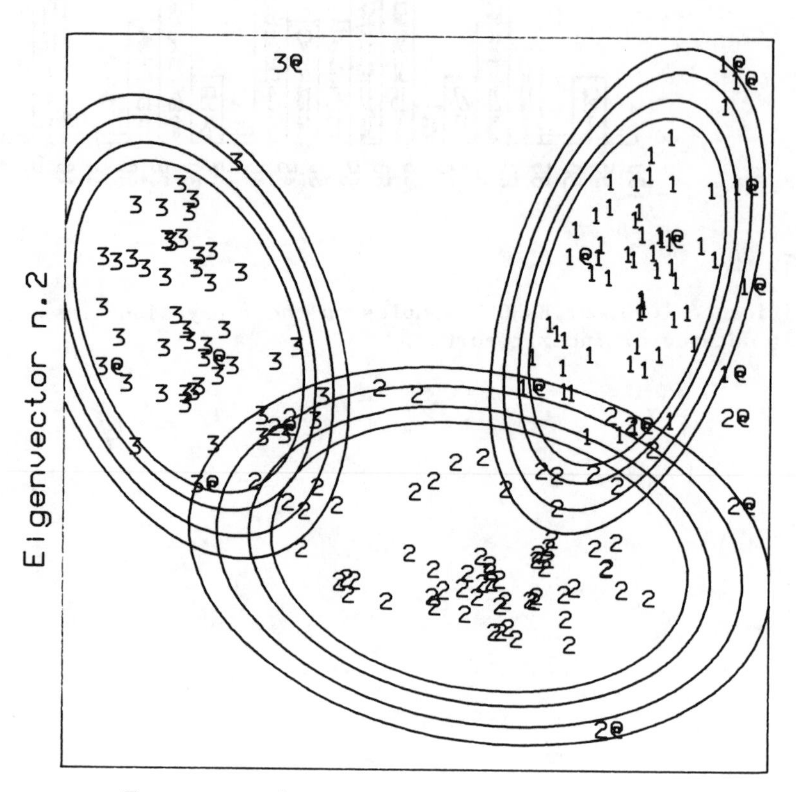

Figure 21.

Projections of confidence hyperellipsoids (70-80-90-95%) confidence level). 1: Barolo; 2: Grignolino; 3: Barbera after outlier deletion (outlier class symbol followed by ASCII character 64).

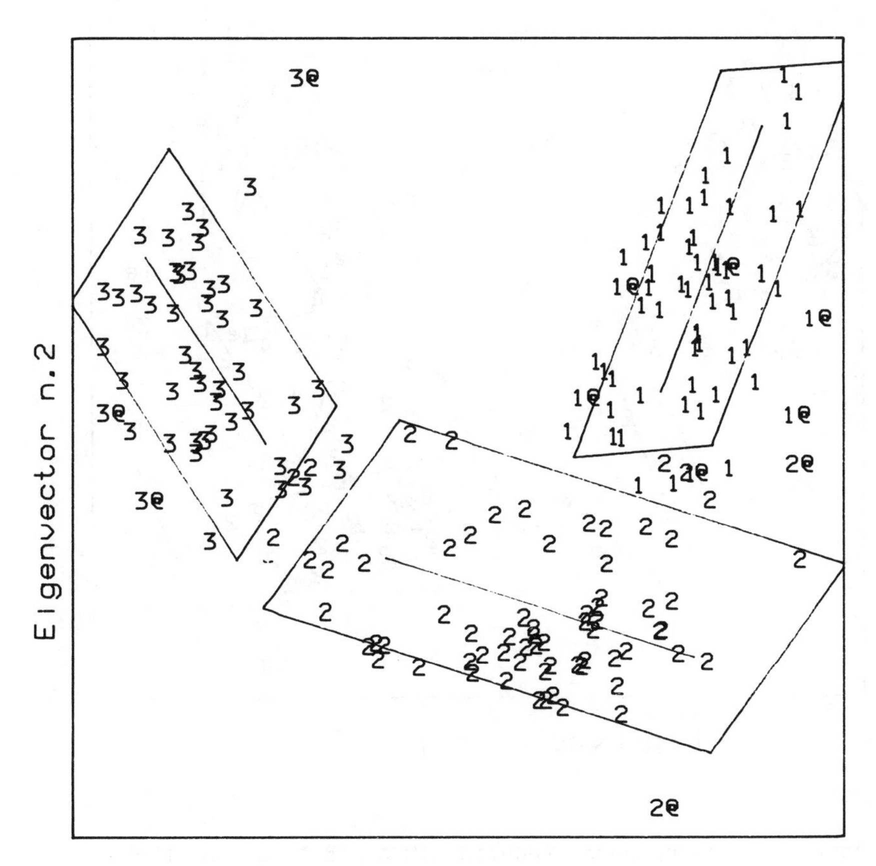

Figure 22.

Projection of 1-component reduced SIMCA models and the 2-component SIMCA base-models after one cycle of outlier deletion. Non-weighted variables.

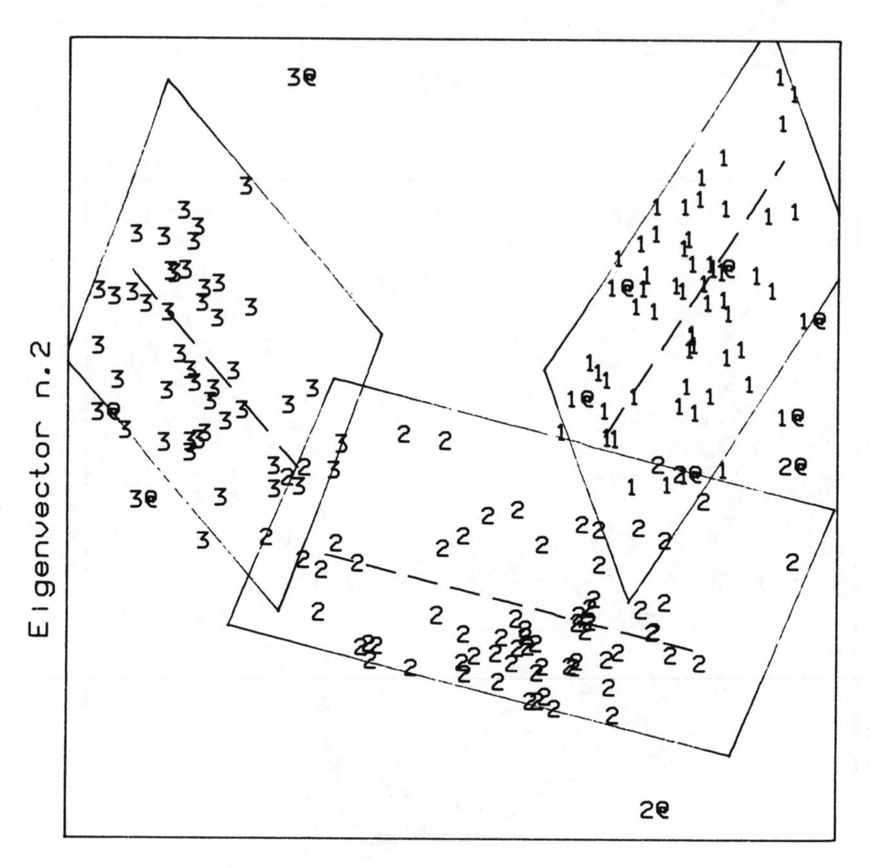

Figure 23.

Projection of 1-component reduced SIMCA models and of the 2-component SIMCA base-models after one cycle of outlier deletion. Variables weighted by their discrimination power.

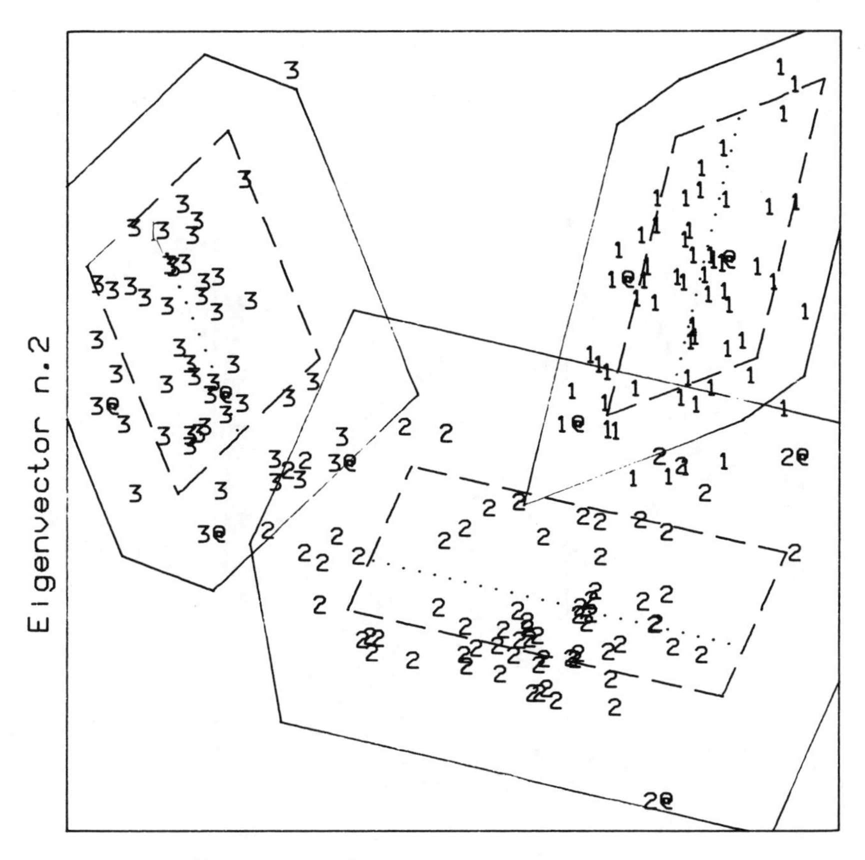

Figure 24.

Projection of 2-component reduced models and 3-component base-models after outlier deletion.

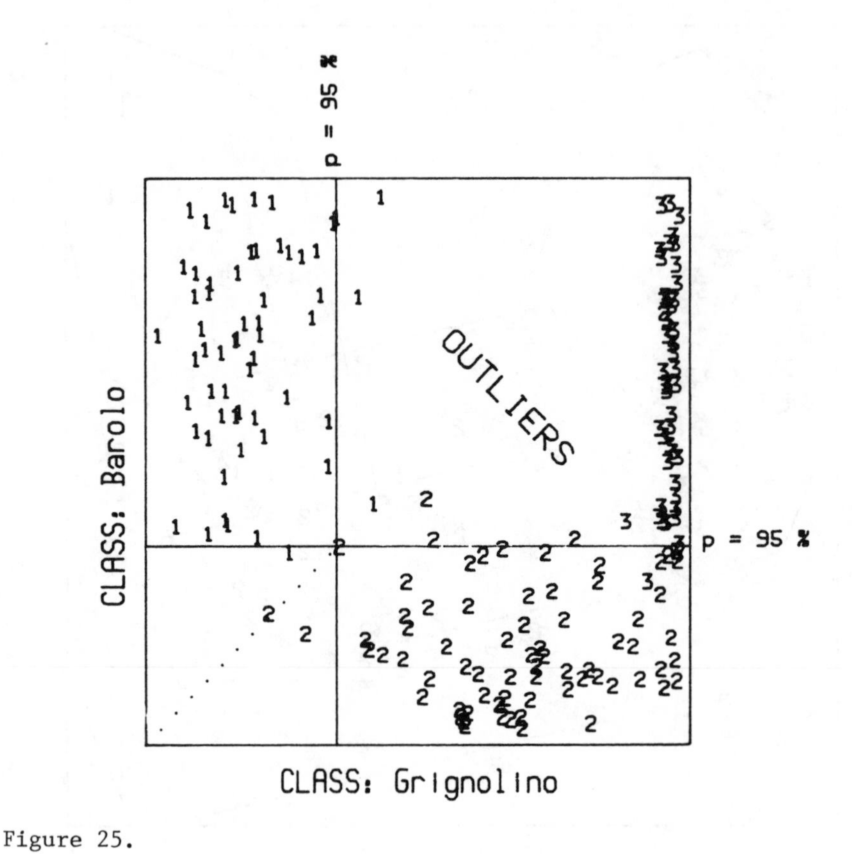

Figure 25.

Coomans diagram with SIMCA-2 components reduced model after outlier deletion.

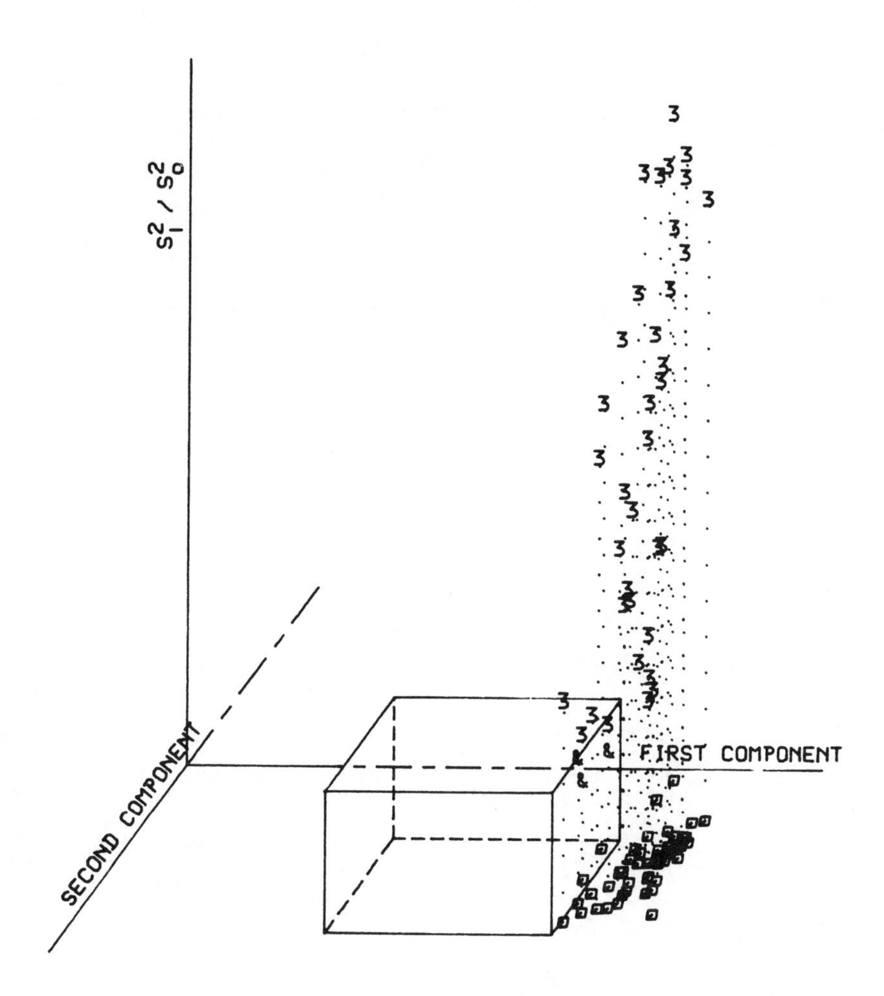

Figure 26.

SIMCA box of Grignolino. Reduced 2-component model Barbera samples with their shadow on the plane of Grignolino model and shows three Barbera samples which fit Grignolino model.

MULTIDIMENSIONAL DATA REPRESENTATION IN MEDICINAL CHEMISTRY

Paul J. Lewi

Research Laboratories
Janssen Pharmaceutica NV
B-2340 Beerse Belgium

Three methods of multivariate data representation are compared, namely Principal components analysis, Spectral map analysis and Factorial analysis of correspondences. It is shown that the three methods possess strong similarities. Yet the data to which they are most applicable can be very different.

INTRODUCTION

In this contribution we describe three related methods of factorial representation and their application to medicinal chemistry, more particularly to pharmacology. A table of pharmacological data describes a number of chemical compounds in terms of measurements made on them. These measurements include physicochemical determinations together with biological observations on tissues, organs or whole animals. By convention, we assume that compounds are arranged row-wise, while measurements appear column-wise in the table.

Tables of pharmacological data represent multidimensional structures which we wish to reduce to a low-dimensional representation without loss of much information. Reduction of the apparent complex dimensionality of a data structure is possible whenever measurements appear to be correlated and whenever groups of compounds possess similar chemical and biological properties. Extraction of orthogonal factors from a variance-covariance matrix and projection of the data onto the

B. R. Kowalski (ed.), Chemometrics. Mathematics and Statistics in Chemistry, 351–376

two or three most significant factors are well-known procedures of multidimensional data reduction. A discussion of the technique of data representation has been presented elsewhere (Lewi, 1982).

A striking property of factorial methods is that they allow to display both compounds and measurements within a common frame of factor axes. There are several factorial methods that can achieve reduction of the dimensionality of a data structure. We will discuss three of them, namely principal components analysis (PCA), spectral map analysis (SMA) and factorial analysis of correspondences (FAC). We will show that there is but little difference between these methods, mainly due to different types of re-expression, centering, standardization and weighting. The methods differ greatly, however, as to the type of data for which they are best suited.

1. PRINCIPAL COMPONENTS ANALYSIS (PCA)

Principal components analysis (PCA) is by far the simplest and most widely used method of factorial representation (Cooley and Lohnes, 1962). It can be defined in terms of a series of elementary operations (Fig.1.a) : (1) optional re-expression of the data, (2) definition of constant weight coefficients, (3) column-centering, (4) column-standardization, (5) weighting of the data, (6) extraction of orthogonal factors (or eigenvectors) from the variance-covariance matrix, and (7) projection of the data table onto the factor matrix. The latter operation yields factor scores for compounds. Scaling of the factor matrix finally produces factor loadings for measurements. The scaling is such that factor scores and factor loadings possess identical factor variances (or associated eigenvalues) along each of the factor axes. As a result, factor scores and loadings can be treated as coordinates of points, representing compounds and measurements, within the same frame of factor axes. The most common frame is a two-dimensional plot of the two most significant factors. Note that the fifth step is not required in standard PCA. The effect of weighting of the data in this case is that of multiplication by a constant. The step is introduced here in order to emphasize analogies between factorial methods of data representation.

1	PRINCIPAL COMPONENTS ANALYSIS	PCA
1.1	RE-EXPRESSION (OPTIONAL) : log, reciprocal, log reciprocal,...	
1.2	WEIGHT COEFFICIENTS (CONSTANT) : $W_{ii} = 1/n$ $W^{*}_{jj} = 1/p$	
1.3	COLUMN-CENTERING : $Y_{ij} = X_{ij} - m_j$ with $m_j = \sum_i W_{ii} X_{ij}$	
1.4	COLUMN-STANDARDIZATION : $Z_{ij} = Y_{ij}/v_j^{1/2}$ with $v_j = \sum_i W_{ii} Y^2_{ij}$	
1.5	WEIGHTING : $Z_{ij} = Z_{ij} W^{*\ 1/2}_{jj}$	
1.6	FACTORIZATION : $C = Z^t.W.Z$ $\Lambda = F^t.C.F$ where $F^t.F = I$	
1.7	FACTOR SCORES AND LOADINGS : $S = Z.F$ $L = W^{*-1/2}.F.\Lambda^{1/2}$	

1.a Schematic of principal components analysis (PCA). Capital letters indicate matrices; lower case letters represent vectors and constants. X, Y, Z refer to the data table in various stages of the analysis. W, W^* are diagonal weighting matrices for row- and column-items respectively. C, Λ represent original and diagonalized variance-covariance matrices. F, S, L denote orthonormal factor, factor score and factor loading matrices. I is a unit matrix. Symbols m, v are used for column-means and column-variances. The number of rows and columns are defined by n, p with corresponding indices i, j. The matrix product is indicated by a solid dot.

By way of illustration of PCA, we have analyzed a table of mixed physicochemical and pharmacological measurements on a series of dibenzothiazepine and dibenzo-oxazepine neuroleptics (Fig.1.b), as reported by Schmutz (1975). Neuroleptics are drugs that are used in psychiatry for the relief of schizophrenia, mania, autism and other mental disorders. In the table, Hammett's constant σ_m expresses the electronic effect caused by various meta-substituents on a molecular structure (indicated by the symbols X and R in the insert of Fig.1.c). A measure of lipophilicity (i.e. the tendency to dissolve in fatty material) is provided by the logarithm of the partition coefficient between octanol and water and is denoted by the symbol π. In studies of structure-activity relationships it is customary to include the parabolic term π^2. Pharmacological observations are expressed as the negative logarithms of the median effective doses required to produce a given biological effect. Catalepsy refers to the ability of rats to maintain abnormal postures for a prolonged time. It is also observed in humans and is considered as an undesirable side-effect of many highly potent and specific antipsychotic drugs. The other observation relates to the inhibition of stereotyped behaviour produced by an injection of apomorphine. Both observations in rats are strongly predictive for antipsychotic effects of neuroleptic compounds in humans.

The objective of PCA here is to relate pharmacological observations on a series of chemical analogs to their structural and physicochemical properties. The COMPONENTS map (Fig.1.c) displays the position of twelve chemical analogs with respect to five standardized measurements. The first and second factors are arranged horizontally and vertically, while the third factor is oriented perpendicularly to the plane of the map. In this representation we distinguish between compounds and measurements by means of circles and squares. The magnitude of the third factor has been coded by varying the thickness of the contour of the symbols. Since all compounds (except one) and all measurements are well-represented by the first two factors, the average contour is indicative for the position within or close to the plane of the COMPONENTS map. A thick outline indicates that the actual position of the compound or measurement projects above the plane of the first two factors. (There are none in our illustration.) A thin contour means that the actual position of

PHYSICOCHEMISTRY AND PHARMACOLOGY OF THIAZEPINES AND OXAZEPINES

A :	SIGMA-M
B :	PI
C :	PI-SQUARED
D :	APOMORPHINE
E :	CATALEPSY

	A	B	C	D	E
1 : O,CH3	¯.070	.487	.237	¯.117	¯.332
2 : O,SCH3	.151	.597	.356	.069	¯.550
3 : O,CL	.373	.744	.554	.707	¯.033
4 : O,CN	.564	¯.322	.104	.680	.428
5 : O,SO2CH3	.595	¯1.279	1.636	.680	¯1.176
6 : O,NO2	.706	.340	.116	.760	.836
7 : S,CH3	¯.116	.524	.275	¯.755	¯1.176
8 : S,SCH3	.151	.597	.356	¯.170	¯1.094
9 : S,CL	.373	.744	.554	.122	¯.305
10 : S,CN	.558	¯.285	.081	.707	.184
11 : S,SO2CH3	.600	¯1.242	1.543	.707	¯1.012
12 : S,NO2	.706	.376	.141	1.239	¯.360

SCHMUTZ, 1975

1.b Physicochemical and pharmacological data on dibenzothiazepine and dibenzo-azepine analogs reported by Schmutz (1975). For explanation see text.

the compound or measurement lies below the plane of the map. (This occurs with the NO_2-substituted thiazepine.)

In the COMPONENTS map of the neuroleptics 96 percent of the total variance in the data (after centering and standardization of the columns) is represented by the first three factors. As a result of very large contributions by the first two factors (51 and 40 percent respectively), the measurements are projected very close to the boundary of a circle around the center of the map. Therefore, we can state that the information in the table is fairly reproduced by the COMPONENTS map which takes account of the two major factors. Note that we express information content in terms of variance of the data after a number of elementary operations such as centering and standardization. When the contribution by the first two factors is high, it makes sense to join the representation of the measurements to the center of the map. This results in a series of axes which are indicated on the COMPONENTS map by dasned lines. Each of these lines can be regarded as a measurement axis in factor space, which means that compounds can be projected upon them. Projections of the compounds upon the apomorphine axis are obtained approximately in the same order as in the corresponding column of the table. For example, NO_2 - substituted compounds rank highest, while CH_3 - derivatives obtain the lowest score for inhibition of apomorphine in rats. The concordance is not perfect because 9 percent of the variance is carried by the third and higher order factors.

On first acquaintance with factorial representations one may be confused by the appearance of multiple axes in one and the same diagram. Of course, one is used to plane Cartesian diagrams in which two and only two measurements are displayed by means of two orthogonal axes. In factorial maps, however, we obtain several axes, with variable angular separations. The cosine of the angle between two well-represented axes in the plane of the map equals the coefficent of correlation between the corresponding measurements.

In the particular COMPONENTS map of neuroleptics we find that inhibition of stereotyped behavior, provoked by apomorphine in rats, is strongly correlated with Hammett's electronic effect of the substituents σ_m. Note that a near-perfect correlation is

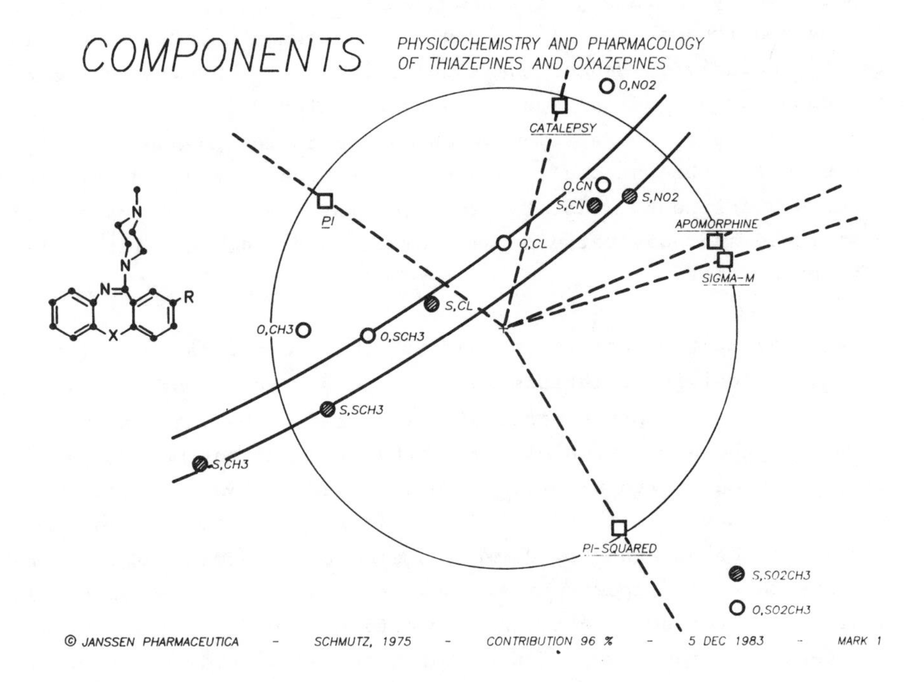

1.c COMPONENTS map obtained by principal components analysis (PCA) of the table in Fig.1.b . Circles denote compounds derived from substitutions on the parent molecule in the insert. The labels on the map refer to substitutions at the X and R positions respectively. Squares represent physicochemical and pharmacological measurements. The position of the symbols is defined by the two most significant factors. Thickness of the symbols is coded according to the third significant factor. 96 percent of the variance (after centering and standardization) is contributed by the three factors that are expressed by the map.

reflected on the map by a very small angular separation between the representative axes. It must be enphasized here that intercorrelations among measurements are restricted to the set of compounds that is being studied. The magnitude of the correlation can be quite different when derived from a different set of compounds. We also observe from the COMPONENTS map that lipophilicity π is positioned almost oppositely to the representation of π^2. The correlation between the latter two is largely negative. Hence, the angular separation between the axes representing π and its square is near to 180 degrees. Note that this is a chance correlation which is due to the presence in the table of two strongly negative π-values produced by the SO_2CH_3-substituents. Cataleptic effect of these neuroleptics is positioned half-way between their electronic (σ_m) and lipophilic (π) properties.

Looking at the representation of the compounds we find a structure-activity relationship in at least ten out of the twelve derivatives. The projections of oxazepines (having an oxygen atom in the X-position of the central ring of the molecule) and of thiazepines (with a sulfur atom at X) form two distinct and slightly curved patterns. The linear part of the patterns appears to be governed by Hammett's electronic effect σ_m and by inhibition of apomorphine. As a result of a systematic difference between cataleptic effect in oxazepines and thiazepines, the two line patterns are separated. This can be confirmed by inspection of the tabulated data. An exception to this structure-activity rule is made by the SO_2CH_3- substituents which clearly violate the overall pattern.

In general, principal components analysis (PCA) is most indicated (a) when one deals with measurements that are expressed in different physical units and (b) when one wants to classify compounds (or any other type of objects or individuals) with respect to the magnitude of their results in various measurements. The latter statement implies that we do not wish to correct for differences in potency between compounds (or importance between objects) before submitting the data to factorial analysis.

2. SPECTRAL MAP ANALYSIS (SMA)

Spectral map analysis (SMA) has been introduced by the author (1976) as a method of factorial representation in medicinal chemistry. SPECTRAMAP is a name used on graphics obtained by means of SMA. Basically, SMA can be defined by the following operations (Fig.2.a) : (1) optional re-expression of the data, (2) definition of weights for compounds and measurements, (3) double centering with respect to both compounds and measurements, (4) global standardization, (5) weighting of the data, (6) factor extraction and (7) calculation of factor scores and factor loadings as described above in the case of PCA.

The objective of SMA is to represent relationships between compounds and measurements, irrespective of the potency of compounds and independently of the sensitivity of the measurements. In pharmacological studies, a compound is said to be more potent when it requires a lesser amount of substance to produce a given observable effect. Similarly, a measurement is considered to be more sensitive when a given effect is observed at lower levels of dosage or concentration of the active compounds.

By way of illustration, we present the SPECTRAMAP derived from a study by Wood e.a. (1981) on receptor binding of narcotic analgesics. Among the latter, we find morphine analogs such as dihydromorphine, synthetic drugs such as lofentanil, and endogenous peptides among which enkephalins and beta-endorphin. The underlying model in receptor binding studies is that narcotic analgesics can attach to diverse binding sites on so-called receptors (i.e. proteins embedded in membranes of brain cells) and thus interfere with the transmission and perception of pain. In receptor binding assays one expresses the affinity of a compound for a radioactively tagged binding site by means of the median inhibitory concentration IC_{50} (usually expressed in nanomoles). The experimental procedure roughly involves binding of a specific radioactively labelled ligand and subsequent displacement of the ligand by increasing concentrations of the test compound. In the example (Fig.2.b), 26 compounds have been tested for their ability to displace four ligands, namely tritiated dihydromorphine (DHM), naloxone (NAL), d-ala d-leu

2	SPECTRAL MAP ANALYSIS	SMA
2.1	RE-EXPRESSION : as in 1.1 $X_{ij} = X_{ij} - \mathrm{Min}(0, \underset{i}{\mathrm{Min}} \underset{j}{\mathrm{Min}} X_{ij})$ $X_{ij} = X_{ij} / \sum_i \sum_j X_{ij}$	
2.2	WEIGHT COEFFICIENTS (FROM MARGINAL TOTALS) : $W_{ii} = \sum_j X_{ij}$ $W^{\star}_{jj} = \sum_i X_{ij}$	
2.3	DOUBLE CENTERING : $Y_{ij} = X_{ij} - (m_i + m^{\star}_j - m^g)$ with $m_i = \sum_j X_{ij} W^{\star}_{jj}$ $m^{\star}_j = \sum_i W_{ii} X_{ij}$ $m^g = \sum_i \sum_j W_{ii} X_{ij} W^{\star}_{jj}$	
2.4	GLOBAL STANDARDIZATION : $Z_{ij} = Y_{ij} / v^{g^{1/2}}$ with $v^g = \sum_i \sum_j W_{ii} Y^2_{ij} W^{\star}_{jj}$	
2.5	WEIGHTING : as in 1.5	
2.6	FACTORIZATION : as in 1.6	
2.7	FACTOR SCORES AND LOADINGS : as in 1.7	

2.a Schematic of spectral map analysis (SMA). Step 1 ensures that data are nonnegative and sum to unity. Index g denotes globalization over all rows and columns. Symbol m is used for row-, column- and global means. v now refers to global variance. All other symbols are defined as in PCA.

enkephalin (ENK) and ethylketocyclazocine (EKC). It is convenient to convert IC_{50} values into equilibrium inhibition constants K_i (by means of the Cheng-Prusoff formula) $K_i = IC_{50}/(1+C/K_D)$ which accounts for the ligand concentration C and the receptor-ligand dissociation constant K_D. Receptor binding studies of narcotic analgesics suggest that there exist several distinct binding sites among which a μ-type site, which is specifically occupied by morphine-like analogs, a δ-type site which is most sensitive to endogenous peptides and various others among which the κ-type site. In practice, drugs can bind to various extents to several receptors and binding sites. From numerical data only, it is rather difficult to associate drugs with binding sites for which they show preferential affinity. SMA is suitable for exploratory analysis of the spectrum of binding affinities of compounds and binding sites.

First, we have to define binding affinity of a compound as the logarithm of the reciprocal inhibition constant K_i. (Note that log reciprocal values can also be expressed as negative log values.) Reciprocal values are appropriate because of the inverse relationship between biological effect and dose (or concentration) of an active compound required to produce that particular effect. Logarithms are mandatory because of the positively skewed distribution of reciprocal IC_{50} and K_i-values. Secondly, we refer to specificity of a compound when we consider its relative affinity for different binding sites. A particular compound, such as morphine, may have high specificity for the DHM-labelled binding site, although its absolute affinity for this binding site is rather small when compared with more potent compounds such as butorphanol and others. In SMA, as opposed to PCA, we are more interested in studying the spectrum of relative specificities, rather than in the grouping by absolute affinities. To this effect we have to make abstraction of the individual differences between absolute affinities (or potencies) of the compounds. In this context, affinity refers to the log reciprocal value of an inhibition constant K_i.

One way to correct for differences of affinity between drugs is to center the rows of the table (after re-expression into log reciprocal values) with respect to the corresponding row-means.

NARCOTIC ANALGESICS IN RECEPTOR BINDING ASSAYS

A : EKC
B : DHM
C : ENK
D : NAL

	A	B	C	D
1 : MORPHINE	696.0	1.3	26.1	3.5
2 : PHENAZOCINE	9.1	.2	2.6	.8
3 : LEVORPHANOL	47.0	.2	4.5	.7
4 : ETORPHINE	4.8	.1	.9	.2
5 : DIHYDROMORPHINE	6012.0	1.0	16.6	1.3
6 : D-ALA,D-LEU-ENKEPHALIN	113.0	8.3	1.3	5.7
7 : MET-ENKEPHALIN	66.0	3.6	2.8	4.6
8 : LEU-ENKEPHALIN	120.0	8.1	3.1	11.6
9 : LEU-ENKEPHALIN,ARG-PHE	4430.0	4.0	3.0	4.2
10 : FK-33824	6329.0	.4	4.9	1.0
11 : LY-127623	3924.0	.7	2.5	1.2
12 : BETA-ENDORPHIN	1835.0	.3	.3	.8
13 : SKF 10074	241.0	1.4	5.0	1.1
14 : KETAZOCINE	7.3	7.9	6.6	12.7
15 : ETHYLKETOCYCLAZOCINE	4.7	4.1	7.6	7.5
16 : MR 2034	5.7	.3	.5	.2
17 : BUPRENORPHINE	2.0	3.1	3.6	1.6
18 : LOFENTANIL	2.7	.7	1.2	.4
19 : BUTORPHANOL	7.7	.3	1.6	.3
20 : CYCLAZOCINE	5.0	.2	1.1	.2
21 : LEVALLORPHAN	4.0	.2	3.3	.7
22 : NALBUPHINE	38.0	2.4	21.2	2.4
23 : PENTAZOCINE	116.0	3.7	19.8	13.5
24 : NALORPHINE	55.0	.7	13.9	3.7
25 : NALOXONE	11.1	1.4	16.5	1.0
26 : NALTREXONE	9.8	.1	5.0	.5

WOOD E.A. 1981

2.b Receptor binding data of narcotic analgesics reported by Wood e.a. (1981). Values represent equilibrium inhibition constants (nanomoles) obtained from rat brain homogenate. For explanation see text.

The effect of row-centering is to assign an average affinity of zero to each compound. Note that column-centering produces a similar effect on the binding sites. Indeed, the average affinity of a receptor site equals zero after reduction of the columns of the table with respect to the corresponding column-means. It can be proved that the order of centering, row- or column-wise, is irrelevant. Double centering of the data table is an essential part of SMA.

In contrast to PCA, we do not column-standardize, since this would destroy the doubly-centered property of the data. Instead, we divide each element in the table by the square root of the global variance. The global variance is defined here as the mean square value computed over all elements of the table, after re-expression and double centering. Factorial decomposition of the variance-covariance matrix of a doubly-centered table leads to a representation in the plane of the two most dominant factors in exactly the same way as described in the case of PCA. Note that the magnitude of the third factor is also coded by means of the variable thickness of the symbols that stand for compounds (circles) and for binding sites (squares).

We have varied the size of the circles according to the strongest affinity of each drug for the various receptors. Likewise, we have scaled the size of the squares proportionally to the strongest affinity of each binding site for the various drugs. These strongest affinities can also be read-off from the table of K_i-values by determination of the minimum values row- and column-wise, followed by a re-expression into log reciprocal values. (Preliminary division of all elements by the largest value in the table avoids the occurrence of negative values after log reciprocal re-expression.)

It can be seen from the SPECTRAMAP that binding sites labelled with DHM and NAL are strongly correlated, at least within this series of 26 compounds (Fig.2.c). In contrast to PCA, correlation between measurements is expressed by the actual distance of their representations in factor space, rather than by their angular distance as seen from the center of the map. It can also be verified that narcotic analgesics bind in nearly the same proportion to DHM- and NAL-binding sites. Because of this,

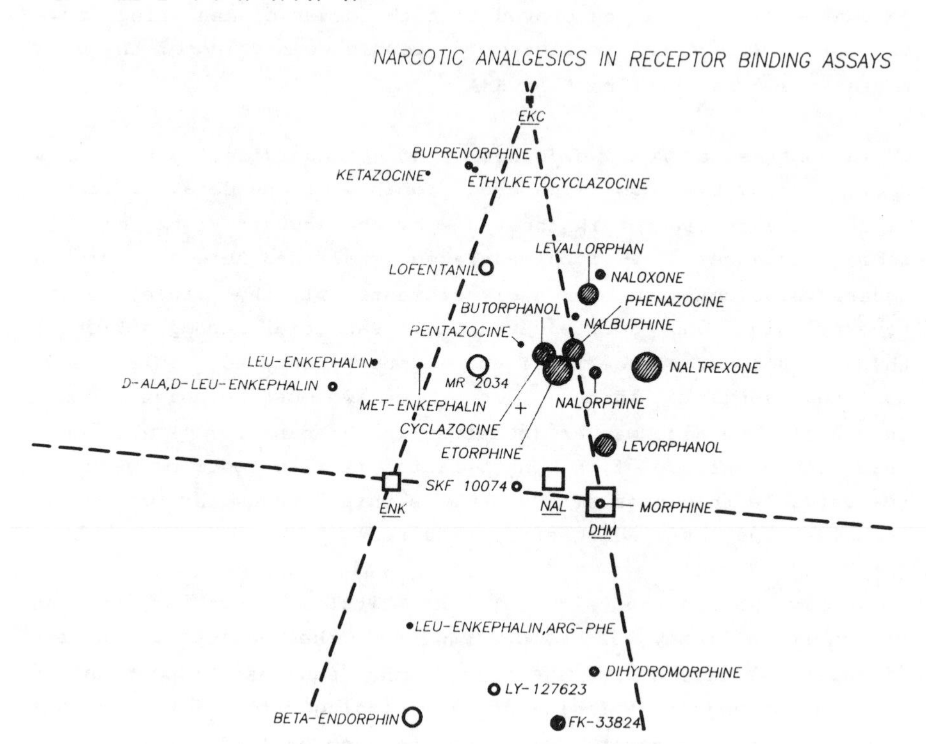

2.c SPECTRAMAP obtained by spectral map analysis (SMA). Circles denote compounds; squares represent radioactive ligands. (See text for meaning of abbreviations.) Size of circles and squares is proportional to the (logarithmic) affinities of compounds and binding sites. Dashed lines define characteristic ratios of affinities for three differentiated binding sites. The barycenter of the map is indicated by a cross. Reproduction is the percentage of global variance (before double centering) explained by differences of affinities between compounds and between observations, and by the first three factors. Other conventions are as indicated in the legend of Fig.1.c .

one may conclude that tritiated dihydromorphine and naloxone label one and the same μ-type site, which is shown to bind specifically to morphine and related alkaloids.

The ENK-site is representative for the δ-type binding in this particular series of analgesic compounds. The specificity of this δ-type site is caused in this study by endogenous peptides, enkephalins and beta-endorphin. Many compounds that bind to the δ-type site also bind to some extent to the μ-type site.

Quite distinct from the μ- and δ-types is the so-called κ-type site which is labelled by EKC. Compounds such as ketazocine and ethylketocylazocine itself bind very specifically to the κ-type site. Their absolute affinities are small. On the other hand, potent compounds, such as beta-endorphin, do not bind at all to the EKC-labelled site.

Looking at the projection of the compounds, we observe several ones that bind to each of the three sites. This group, which includes cyclazocine and etorphine, is projected in the neighbourhood of the barycenter of the map. In SMA the barycenter of the map is at the same time the center of gravity of the data structure as represented by the compounds and by the measurements. As a rule, compounds that possess little selectivity for any of the binding sites are positioned at or near to the barycenter. The same statement applies to binding sites that show little selectivity for the compounds. They too will find their position close to the barycenter, which can be considered as the neutral or indifferent point of the map. The barycenter of the data structure is identified by a small cross on the SPECTRAMAP.

In order to facilitate interpretation of the SPECTRAMAP we have joined three outstanding poles of the map by dashed lines. We define a pole as the projection in the factorial plane of a particular binding site which is distinct from the others, and which at the same time is located at some distance from the barycenter. The three poles that we have selected correspond with the μ-, δ- and κ-type binding sites which we have discussed above. Each of the dashed lines that we have thus constructed represents an axis in the plane of the map. The horizontal axis is related to the ratio of K_i-values obtained from the combined DHM- and NAL-binding with respect to ENK-binding. Specific μ-type compounds (such as naltrexone) are projected on the right side of the horizontal axis, while more specific δ-type compounds (among which the enkephalins) find their projections at the opposite side of the same axis. The ranking of ratios of K_i-values, as determined from the columns labelled DHM and ENK in the table, appears approximately in the same order as the projections of the compounds on the horizontal axis of the SPECTRAMAP. Similar rankings can be constructed for the EKC/DHM and EKC/ENK ratios from projection of the compounds onto the dashed lines that join the μ-, δ- and κ-poles of the map.

We have come across an important difference between SMA and PCA. In the former case, axes can be drawn between representations of measurements, while in the latter case axes always extend from the barycenter of the map toward representations of the measurements. (It must be understood that in SMA, as well as in PCA, the origin of the factor space coincides with the barycenter of the configuration of compounds. In SMA we also obtain the origin at the barycenter of the configuration of measurements.) SMA allows to compare distances between measurements in addition to distances between compounds. Consequently, in SMA one may apply point clustering techniques to measurements as well as to compounds. In CPA, on the contrary, one can only cluster compounds, since correlations between measurements are expressed by angular rather than by point-to-point distances. There is no formal distinction in SMA between compounds and measurements. Indeed, if the original table is submitted to analysis by SMA in its transposed form, we would obtain exactly the same arrangement of compounds and measurements in the SPECTRAMAP. Only the relation of circles and squares to compounds and measurements would be transposed.

The reproduction of information by SPECTRAMAP includes a far greater proportion of the global variance in the data table than is possible with COMPONENTS. By means of SMA we can account for the variance of affinities among compounds (26 percent), for the variance of affinities between labelled binding sites (47 percent) and finally for the covariance between compounds and binding sites (27 percent of the global variance before correction for differences in affinities). The variance of affinities is visualized on a SPECTRAMAP by means of the variation in the sizes of the symbols that represent compounds (circles) and measurements (squares), as has been explained above.

A distinctive characteristic of SMA is the application of variable weighting to compounds and measurements in the calculation of means, variances and covariances. The underlying idea of variable weighting is to attach greater importance in the analysis to compounds that show high affinity to one or more of the binding sites. Vice-versa, a larger weight is assigned to binding sites that have strong affinity to at least some of the compounds that can bind to them. Usually, variable weight coefficients can be derived in SMA from the marginal totals of the data table (usually after re-expression, e.g. by negative logarithms). In the case of pharmacological data it is preferable to compute weight coefficients for compounds and binding sites according to their strongest affinity, rather than on the basis of their average affinity (as would result from the use of marginal totals). The reason is that a highly potent compound that binds selectively to only one site (such as dihydromorphine), may obtain a low average affinity and, hence, a low weight in the analysis due to its inability to bind to the other sites. As a general rule, the weight coefficients for compounds and for measurements must sum up to unity.

3. Factorial analysis of correspondences (FAC)

Factorial analysis of correspondences (FAC) has been introduced as a method of factorial representation of multivariate data by Benzécri and a group of french statisticians (1973). The method has been designed primarily (although not exclusively) for the analysis of tables of counts and frequencies, such as occur in two-way tabulations of census and survey data. Such tables are also referred to as contingency tables. FAC can be defined by means of the following sequence of operations (Fig.3.a) : (1) re-expression to unit sum over rows and columns, (2) definition of weights for row- and column-items from marginal totals of the table, (3) subtraction and subsequent division of each element in the table by its expected value,

3	FACTORIAL ANALYSIS OF CORRESPONDENCES	FAC
3.1	RE-EXPRESSION : as in 2.1	
3.2	WEIGHT COEFFICIENTS : as in 2.2	
3.3	SUBTRACTION-DIVISION BY EXPECTATION : $Y_{ij} = (X_{ij} - W_{ii}W^{\star}_{jj})/W_{ii}W^{\star}_{jj}$	
3.4	GLOBAL STANDARDIZATION : as in 2.4	
3.5	WEIGHTING : as in 1.5 and 2.5	
3.6	FACTORIZATION : as in 1.6 and 2.6	
3.7	FACTOR SCORES AND LOADINGS : as in 1.7 and 2.7	

3.a Schematic of factorial analysis of correspondences (FAC). Only step 3 is distinct from the schematic in Fig.2.a .

(4) global standardization, (5) weighting of the data, (6) factorization of the variance-covariance matrix and (7) projection into a reduced space of most important factors, yielding factor scores and loadings, such as has been described in the previous sections on spectral map analysis (SMA) and principal components analysis (PCA).

In FAC it is customary to divide initially each element of the contingency table by the gross total over all elements in the table. This way, weights for row- and column-items can be simply expressed as the marginal row- and column-totals. The requirement that weight coefficients for row- and column-items sum up to unity is then automatically satisfied. Furthermore, expected values for each element in a contingency table can thus be expressed as the product of the corresponding row- and column-weights. From a statistician's point of view, marginal totals of a contingency table define the probability densities for row- and column-items of the table. Factorial analysis of correspondence (FAC) makes consistent use of this property.

Means, variances and covariances are the result of weighting in FAC (as well as in SMA). For example, the global variance is computed in FAC as the sum of the squared data (after subtraction-division by their expectation) multiplied by the corresponding row- and column-weights. It can be shown that variance defined in FAC takes the form of a chi-square statistic. For this reason, the metric used for representing row- and column-items in FAC is said to be an euclidean metric of chi-square.

A distinctive characteristic of FAC lies in the subtraction-division by expected values of the contingency table. This can be understood as a correction for differences in importance between row- and column-items, as expressed by the marginal totals of the table. Indeed, after subtraction-division by expectations, (weighted) mean values are identically equal to zero, whether computed row-wise, column-wise or globally over all elements of the table. In this respect, we find a close analogy with the result of double-centering which we have described above in the discussion of SMA. Furthermore, due to the symmetry of operations in FAC, this method too is invariant under a transposition of rows and columns in the data table.

For the purpose of illustration, we have analyzed a table of pharmacological data which describes 13 neuroleptics and 13 morphinomimetics in six elementary screening tests in rats (Niemegeers, 1974). In this preliminary screening assay, a fixed dose of each compound has been administered to six animals. Normally, a considerable number of behavioral observations are scored for presence or absence during a three-hour session. For didactic reasons we have selected six observations that discriminate least between the two classes of compounds. These

NEUROLEPTICS AND MORPHINOMIMETICS IN SCREENING TESTS ON RATS

A : PROSTRATION
B : TEMPERATURE INCR.
C : TEMPERATURE DECR.
D : PUPIL DIAMETER INCR.
E : PUPIL DIAMETER DECR.
F : PALPEBRAL OPENING DECR.

	A	B	C	D	E	F	TT
1 : PROPOXYPHENE	0	3	1	3	0	0	7
2 : MORPHINE	0	0	4	2	0	2	8
3 : METHADONE	0	0	6	6	0	3	15
4 : PETHIDINE	0	1	4	6	0	2	13
5 : CODEINE	0	4	0	3	0	1	8
6 : DEXTROMORAMIDE	0	0	6	4	0	3	13
7 : ANILERIDINE	0	2	3	4	0	2	11
8 : PHENOPERIDINE	0	2	5	2	0	3	12
9 : ETONITAZINE	0	0	6	6	0	0	12
10 : PHENAZOCINE	0	2	4	4	0	2	12
11 : PIRITRAMIDE	0	2	4	4	0	3	13
12 : FENTANYL	0	1	6	6	0	3	16
13 : BEZITRAMIDE	0	0	2	2	0	1	5
14 : CHLORPROMAZINE	6	0	6	5	0	6	23
15 : PERPHENAZINE	6	0	6	0	0	6	18
16 : HALOPERIDOL	0	0	6	0	3	6	15
17 : TRIFLUOPERAZINE	6	0	3	0	0	6	15
18 : CHLORPROTHIXENE	5	0	6	6	0	6	23
19 : SPIRILENE	0	0	3	0	0	5	8
20 : PIMOZIDE	0	0	5	0	1	5	11
21 : PIPAMPERONE	2	0	6	0	1	6	15
22 : SULPIRIDE	0	0	0	0	1	2	3
23 : BENPERIDOL	4	0	6	6	0	6	22
24 : SPIPERONE	0	3	0	3	0	5	11
25 : OXYPERTINE	6	0	6	0	1	6	19
26 : THIORIDAZINE	1	0	6	6	0	6	19
TT : TABLE TOTAL	36	20	110	78	7	96	347

NIEMEGEERS C. , 1974

3.b Pharmacological screening data of neuroleptics and morphinomimetics (Niemegeers, 1974). Data represent the number of animals out of six that showed a particular behavioral or physiological symptom after administration of a compound.

observations include prostration, pupil diameter increase and decrease, temperature increase and decrease, and decrease of palpebral opening. Each number in this contingency table expresses the number of rats (out of six) that showed a particular behavioral symptom after treatment with one of the 26 compounds.

The projection in the plane of the two most important factors reveals the pattern of compounds and of screening observations, together with their specificities (Fig.3.c). It is readily seen on the map of CORRESPONDENCES that morphinomimetics produce a selective effect on temperature increase and on pupil diameter increase, while prostration, pupil diameter decrease and decrease of palpebral opening are more frequently observed with neuroleptics. Temperature decrease is about equally frequent in all the compounds. This is why its representation in the map of CORRESPONDENCES is very close to the barycenter (which also happens to be the origin of the factor space). The two classes of compounds appear to be almost separable by the six behavioral observations with the exception of the neuroleptic spiperone which is projected among the group of morphinomimetics.

Factorial representation of compounds and observations by means of a map of CORRESPONDENCES follows the same conventions as adopted for SPECTRAMAP. On the CORRESPONDENCES map, symbols representing compounds and observations are given different sizes according to their respective weights in the analysis. These weights are proportional to the marginal totals (or marginal probabilities) of the contingency table. Here too, we varied the thickness of the contours in order to signal compounds and observations that are located above or below the plane of the map, by using thick and thin outlines respectively. The latter emphasize the contribution by the third most important factor. For example, codeine and temperature increase are represented above the plane . Similarly, etonitazine is seen to lie below the plane of the map. A graphic feature, not discussed previously, appears on the CORRESPONDENCES map and affects the symbols representing the neuroleptics sulpiride and spirilene.

Broken contours signal that the compounds contribute to higher order factors and their location in a three-factor space does not

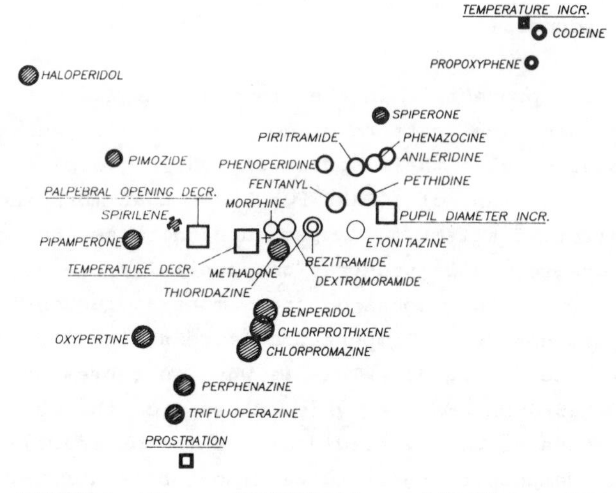

3.c CORRESPONDENCES map obtained by factorial analysis of correspondences (FAC). Contribution equals the percentage of the global variance (after subtraction-division by expectations) or chi-square that is accounted for by the three most important factors. Other conventions are described in the legends of Fig.1.c and 2.c .

explain exhaustively the numerical information in the table. (Of course, broken contours may also occur in symbols representing measurements.)

In the CORRESPONDENCES map of our example, we find that the first three factors contribute 91 percent of the global variance of the data after subtraction-division by expectations. At present, we have not discovered a way in FAC to combine factor variances with the variance of expected values. Such a pooling of variances has been possible in SMA which involves the simpler operation of double centering by subtraction of expected values. For this reason, contributions derived from FAC refer to a fraction of the global variance after subtraction-division by expectations. In SMA, on the other hand, the term reproduction has been used for the fraction of the global variance before double centering by expected values.

DISCUSSION

In our description of multivariate data representation we have selected three factorial methods : principal components analysis (PCA), spectral map analysis (SMA) and factorial analysis of correspondences (FAC). These three methods make use of factor extraction and projection of the data structure on a low-dimensional space spanned by orthogonal factors. They are distinct by the operations that are applied to the table prior to factorization. Distinctive operations include : definition of weight coefficients for row- and column-items, subtraction of row-, column- or global means, division by row- and column-totals, division by column- or global standard deviations and weighting of the data according to the relative importance of row- and column-items.

In standard PCA, data are column-centered, column-standardized and weights assigned to row- and column-items are constant. PCA does correct for differences of importance among column-items (which, by convention, represent measurements). On the other hand, PCA does not modify the representation of row-items (objects). Tables that are submitted to PCA frequently have column-items expressed in different physical units. Also, one assumes in this case that there is no significant 'size' component present in the row-items of the table.

Because of the different treatment of row- and column-items in PCA, the result of PCA is not invariant under transposition of the data table. Axes drawn in a plane spanned by two factors, always start from the origin of the map and extend toward the representation of column-items. Only row-items can be clustered within the factor plane. Axes can only be constructed through column-items. The roles of rows and columns cannot be interchanged as it is the case with the other two methods.

When data are expressed in the same units, when they represent frequencies, or when a significant 'size' component is present in the row-items of a table, then SMA and FAC appear to be more appropriate than PCA. The result of SMA and FAC are invariant with respect to transposition of rows and columns in a table. Clusters can be formed from row-items as well as from column-items. Axes can be drawn in a factor plane by joining any pair of row-items or any pair of column-items. In SMA there is a direct relationship between euclidean distances measured along these axes and the data in the table. In FAC this relation is less obvious.

The barycenter of the map obtained with SMA and FAC represents a neutral or 'grey' point on the map. In applications to medicinal chemistry, this point can be regarded as the representation of a hypothetical compound that performs equally on account of all measurements. Vice-versa, the barycenter can also be viewed as representing a hypothetical measurement that would not favor any of the compounds.

More specific or 'colored' items are projected toward the border of the maps. We observed that row- and column-items that add up to small marginal frequencies are positioned farther away from the barycenter in FAC than this is the case in SMA.

There is a close analogy between the standard color diagram or so-called 'chromaticity' diagram and the mappings obtained from SMA and FAC. Indeed, colors can be expressed by their brightness, hue and degree of saturation. In the color diagram each color is represented as a function of hue and saturation, irrespective of its brightness. Highly saturated or 'pure' colors are displayed near the border of a horse shoe-like contour representing the spectral colors. Moving toward the center of the diagram one finds increasingly unsaturated and mixed colors, and eventually a shade of grey at the center. Complementary colors, i.e. colors that add up to grey when mixed together, are located on axes through the origin.

What makes FAC different from SMA is the special euclidean metric of chi-square used in FAC for representing distances between row-items and between column-items. In FAC, variance takes the form of a chi-square value. Because of this, FAC is appropriate for contingency tables (i.e. tables in which the items represent counts or frequencies). The latter often arise in surveys where a population is subdivided according to two categories (e.g. population by age and by geographical location). In a preliminary step of FAC each item of the table is subtracted and divided by its expected value. When marginal totals of a table are regarded as probabilities, then an expected frequency is simply the product of the corresponding row- and column-totals.

Weighting schemes in SMA can be varied in order to enhance a particular aspect of the information in a table. FAC defines weight coefficients for row- and column-items proportionally to marginal totals of the table, in order to preserve the distance metric under a law of chi-square. In SMA weighting can be based on marginal totals, maximal values by rows and columns, or on externally defined weight coefficients. In the simplest case of SMA, weight coefficients are assumed to be constant and distances are defined by means of a simple euclidean metric in a doubly-centered data space. In pharmacological tables we prefer to weigh compounds and observations according to the maximal activity in the corresponding rows and columns.

Provided that data are expressed in the same units, FAC appears also to be applicable to tables that are not proper contingency tables. In the case of the receptor binding data, however, this would necessitate log reciprocal re-expression and conversion to nonnegative values. Likewise, SMA can be applied to contingency tables, with the exception of sparse and binary coded tables. As stated in the introduction, the choice of a method is dictated by the properties of the data and by the objective of the analysis.

CONCLUSION

Each of the three methods of data representation which we described above, (PCA, SMA and FAC) allows to visualize part of the information hidden in a table of numbers. None of them can be viewed as 'better' than the others. For each of them, however, there exists a particular type of tabulated data for which one method may be preferred. The factorial methods which we have discussed are primarily for exploratory use. Their effectiveness depends to a large extent on the properties of the data and on the inventiveness of the analyst. It is convenient to make a distinction between methods that treat row- and column-items alike (SMA and FAC) and those that don't (PCA).

Factorial methods of data representation are attractive because they are graphic. Geometrical patterns form a useful complement to numerical tables. Mappings of the kind described above may help in better understanding the information that is contained in a table. They do not furnish a proof of any kind and, hence, conclusions drawn from them must be validated.

There is a place for powerful, flexible and reliable methods of data representation for the purpose of graphic documentation and communication, not only in medicinal chemistry but also in many fields of science, technology and business. It will take some time, however, before the graphic language of factorial mappings will be acquainted by the public at large.

REFERENCES

Benzécri J.-P., L'analyse des donées, Vol. II, L'analyse des correspondances, Dunod, Paris, 1973.

Cooley W.W. and Lohnes P.R., Multivariate procedures for the behavioral sciences. J. Wiley, New York NY, 1962.

Lewi P.J., Multivariate data analysis in industrial practice. Research Studies Press (J. Wiley), Chichester Eng., 1982.

Lewi P.J., Spectral mapping, a technique for classifying biological activity profiles of chemical compounds. Arzneim. Forsch. (Drug Res.), 26, 1295-1300, 1976.

Niemegeers C.J.E., Pharmacology Dept., Janssen Pharmaceutica, Unpublished results, 1974.

Schmutz J., Neuroleptic piperazinyl-dibenzoazepines. Arzneim. Forsch. (Drug Res.), 25, 712-719, 1975.

Wood P.L., Charleson S.E., Lane D. and Hudgin R.L., Multiple opiate receptors : differential binding of μ, κ and δ agonists. Neuropharmacol., 20, 1215-1220, 1981.

ANALYSIS OF VARIANCE AND LINEAR MODELS

Giovanni LATORRE

University of Calabria, Cosenza, Italy

Both the methodologies, the Regression Analysis and the Analysis of Variance, can be seen as particular cases of the methodology of the Linear Models. More specifically, the former constitutes the Full-Rank Case and the latter the Not-Full-Rank Case. Furthermore this general frame-work is very useful in order to understand the common philosophy which underlies the different specifications of the Analysis of Variance for the numerous experimental designs.

1. INTRODUCTION

The analysis of variance is one of the best known statistical methods which is particularly important in the analysis of experimental data. The main purpose of the ANOVA consists, firstly, in partitioning the total variability of the response variable into the different sources of variability such as experimental factors, interactions of factors and experimental error (or chance) and, secondly, in testing the significance of the contributions of each source to detect what are the factors or interaction of factors, if any, which strongly affect the response.

This methodology is often presented in a fragmentary way, which makes the ANOVA strongly dependent on the type of design one is using.

This usually confuses the user, and on the other hand does not provide a method to the user to make his own ANOVA for his peculiar design.

The lack of emphasis on a general approach to ANOVA also causes

B. R. Kowalski (ed.), Chemometrics. Mathematics and Statistics in Chemistry, 377–391.

problems from the computational viewpoint because the user very often finds difficult to understand whether the ANOVA program described in the obscure manual of his statistical package is the right one for his design.
What we will show here is the general theory underlying the ANOVA methodology which can be applied to any kind of design.
We will see that ANOVA has a lot to do with regression and linear models and we will finally show that as long as one is able to describe his own design and has access to a multiple regression program he will also be able to make an ANOVA even for data whose design is not standard, i.e. incomplete or unbalanced.
In order to do this, firstly, we will review the methodology of the multiple regression, which we will also denominate Linear models in the full-rank case.
Secondly, we will see that Linear models can also be used for the analysis of experimental data, and we will denominate this methodology as: Linear models in the not-full-rank case.

2. FULL RANK CASE.

In the usual regression problem the data are formed by n observations on the response variable y and on the p explanatory variables $X_1, \ldots, X_p$.
A linear relation between y and $(X_1, \ldots, X_p)$ is represented by the model:

$$y = \beta_o + \beta_1 X_1 + \ldots + \beta_p X_p + \varepsilon \tag{2.1}$$

where ε is the error disturbance.
Under (2.1) the data have to satisfy the following n equations:

$$y_i = \beta_o + \beta_1 x_{1i} + \ldots + \beta_p x_{pi} + \varepsilon_i \quad (i=1,n) \tag{2.2}$$

which can be written in matrix terms:

$$\underset{n\times 1}{y} = \underset{n\times(p+1)}{X} \cdot \underset{(p+1)\times 1}{\underset{\sim}{\beta}} + \underset{(n+1)}{\underset{\sim}{\varepsilon}} \tag{2.3}$$

where:

$$\underset{\sim}{y} = \begin{pmatrix} y_1 \\ \vdots \\ y_n \end{pmatrix}, \quad X = \begin{pmatrix} 1 & x_{11} & & x_{p1} \\ \vdots & \vdots & & \vdots \\ 1 & x_{1n} & \ldots & x_{pn} \end{pmatrix}, \quad \underset{\sim}{\beta} = \begin{pmatrix} \beta_o \\ \vdots \\ \beta_p \end{pmatrix}, \quad \underset{\sim}{\varepsilon} = \begin{pmatrix} \varepsilon_1 \\ \vdots \\ \varepsilon_n \end{pmatrix}$$

The error disturbance is assumed to satisfy:

$$\varepsilon_i \sim N(0, \sigma^2) \text{ and } COV(\varepsilon_i, \varepsilon_j) = 0 \text{ for all } i \neq j. \qquad (2.4)$$

Finally, the matrix X is assumed to be of full-rank, i.e. $r(X) = r = p+1$.

2.1 Least squares estimation of the parameters.

The least squares estimator of $\underset{\sim}{\beta}$ has to minimize the function:

$$S(\underset{\sim}{\beta})=\underset{\sim}{\varepsilon}'\underset{\sim}{\varepsilon}=(\underset{\sim}{y}-X\underset{\sim}{\beta})'(\underset{\sim}{y}-X\underset{\sim}{\beta})=\underset{\sim}{y}'\underset{\sim}{y}-2\underset{\sim}{\beta}'\ X'\underset{\sim}{y}+\underset{\sim}{\beta}'X'X\underset{\sim}{\beta}$$

By differentiating with respect to $\underset{\sim}{\beta}$ we obtain:

$$\delta S(\underset{\sim}{\beta})/\delta\underset{\sim}{\beta} = -2X'\underset{\sim}{y} + 2X'X\underset{\sim}{\beta}$$

by equating to $\underset{\sim}{0}$ we obtain the normal equations:

$$X'X\ \hat{\underset{\sim}{\beta}} = X'\underset{\sim}{y} \qquad (2.1.1)$$

Since X is full rank, so is (X'X), thus its inverse exists and we can easily solve (2.1.1) by premultiplying by $(X'X)^{-1}$, so that one gets:

$$\underset{\sim}{\beta} = (X'X)^{-1}X'\underset{\sim}{y} \qquad (2.1.2)$$

which is the unique solution of the normal equations.
The fitted model is given by:

$$\hat{\underset{\sim}{y}} = X\ \hat{\underset{\sim}{\beta}} \qquad (2.1.3)$$

Notice that:

$$\hat{\underset{\sim}{y}} = X(X'X)^{-1}\ X'\underset{\sim}{y} = P\underset{\sim}{y}$$

where P is an idempotent matrix, i.e.:

$$P^2 = \left[X(X'X)^{-1}\ X'\right]\left[X(X'X)^{-1}\ X'\right] = X(X'X)^{-1}\ (X'X)\ \cdot$$
$$\cdot\ (X'X)^{-1}\ X' = X(X'X)^{-1}\ X' = P$$

hence:

$$\underset{\sim}{y}'\hat{\underset{\sim}{y}} = \underset{\sim}{y}'\ P\underset{\sim}{y} = \underset{\sim}{y}'\ P^2\underset{\sim}{y} = (P\ \underset{\sim}{y})'\ (P\ \underset{\sim}{y}) = \hat{\underset{\sim}{y}}'\hat{\underset{\sim}{y}}. \qquad (2.1.4)$$

A final remark is the following:

$$\sum_{l=1}^{n} y_i = \sum_{i=1}^{n} \hat{y}_i\ ,$$

that is both the y_i's and the $\hat{y}_i$'s have the same average $\overline{y}$ in fact from (2.1.1) we have:

$$X'\hat{\underset{\sim}{y}} = X'\underset{\sim}{y}$$

which are n equations, the first of those equations is:

$$(11\cdots\cdots 1)\,\hat{\underset{\sim}{y}} = (11\cdots\cdots 1)\,\underset{\sim}{y}$$

i.e.:

$$\sum_{i=1}^{n} \hat{y}_i = \sum_{i=1}^{n} y_i.$$

2.2 Partitioning of Sums of Squares.

Having estimated the parameters one wants to know whether the model (2.1) explains a significant portion of the variation of the y-variable.
Firstly, let's notice that the minimum of the quadratic form $S(\underset{\sim}{\beta})$ is given by:

$$\begin{aligned}0 &\leq (\underset{\sim}{y} - X\hat{\underset{\sim}{\beta}})'(\underset{\sim}{y} - X\hat{\underset{\sim}{\beta}}) = (\underset{\sim}{y} - \hat{\underset{\sim}{y}})'(\underset{\sim}{y} - \hat{\underset{\sim}{y}}) = \\ &= \underset{\sim}{y}'\underset{\sim}{y} - \underset{\sim}{y}'\hat{\underset{\sim}{y}} - \hat{\underset{\sim}{y}}'\underset{\sim}{y} + \hat{\underset{\sim}{y}}'\hat{\underset{\sim}{y}} = \underset{\sim}{y}'\underset{\sim}{y} - \hat{\underset{\sim}{y}}'\hat{\underset{\sim}{y}}\end{aligned}$$

by (2.1.4), hence:

$$\underset{\sim}{y}'\underset{\sim}{y} \geq \hat{\underset{\sim}{y}}'\hat{\underset{\sim}{y}}.$$

We can measure the variation of the y_i's by:

$$\sum_{i=1}^{n} (y_i - \overline{y})^2 = \sum_{i=1}^{n} y_i^2 - n\,\overline{y}^2 = \underset{\sim}{y}'\underset{\sim}{y} - n\,\overline{y}^2 \qquad (2.2.1)$$

Since the average of the $\hat{y}_i$ is also $\overline{y}$, the corresponding measure of variation due to the model is:

$$\sum_{i=1}^{n} (\hat{y}_i - \overline{y})^2 = \sum_{i=1}^{n} \hat{y}_i^2 - n\overline{y}^2 = \hat{\underset{\sim}{y}}'\hat{\underset{\sim}{y}} - n\overline{y}^2 \qquad (2.2.2)$$

The unexplained variation (or variation due to error) is:

$$(\underset{\sim}{y}'\underset{\sim}{y} - n\,\overline{y}^2) - (\hat{\underset{\sim}{y}}'\hat{\underset{\sim}{y}} - n\,\overline{y}^2) = \underset{\sim}{y}'\underset{\sim}{y} - \hat{\underset{\sim}{y}}\hat{\underset{\sim}{y}} \qquad (2.2.3)$$

Intuitively, the greater the variation (2.2.2) due to the model the better is the model itself.
We can now define:

Sum of Squares due to the model: SSM = $\hat{y}'\hat{y}$
Sum of Squares due to the error: SSE = $y'y - \hat{y}'\hat{y}$
Total Sum of Squares : SST = $y'y$
where:

$$SST = SSM + SSE.$$

A first rough measure of the performance of the model is given by:

$$R^2 = (\hat{y}'\hat{y} - n\bar{y}^2)/(y'y - n\bar{y}^2)=(SSM-n\bar{y}^2)/(SST-n\bar{y}^2) \quad (2.2.4)$$

such that $0 \leq R^2 \leq 1$, where R^2 intuitively represents the fraction of SST which is accounted for by fitting the model. A drawback of R^2 is that its value can always be increased by adding new variables to the model.
A better way to judge the performance of the model consists in testing whether $(SSM - n\bar{y}^2)$, the adjusted model sum of squares, is significantly large.
In fact, it can be shown (Searle, p. 101) that an appropriate reference distribution for the ratio:

$$f = \left[(SSM - n\bar{y}^2)/(r-1)\right]/\left[(SSE/(n-r)\right] \quad (2.2.5)$$

is given by the F-distribution with (r-1) and (n-r) degrees of freedom (d.f.).
Hence, in order to test whether the model (2.1) "accounts for a significant portion of the total variations of y...." (Searle, p. 104) one has to verify that:

$$f > F\ (\alpha;\ r-1,\ n-r)$$

where α is the pre-selected level of significance and $F(\alpha;r-1,n-r)$ is the tabulated threshold value of the F-distribution with (r-1) and (n-r) d.f. corresponding to α.
The above results can be summarized in a table. Before doing this let's notice that:

$$SSM = \hat{y}'\hat{y} = \hat{\beta}'X'X\hat{\beta} = \hat{\beta}'X'X\left[(X'X)^{-1}\ X'y\right] = \hat{\beta}'X'y.$$

Moreover it can be shown (Searle, p. 93) that:

$$E\left[SSE/(n-r)\right]= E\left[(y'y - \hat{\beta}'X'y)/(n-r)\right] = \sigma^2$$

i.e.

$$\hat{\sigma}^2 = SSE/(n-r)$$

is an unbiased estimator of σ^2.
The following ANOVA-table summarizes all the operations and

computations needed to test the significance of the model:

Source of variation	Sums of Squares	d.f.	Mean Squares	f-ratio
Model	$\hat{\beta}'X'y-n\bar{y}^2$	r-1	$\frac{\hat{\beta}'X'y - n\bar{y}^2}{r-1}$	$\frac{\hat{\beta}'X'y - n\bar{y}^2}{\hat{\sigma}^2(r-1)}$
Error	$y'y-\hat{\beta}'X'y$	n-r	$\hat{\sigma}^2$	-
Total	$y'y-n\bar{y}^2$	n-1	-	-

$r = r(x) = p+1.$

Table (2.2.1): ANOVA table.

2.3 Reduction Sum of Squares.

Having tested the significance of the model (2.1) one, typically, wants to investigate whether a subset of the p originally considered explanatory variables has contributed to the above significance. Let's partition the matrix X in two submatrices:

$$\underset{nx(p+1)}{X} = (\underset{nx(p_1+1)}{X_1} , \underset{nx(p-p_1)}{X_2})$$

and let X_2 be the matrix of the subset of variables whose contribution is under analysis. Let's also denominate

$$y = X\beta + \varepsilon = X_1\beta_1 + X_2\beta_2 + \varepsilon \qquad (2.3.1)$$

the "Full Model", and

$$y = X_1\beta_1 + \varepsilon_1 \qquad (2.3.2)$$

the "Reduced Model", where $\beta' = (\beta_1' , \beta_2')$, and

$$r_1 = r(X_1) = p_1 + 1.$$

Let's also indicate the SSM of the Full Model by:

$$R(\beta) = \hat{\beta}' X' y \qquad (2.3.3)$$

and the SSM of the Reduced Model by:

$$R(\beta_1) = \hat{\beta}_1' X' y \qquad (2.3.4)$$

We can now define:

$$R(\underset{\sim}{\beta}_2 \mid \underset{\sim}{\beta}_1) = R(\underset{\sim}{\beta}) - R(\underset{\sim}{\beta}_1) \qquad (2.3.5)$$

as the Reduction sum of squares, measuring the contribution of the subset of variables X_2 to the SSM of the Full Model. Common sense logic suggests that if one wants to judge the contribution of the subset of variables X_2 to the Full Model, then he has to test whether $R(\underset{\sim}{\beta}_2|\underset{\sim}{\beta}_1)$ is significantly large or not. It can be shown (Searle, p. 247) that an appropriate reference distribution for the ratio

$$f_d = \left[R(\underset{\sim}{\beta}_2|\underset{\sim}{\beta}_1)/(r-r_1)\right]/\hat{\sigma}^2 \qquad (2.3.6)$$

is given by the F-distribution with $(r-r_1)$ and $(n-r)$ d.f. Hence, in order to test whether the subset of variables X_2 provides a significant contribution to the Full Model one has to verify that:

$$f_d > F(\alpha;\ r-r_1,\ n-r).$$

A summary of the above computations is provided by the following table:

Source of variation	Sums of Squares	d.f.	Mean Squares	f-ratios
Full Model	$R(\underset{\sim}{\beta})-n\bar{y}^2=\hat{\underset{\sim}{\beta}}X'\underset{\sim}{y}-n\bar{y}^2$	$r-1$	$\frac{R(\underset{\sim}{\beta})-n\bar{y}^2}{r-1}$	$f=\frac{R(\underset{\sim}{\beta})-n\bar{y}^2}{(r-1)\hat{\sigma}^2}$
Reduced Model	$R(\underset{\sim}{\beta}_1)-n\bar{y}^2=\hat{\beta}_1X'\underset{\sim}{y}-n\bar{y}^2$	r_1-1	-	-
Reduction	$R(\underset{\sim}{\beta}_2\mid\underset{\sim}{\beta}_1)=R(\underset{\sim}{\beta})-R(\underset{\sim}{\beta}_1)$	$r-r_1$	$\frac{R(\beta_2\mid\beta_1)}{r-r_1}$	$f_d=\frac{R(\beta_2\mid\beta_1)}{(r-r_1)\hat{\sigma}^2}$
Error	$\underset{\sim}{y}'\underset{\sim}{y}-\hat{\underset{\sim}{\beta}}'X'\underset{\sim}{y}$	$n-r$	$\hat{\sigma}^2$	-
Total	$\underset{\sim}{y}'\underset{\sim}{y}-n\bar{y}^2$	$n-1$	-	-

$r = r(X) = p+1, \quad r_1 = r(X_1) = p_1+1.$

Table (2.3.1): ANOVA table.

3. MODELS OF NOT-FULL-RANK.

The methodology of the linear models can be applied also to the analysis of experimental data. In this case the element y_i of the vector $\underset{\sim}{y}$ is the outcome of the i-th experiment. The row vector

$(x_{i1}, x_{i2}, \ldots, x_{ip})$, i.e. i-th row of the matrix X, describes the presence, $x_{ij}=1$, or the absence, $x_{ij}=0$, of the j-th experimental factor in the i-th experiment. For this reason X is usually called the "design" matrix. Finally, the element β_j of the vector β represents the effect of the j-th experimental factor on the variable y.
For example, let y be the yeld in a chemical reaction. The first experimental factor is A, for instance the pressure, with levels A_1, A_2 and A_3. The second experimental factor is B, which consists of two alternative methods, B_1 and B_2, in order to obtain the yield. The whole experiment is formed, then, by six experimental conditions and six experimental outcomes:

B \ A	A_1	A_2	A_3
B_1	y_{11}	y_{12}	y_{13}
B_2	y_{21}	y_{22}	y_{23}

We shall indicate the effects of A by: π_1, π_2, π_3 and the effects of B by: τ_1, τ_2. If we make the assumption of a linear relation between yeld and effects we can write:

$$\begin{aligned} y_{11} &= \mu + \tau_1 + \pi_1 + \varepsilon_{11} \\ y_{12} &= \mu + \tau_1 + \pi_2 + \varepsilon_{12} \\ y_{13} &= \mu + \tau_1 + \pi_3 + \varepsilon_{13} \\ y_{21} &= \mu + \tau_2 + \pi_1 + \varepsilon_{21} \\ y_{22} &= \mu + \tau_2 + \pi_2 + \varepsilon_{22} \\ y_{23} &= \mu + \tau_2 + \pi_3 + \varepsilon_{23} \end{aligned} \tag{3.1}$$

where μ is a mean effect and ε_{ij} are the standard normal independent random noises.
In matrix terms (3.1) can be written as:

$$\underset{\sim}{y} = X \underset{\sim}{\beta} + \underset{\sim}{\varepsilon} \tag{3.2}$$

where:

$$\underset{\sim}{y} = \begin{pmatrix} y_{11} \\ \vdots \\ y_{23} \end{pmatrix}, \quad X = \begin{matrix} (\mu) & (\tau_1) & (\tau_2) & (\pi_1) & (\pi_2) & (\pi_3) \end{matrix} \begin{pmatrix} 1 & 1 & 0 & 1 & 0 & 0 \\ 1 & 1 & 0 & 0 & 1 & 0 \\ 1 & 1 & 0 & 0 & 0 & 1 \\ 1 & 0 & 1 & 1 & 0 & 0 \\ 1 & 0 & 1 & 0 & 1 & 0 \\ 1 & 0 & 1 & 0 & 0 & 1 \end{pmatrix}, \quad \underset{\sim}{\varepsilon} = \begin{pmatrix} \varepsilon_{11} \\ \vdots \\ \varepsilon_{23} \end{pmatrix}$$

and $r(X)=r < p+1$.
The only difference between (3.2) and (2.3) is that in (3.2) X

is not of full-rank; in fact, since $(\mu)=(\tau_1)+(\tau_2)=(\pi_1)+(\pi_2)+(\pi_3)$, $r(x) = 6-2 = 4$.
Since $r(X'X) = r(X) = r < p+1$, the inverse of $(X'X)$ no longer exists and we cannot solve the normal equations:

$$(X'X)\,\hat{\beta} = X'\,y \tag{3.3}$$

In order to overcome above difficulties let's define G a generalized inverse (g.i.) of (X'X) if:

$$(X'X)\,G(X'X) = (X'X) \tag{3.4}$$

G is a matrix not uniquely defined, which satisfies, also, the following properties:

(i) $X\,G\,X'$ is invariant to the choice of G (Searle, p. 20)(3.5) (i.e. it is uniquely defined);

(ii) $X\,G\,X'\,X = X$ and $X'\,X\,G\,X' = X'$ (Searle, p. 12); (3.6)

(iii) finally, if X_* is a full-rank submatrix of X, i.e. $X = (X_*, X_\circ)$ and $r(X_*)=r(X)=r$, then a g.i. of

$$(X'X) = \begin{pmatrix} X_*'X_* & X_*'X_\circ \\ X_\circ'X_* & X_\circ'X_\circ \end{pmatrix}$$

is given by:

$$G = \begin{pmatrix} (X_*'X_*)^{-1} & 0 \\ 0 & 0 \end{pmatrix} \tag{3.7}$$

(Searle, p. 5).

3.1 Analysis of variance.

The above defined g.i. will be now used to find a solution of the normal equation (3.3). If G is a g.i. of X'X, then a solution of (3.3) is given by:

$$\hat{\beta}= G\,X'\,y. \tag{3.1.1}$$

(3.1.1) is one of the infinite solutions which satisfies (3.3), in fact:

$$(X'X)\,\hat{\beta} = X'X\,G\,X'\,y = X'\,y.$$

The lack of uniqueness of the solution (3.2.1) is not a problem in the not-full-rank case, because the emphasis is not on estimation of parameters, but in testing hypotheses.
In particular, all the relevant sums of squares which we use to test the performance of the model (3.2) by means of the Analysis of Variance, are uniquely defined. In fact:

(i) $\hat{y} = X\,\hat{\beta} = X\,G\,X'\,y$
is uniquely defined, by (3.5);

(ii) $SSM = R(\beta) = \hat{\beta}\,X'\,y = y'\,X\,G\,X'\,y$ (3.1.2)
is uniquely defined, by (3.1.5.);

(iii) $SSE = y'y - y'\,X\,G\,X'\,y$ (3.1.3)
is uniquely defined.

The above results make it possible to apply the same methodogy, which has already been shown for the full-rank case, to the not full-rank case.
In fact in order to verify the performance of the model (3.2.1), one can test whether $(SSM - n\,\bar{y}^2)$ is significantly large by comparing the ratio

$$f = [(SSM - n\,\bar{y}^2)/(r-1)]/[SSE/(n-r)]$$

with:

$$F\ (\alpha;\ r-1,\ n-r).$$

Once having tested the significance of the full-model (3.2) as in the full-rank case, one may want to verify whether a subset of factors (or interactions of factors) has contributed to the significance of (3.2).
If we denote by the submatrix X_2 of X the above subset, then:

$$y = X\,\beta + \varepsilon = X_1\,\beta_1 + X_2\,\beta_2 + \varepsilon \quad (3.1.4)$$

is the full-model, and

$$y = X_1\,\beta_1 + \varepsilon_1 \quad (3.1.5)$$

is the reduced model, with:

$$r\left(\underset{n\times(p_1+1)}{X_1}\right) = r_1 \leq p_1 + 1.$$

Moreover, let G_1 be a g.i. of $(X_1'\,X_1)$. Then, in analogy to (3.1.2), the model sum of squares of the reduced model (3.1.5) is

$$R(\beta_1) = \hat{\beta}_1\,X_1'\,y = y'\,X_1\,G_1\,X_1'\,y,$$

and the reduction sum of squares is:

$$R(\beta_2|\beta_1) = R(\beta) - R(\beta_1).$$

Finally, in order to test whether $R(\beta_2|\beta_1)$ is significantly large, one has to compare:

$$f_d = [R(\beta_2|\beta_1)/(r-r_1)]/\hat{\sigma}^2$$

with $F(\alpha; r-r_1, n-r)$.
The following ANOVA table summarizes all the necessary computations:

Source of variation	Sums of Squares	d.f.	Mean Squares	f-ratios
Full Model	$R(\beta)-n\bar{y}^2=y'XGX'y-n\bar{y}^2$	$r-1$	$\frac{R(\beta)-n\bar{y}^2}{r-1}$	$f=\frac{R(\beta)-n\bar{y}^2}{(r-1)\hat{\sigma}^2}$
Reduced Model	$R(\beta_1)-n\bar{y}^2=y'X_1G_1X_1'y$ $-n\bar{y}^2r_1$	r_1-1	-	-
Reduction	$R(\beta_2\|\beta_1)=R(\beta)-R(\beta_1)$	$r-r_1$	$\frac{R(\beta_2\|\beta_1)}{r-r_1}$	$f_d=\frac{R(\beta_2\|\beta_1)}{(r-r_1)\hat{\sigma}^2}$
Error	$y'y - y'XGX'y$	$n-r$	$\hat{\sigma}^2$	-
Total	$y'y - n\bar{y}^2$	$n-1$	-	-

$r(X) = r < p+1;\ r(X_1) = r_1 \leq p_1 + 1.$

Table (3.1.1): ANOVA table.

3.2 Computational aspects.

Given the full model:

$$y = X_1\beta_1 + X_2\beta_2 + \varepsilon$$

from table (3.1.1) we know that in order to test the two relevant hypotheses:
1) H = {the full model provides a significant explanation of the total variation of y};
2) H_d = {the subset of variables (in the full-rank case) or factors and interactions (in the not-full-rank case) identified by X_2 provides a significant contribution to the full model};

one needs to compute the quantities: $R(\beta)$, $\hat{\sigma}^2$, $R(\beta_1)$ and $R(\beta_2|\beta_1)= R(\beta) - R(\beta_1)$.
In the full-rank case those quantities can be readily obtained by a multiple regression program. In fact we would need only two computer runs. In the first one we would provide X as the matrix of explanatory variables and we would obtain the quantities $R(\beta)$ and $\hat{\sigma}^2$ in the full-model analysis of variance table. In the second run we would provide the matrix X_1 and we would obtain $R(\beta_1)$ and then we would calculate $R(\beta_2|\beta_1)$. So that all the entries of table (2.3.1) would be known and H and H_d could be tested.
In the not-full-rank case we cannot directly use a regression

program because X and, in general, also X_1 are not-full-rank matrices. But by using (3.7) we can still use the computational procedure described above.
Let's partition the matrices X and X_1 in such a way that one submatrix is of full rank, i.e.:

$$\underset{n\times(p+1)}{X} = (\underset{n\times r}{X_*}, \underset{n\times(p+1-r)}{X_\circ}), \quad r(\underset{n\times r}{X_*}) = r(X) = r < p+1$$

$$\underset{n\times(p_1+1)}{X_1} = (\underset{n\times r_1}{X_{1*}}, \underset{n\times(p_1+1-r_1)}{X_{1\circ}}), \quad r(\underset{n\times r_1}{X_{1*}}) = r(X_1) = r_1 \leq p_1+1$$

Then by (3.7) the following matrices are g.i. of, respectively, $(X'X)$ and $(X_1'X_1)$:

$$G = \begin{pmatrix} (X_*'X_*)^{-1} & 0 \\ 0 & 0 \end{pmatrix}, \quad G_1 = \begin{pmatrix} (X_{1*}'X_{1*})^{-1} & 0 \\ 0 & 0 \end{pmatrix}$$

Now we notice that:

$$R(\underset{\sim}{\beta}) = \underset{\sim}{y}'\, X\, G\, X'\, \underset{\sim}{y} = \underset{\sim}{y}'\, (X_*, X_\circ)\, G \begin{pmatrix} X_*' \\ X_\circ' \end{pmatrix} y =$$

$$= \underset{\sim}{y}'\, X_*\, (X_*'\, X_*)^{-1}\, X_*'\, \underset{\sim}{y} = R(\underset{\sim}{\beta}_*) \tag{3.3.1}$$

where $R(\underset{\sim}{\beta}_*)$ is the model sum of squares of the full-rank model:

$$y = X_*\, \underset{\sim}{\beta}_* + \underset{\sim}{\varepsilon}_* \tag{3.3.2}$$

In the same way:

$$R(\underset{\sim}{\beta}_1) = \underset{\sim}{y}'\, X_1\, G_1\, X_1' = \underset{\sim}{y}'\, (X_{1*}, X_{1\circ})\, G_1 \begin{pmatrix} X_{1*}' \\ X_{1\circ}' \end{pmatrix} =$$

$$= \underset{\sim}{y}'\, X_{1*}\, (X_{1*}'\, X_{1*})^{-1}\, X_{1*}\, \underset{\sim}{y} = R(\underset{\sim}{\beta}_{1*}) \tag{3.3.3}$$

Where $R(\underset{\sim}{\beta}_{1*})$ is the model sum of squares of the full-rank-model:

$$\underset{\sim}{y} = X_{1*}\, \underset{\sim}{\beta}_{1*} + \underset{\sim}{\varepsilon}_{1*} \tag{3.3.4}$$

In conclusion in the not-full-rank case the entries of table (2.3.1), which can be used to test the hypotheses H and H_d, can be obtained by the following procedure:
(i) find the ranks r and r_1 of X and X_1,
(ii) find the full-rank submatrices X_* and X_{1*} of X and X_1,
(iii) fit the model (3.3.2) and obtain $R(\underset{\sim}{\beta}_*) = R(\underset{\sim}{\beta})$ and $\hat{\sigma}^2$,
(iv) fit the model (3.3.4) and obtain $R(\underset{\sim}{\beta}_{1*}) = R(\underset{\sim}{\beta}_1)$,
(v) calculate $R(\underset{\sim}{\beta}_2 | \underset{\sim}{\beta}_1) = R(\underset{\sim}{\beta}) - R(\underset{\sim}{\beta}_1)$.

4. GENERALIZATION.

The generalization of the testing procedure illustrated above to more than one subset of factors and interactions is very straightforward. In fact let's partition the design matrix X in k submatrices, each of them representing a subset of variables or factors and interactions, whose individual contribution to the full model we want to ascertain.
We will have:

$$X = (X_1, X_2, \ldots, X_k)$$

The full model is:

$$\underset{\sim}{y} = X_1 \underset{\sim}{\beta}_1 + \ldots + X_k \underset{\sim}{\beta}_k + \underset{\sim}{\varepsilon}^* = X \underset{\sim}{\beta} + \underset{\sim}{\varepsilon}^* \qquad (4.1)$$

and the reduced models are:

$$\left\{\begin{array}{l} \underset{\sim}{y} = X_1 \underset{\sim}{\beta}_1 + \ldots + X_{k-1} \underset{\sim}{\beta}_{k-1} + \underset{\sim}{\varepsilon}^*_{k-1} = X^*_{k-1} \underset{\sim}{\beta}^*_{k-1} + \underset{\sim}{\varepsilon}^*_{k-1} \\ \vdots \\ \underset{\sim}{y} = X_1 \underset{\sim}{\beta}_1 + X_2 \underset{\sim}{\beta}_2 + \underset{\sim}{\varepsilon}^*_2 = X^*_2 \underset{\sim}{\beta}^*_2 + \underset{\sim}{\varepsilon}^*_k \end{array}\right. \qquad (4.2)$$

$$\underset{\sim}{y} = X_1 \underset{\sim}{\beta}_1 + \underset{\sim}{\varepsilon}_1 = X^*_1 \underset{\sim}{\beta}^*_1 + \underset{\sim}{\varepsilon}^*_1$$

The corresponding model sums of squares are:

$$\left\{\begin{array}{l} R(\underset{\sim}{\beta}^*_k) = R(\underset{\sim}{\beta}_1, \ldots, \underset{\sim}{\beta}_k) \\ R(\underset{\sim}{\beta}^*_{k-1}) = R(\underset{\sim}{\beta}_1, \ldots, \underset{\sim}{\beta}_{k-1}) \\ \vdots \\ R(\underset{\sim}{\beta}^*_2) = R(\underset{\sim}{\beta}_1, \underset{\sim}{\beta}_2) \\ R(\underset{\sim}{\beta}^*_1) = R(\underset{\sim}{\beta}_1) \end{array}\right.$$

and the set of the possible reductions sums of squares are:

$$\begin{cases} R(\underset{\sim}{\beta}_k \mid \underset{\sim}{\beta}^*_{k-1}) = R(\underset{\sim}{\beta}^*_k) - R(\underset{\sim}{\beta}^*_{k-1}) \\ R(\underset{\sim}{\beta}_{k-1} \mid \underset{\sim}{\beta}^*_{k-2}) = R(\underset{\sim}{\beta}^*_{k-1}) - R(\underset{\sim}{\beta}^*_{k-2}) \\ \vdots \\ R(\underset{\sim}{\beta}_2 \mid \underset{\sim}{\beta}^*_1) = R(\underset{\sim}{\beta}^*_2) - R(\underset{\sim}{\beta}^*_1) \end{cases}$$

Naturally:

$$R(\underset{\sim}{\beta}) = R(\underset{\sim}{\beta}^*_k) = R(\underset{\sim}{\beta}^*_1) + R(\underset{\sim}{\beta}_2 \mid \underset{\sim}{\beta}^*_1) + \ldots + R(\underset{\sim}{\beta}_k \mid \underset{\sim}{\beta}^*_{k-1}).$$

Since all the models (4.2) are hierarchical, in the sense that each one is a submodel of the previous one, it can be shown (Seber, (1980) p. 41) that in order to test the significance of the contribution of X_j to the j-th reduced model one can use the statistcs

$$f_j = R(\underset{\sim}{\beta}_j \mid \underset{\sim}{\beta}^*_{j-1}) / \left[(r_j - r_{j-1})\, \hat{\sigma}^2 \right]$$

and compare it with the threshold value

$$F\,(\alpha;\ r_j - r_{j-1},\ n-r)$$

where $r = r(X)$, $r_j = r(X_j)$, $r_{j-1} = r(X_{j-1})$

As, usual, in the ANOVA table one can summarize all the relevant computations.

Source of variation	Sums os Squares	d.f.	Mean Squares	f-ratios
Full Model	$R(\underset{\sim}{\beta}) - n\bar{y}^2$	$r-1$	$\frac{R(\underset{\sim}{\beta}) - n\bar{y}}{r-1}$	$f = \frac{R(\underset{\sim}{\beta}) - n\bar{y}}{\hat{\sigma}^2(r-1)}$
	$R(\underset{\sim}{\beta}_1) - n\bar{y}^2$	$r_1 - 1$	$\frac{R(\underset{\sim}{\beta}_1) - n\bar{y}^2}{r_1 - 1}$	$f_1 = \frac{R(\underset{\sim}{\beta}_1) - n\bar{y}^2}{\hat{\sigma}^2(r_1 - 1)}$
	$R(\beta_2 \mid \beta_1)$	$r_2 - r_1$	$\frac{R(\underset{\sim}{\beta}_2 \mid \underset{\sim}{\beta}_1)}{(r_2 - r_1)}$	$f_2 = \frac{R(\underset{\sim}{\beta}_2 \mid \underset{\sim}{\beta}_1)}{\hat{\sigma}^2(r_2 - r_1)}$
Reductions	$\vdots$	$\vdots$	$\vdots$	$\vdots$
	$R(\underset{\sim}{\beta}_k \mid \underset{\sim}{\beta}^*_{k-1})$	$r - r_{k-1}$	$\frac{R(\underset{\sim}{\beta}_k \mid \underset{\sim}{\beta}^*_{k-1})}{(r - r_{k-1})}$	$f_k = \frac{R(\underset{\sim}{\beta}_k \mid \underset{\sim}{\beta}^*_{k-1})}{\hat{\sigma}^2(r - r_{k-1})}$
Error	$\underset{\sim}{y}'\underset{\sim}{y} - R(\underset{\sim}{\beta})$	$n-r$	$\hat{\sigma}^2$	-
Total	$\underset{\sim}{y}'\underset{\sim}{y} - n\bar{y}^2$	$n-1$	-	-

$r = r(X) < p+1$; $r_j = r(X_j)$.

Table (4.1): ANOVA table.

The above table in general, is not unique, because it depends on the order in which the k sub-groups X_j of factors are fitted in the reduced models.
This fact raises a serious problem of interpretation of tests of significance referring to the same subgroup, say X_j, which might lead to different conclusions in different orderings (Searle, chap. 7. e.).
However it can be shown (Seber, (1980) p. 42) that the ANOVA table is unique for all the designs which are complete (i.e. the experimental conditions are given by all the possible combinations of the levels of each considered factor) and balanced (i.e. with the same number of observations for each experimental condition). In fact in this situation all the possible reductions sums of squares of the same subgroup, say X_j, are the same, i.e.

$$R(\underset{\sim}{\beta}_j|\underset{\sim}{\beta}_1) = R(\underset{\sim}{\beta}_j|\underset{\sim}{\beta}_1, \underset{\sim}{\beta}_2) = \dots = R(\underset{\sim}{\beta}_j|\underset{\sim}{\beta}_1, \dots, \underset{\sim}{\beta}_{j-1}).$$

Some incomplete but balanced designs, like Latin-squares, Graeco-Latin squares and Hyper-Graeco-Latin-squares designs characterized by some orthogonal structures, enjoy the above properties.

REFERENCES.

1) Box, G.E.P., Hunter, W.G., Hunter, J.S. 1978, "Statistics for experimenters", John Wiley, New York.
2) Dixon, W.J. et al., "BMDP Statistical Software 1981", 1981, University of California Press, Berkeley.
3) Draper, N.R., Smith, H. 1981, "Applied regression analysis", John Wiley, New York.
4) John, P.W.M. 1971, "Statistical design and analysis of experiments", Mac Millan, New York.
5) Miller, R.G.Jr. 1977, "Developments in multiple comparisons", J. Am. Stat. Ass. 72, pp. 779-789.
6) Searle, S.R. 1971, "Linear models", John Wiley, New York.
7) Seber, G.A.F. 1980, "The linear hypothesis: A general theory", Mac Millan, New York.
8) Speed, F.M., Hocking, R.R. 1976, "The use of the R( )-notation with unbalanced data", The Am. Stat. 30, pp. 30-33.
9) Speed, F.M., Hocking, R.R., Hackney, O.P. 1978, "Methods of analysis of linear models with unbalanced data", J. Am. Stat. Ass. 73, pp. 105-112.

CLUSTER ANALYSIS

Prof. Dr. L. Kaufman en Prof. D.L. Massart

Vrije Universiteit Brussel
Centrum voor Statistiek en Operationeel Onderzoek

INTRODUCTION

One of the basic problems in the interpretation of data is finding structure in large sets of empirically gathered information. An important and much used way of solving this problem is to group the objects being studied in a number of sets or subgroups. A general expression for a subgroup of a set of objects is a <u>cluster</u>.

In the process of finding clusters one is looking for a "natural" structure of the data set: this structure must make it possible to explain the cohesion of the objects of each cluster. A generally acceptable definition of natural structure is not simple to find and this is one of the reasons for the large number of clustering methods to be found in the literature. However, most of the methods are based on a simple basic idea: a measure of the quality of a grouping is found by comparing the "similarities" between the different clusters with the "similarities" between the objects within each of the clusters. It is clear that this idea leaves room for many formal methods.

Before selecting a technique to solve a classification problem the particular requirements of the situation must be carefully examined. Here are some of the important questions to be studied:

1) How many clusters are there? Is this number known?
2) How is the similarity between objects and between clusters

B. R. Kowalski (ed.), Chemometrics. Mathematics and Statistics in Chemistry, 393–401.

defined?

3) May clusters overlap?

4) Is it important to find a hierarchy between objects and clusters?

5) Is one looking for representative objects in the clusters?

Studying these questions has led to many significant results both theoretical and practical. It is not possible even to attempt a complete discussion of these results. This paper is limited to a short description of some important or very often applied clustering techniques which are illustrated with small examples.

The following notations and definitions will be used:

- the population under study consists of n elements; it will be denoted by P = {1,2,...,n}
- the n elements are to be divided into k (non-empty and non-overlapping) clusters and k can be either known or unknown; in some methods a value of k can be determined which corresponds to a natural or optimal value. In other methods k is given some or all values between 1 and n
- each element is characterized by the values of p variables
- $x_{i,j}$ is defined as the value of the jth variable (j=1,2,...,p) for the ith element (i=1,2,...,n). This makes it possible to arrange the measurement values into a matrix in the following way:

		variables				
		1	2	3	4	p
objects	1	x_{11}	x_{12}	x_{13}		x_{1p}
	2	x_{21}	x_{22}			
	⋮					
	n	x_{n1}				x_{np}

One of the possible approaches for clustering a set of objects could be to make a list of all possible clusterings and to select the most appealing one. Unfortunately even for relatively small problems the number of clusterings is prohibitively large. For example for 25 objects there are over 4 x 10^{18} possible clusterings. Therefore it is quite clear that some kind of systematic procedure must be used to cluster a set of objects.

An important concept associated with clustering is that of distance or similarity. If i and k are two of the objects of the set which is to be clustered it is necessary to quantify their relative position: in other words, based upon the measurements of the p variables a similarity or distance between any pair of objects must be computed. The distance between objects i and k is often written as d_{ik}. The values d_{ik} do not necessarily satisfy all of the conditions which are required for the mathematical concepts of distance. In cluster analysis one is usually satisfied with the following assumptions:

- $d_{ik} \geq 0$
- $d_{ii} = 0$
- $d_{ik} = d_{ki}$

if $d_{ik} < d_{is}$ it is preferable for objects i and k to be together than objects i and s.

There are many ways to define a distance function on a set of objects. The most commonly used distance is the Euclidean distance defined by:

$$d_{ik} = \sqrt{\sum_{j=1}^{p} (x_{ij}-x_{kj})^2}$$

This distance is the closest to the intuitive and classical geometrical concept shown in a graphical representation

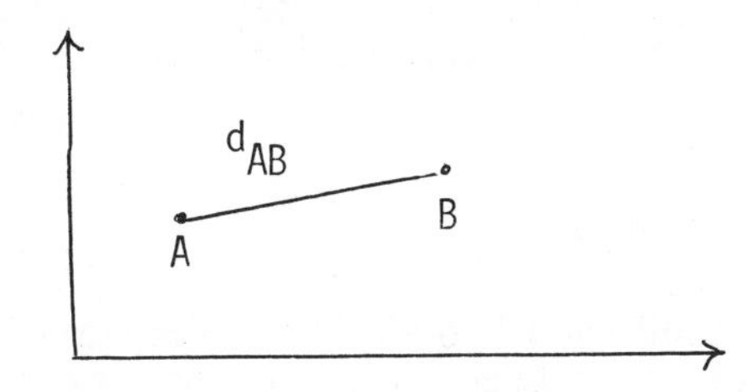

However, in many situations this distance does not represent the actual similarity between points A and B. For example if A and B are locations on a map and the similarity must represent the length of a path between A and B it is better to use a road map to find an adequate measure. The selection of a measure of similarity depends upon what the user considers to be similar. Consider the following example: the following table gives the retention indices of four substances on 3 imaginary stationary phases:

SF	SUBSTANCES 1	2	3	4
1	100	130	150	160
2	150	160	150	100
3	150	170	185	200

Considering Euclidean distance between stationary phases we find that the distances d_{21} and d_{31} are almost the same. d_{21} is approximately 84 and d_{31} is 83. Let us now consider another measure: the correlation coëfficient. Stationary phases 1 and 3 are very strongly correlated: large values for SF1 correspond to large values for SF3. The correlation r_{13} between SF1 and 3 is almost equal to 1. Stationary phases 1 and 2 are very weakly correlated; r_{12} is close to zero.

According to distance measurement SF1 is about as similar with SF2 as it is with SF3. According to the correlation coefficient SF1 is much more similar to SF3 than with SF2. Taking a close look at the data you will find that SF3 is much more polar than SF1 but that they have the same type of specific interactions with the 4 substances, while SF2 and SF1 differ because substances 1 and 4 interact differently for the two phases. Clearly the user of cluster analysis has to decide which measure to use. A final remark: when using the correlation coefficient one usually considers positive and negative correlations to be equivalent. A measure used is for example

$$d_{ik} = \sqrt{1 - r_{ik}^2} \; .$$

We can now start the discussion of clustering methods with an example from a group of techniques called hierarchical clustering techniques.

1. HIERARCHICAL CLUSTERING METHODS

A distance measure is used by a hierarchical clustering method to build a tree of clusters such as in the following example:

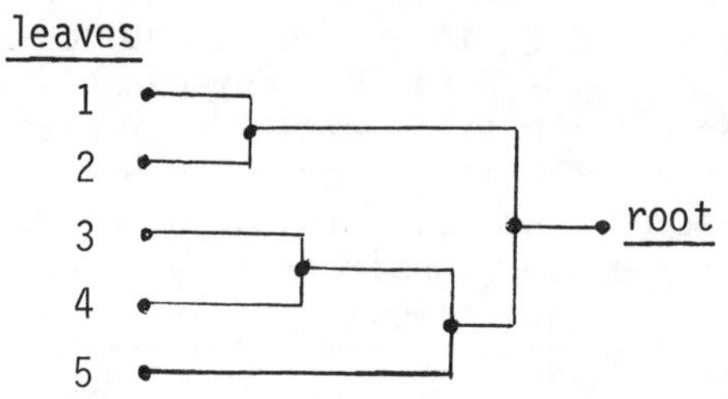

The leaves on the left represent the individual elements while the root on the right corresponds to a single cluster grouping all elements. Going from left to right in the tree the clusters become larger until all elements fall into a single cluster: hierarchical methods which start with individual elements and end with a single cluster are called agglomerative. At each step of such an algorithm two clusters are grouped to form a single cluster. The methods that move from right to left by dividing clusters into two (or more) subclusters are called divisive. When the tree of a hierarchical method has been completely constructed one can choose between several clusterings. In the example above one can obtain the following clusterings

Number of clusters	Clusters
5	(1),(2),(3),(4),(5)
4	(1,2),(3),(4),(5)
3	(1,2),(3,4),(5)
2	(1,2),(3,4,5)
1	(1,2,3,4,5)

It should be remarked that when one goes from 5 clusters to a single cluster in an agglomerative way, two elements which are grouped into a cluster stay together permanently. Similarly in a divisive method when two elements are separated in an early stage they remain in separate clusters; it is even possible to treat two separate clusters separately. These remarks contain the strength and weakness of hierarchical clustering.
- by considering each decision, even the earliest, as permanent, the number of possible clusterings is drastically reduced; this makes it possible to find a "good" clustering with a reasonable amount of calculations;
- on the other hand if an early decision is wrong there is no possibility of a later correction.

We will now consider an agglomerative and a divisive method in some more detail.

An agglomerative method: single linkage

Most agglomerative methods can be considered as special cases of the general scheme described below. d_{ik} is the distance between elements i and k; as the distances are symmetrical ($d_{ik}=d_{ki}$) they can be represented by the following half matrix:

$$\begin{array}{|lllll}
d_{21} & & & & \\
d_{31} & d_{32} & & & \\
d_{41} & d_{42} & d_{43} & & \\
\vdots & & & & \\
d_{m1} & d_{m2} & \cdots & \cdots & d_{mm-1} \\
\hline
\end{array}$$

The general scheme only makes use of the distances d_{ik}. It consists of the following steps:
1) Each element belongs to a separate cluster. The clusters are numbered from 1 to n. The matrix now represents distances between clusters.
2) Search in the matrix for the two clusters closest to each other. Call these clusters p and q ($p > q$) and their distance d_{pq}. The number of clusters is reduced by one and the new cluster is labelled q.
3) The elements of the distance matrix corresponding to the new cluster q must be changed. The elements corresponding to

cluster p are deleted.
4) Steps 2 and 3 are repeated until a single cluster remains.

Agglomerative methods differ by the way that in step 3 the new distances are defined. In single linkage it is the smallest distance between a cluster and the 2 clusters being merged. For example if clusters p and q are being grouped and if $d_{p1}=5$ and $d_{q1}=7$ then the new distance

$$d_{(p,q)1} = d'_{q1} = 5$$
$$= \min (d_{p1}, d_{q1}).$$

The method is characterized by the property that the distance between two clusters is given by the smallest distance between elements of the two clusters. To illustrate the single linkage method we will consider a small geographical example. The problem is to divide 6 touristic attraction points into two clusters. Using the coordinates of the six cities Berlin, Cairo, Istanbul, London, Paris and Rome the following distance matrix is constructed.

	B	C	I	L	P	R
Berlin	-					
Cairo	29	-				
Istambul	18	13	-			
London	10	35	26	-		
Paris	8	32	22	3	-	
Rome	11	21	14	14	11	-

At first, each city is placed in a separate cluster (step 1). We then search for the two closest cities. They are Paris and London. The new clusters are (B), (C), (I), (L,P), (R).
In the next step the distance matrix must be updated. The distances between clusters (B), (C), (I) and (R) remain unchanged. The new distances between cluster (L,P) and the other clusters are calculated in the following way:

$$d_{LP,B} = \min (d_{LB}, d_{PB}) = \min (10,8) = 8$$
$$d_{LP,C} = \min (35,32) = 32$$
$$d_{LP,I} = \min (26,22) = 22$$
$$d_{LP,R} = \min (14,11) = 11$$

The distance matrix now becomes:

	B	C	I	L,P	R
Berlin	-				
Cairo	29	-			
Istambul	18	13	-		
London-Paris	8	32	22	-	
Rome	11	21	14	11	-

The procedure is continued until all objects belong to a single cluster. To analyse the results we can consider all clusterings obtained during the algorithm:

(B,C,I,L,P,R)
(B,L,P,R), (C,I)
(B,L,P,R), (C), (I)
(B,L,P),(R),(C),(I)
(B),(L,P),(R),(C),(I)
(B),(L),(P),(R),(C),(I).

and select the one that seems adequate. The choice can for example depend on the number of clusters. In the example if 2 clusters are wanted they are (B,L,P,R) and (C,I). Another representation shows the grouping together of clusters to form larger clusters. It is called a dendrogram.

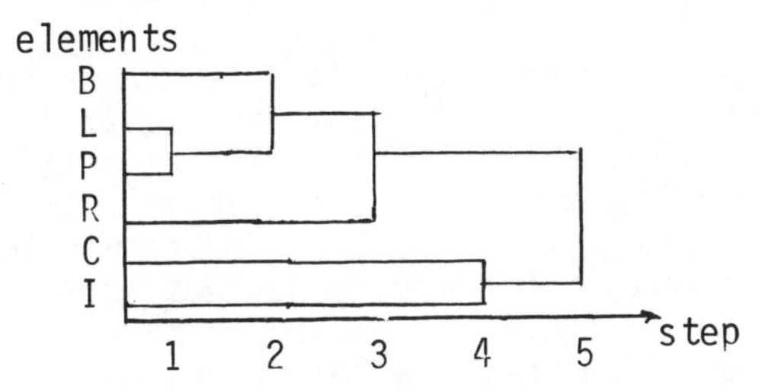

Also the distances can be drawn instead of the step number on the horizontal axis. Many more agglomerative hierarchical methods have been proposed. For example, instead of taking the new distance as: $d_{(p,q),i} = \min(d_{pi}, d_{qi})$, one could consider

$$d_{(p,q),i} = \max(d_{pi}, d_{qi}) \text{ or } d_{(p,q),i} = \frac{d_{pi}+d_{qi}}{2}.$$

These methods are called complete linkage and average linkage, and have also been applied very often.

A divisive method; the Macnaughton method

In this method each cluster of elements considered is divided into two subsets and this procedure continues until every remaining cluster consists of a single element. A cluster is divided into two subsets by means of the construction of a splinter set. This set is started by taking the element for which the average distance to the other elements is maximum. Subsequently another element is a candidate for joining the splinter set if its average distance to the splinter set is smaller than its average distance to the remaining elements. The element is then chosen for which the difference between these two average distances is maximum, and this element joins the splinter set. When no candidates remain, the splitting is carried out. Subsequently each of the two clusters is considered separately and each is divided into two subclusters in the same way. Details of the mathematics involved can be found in ref. 5.

2. A NON-HIERARCHICAL METHOD: K-MEDIAN

Non-hierarchical or partitioning clustering methods search for a division of the set of objects into a number k of clusters such that the elements of the same cluster are close to each other and the different clusters are well separated. Because the k clusters are generated simultaneously the classification is called non-hierarchical.

The number of clusters k can be given or unknown. When it is unknown the algorithm can be repeated for several values of k and many methods make it possible to select a value of k.

In many problems it is important to select a representative object in each cluster. This is the case for several applications in chemistry and therefore such a method will be outlined. To visualize the clustering problem, we again consider a geographical setting. Suppose a company owns 10 supermarkets located at points A,B,C,..., up to J. The company decides to locate two warehouses from which its supermarkets are to be supplied, and for reasons of organization they are to be located in two of the 10 locations right next to a supermarket.

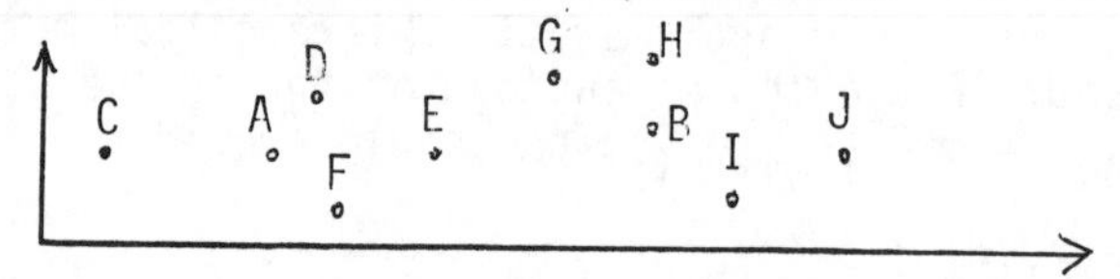

The basic idea is to locate the warehouses in such a way that the total transportation and building costs should be minimized. Without going into an economic analysis a good approximation is found by minimizing the distances from each of the supermarkets to the nearest selected warehouse. There are two ways of solving this problem:

- <u>either</u> consider each partition into two clusters, determine the point in each cluster which minimizes the distances to the other points (this point is called the median) and select the partition for which the sum of distances is minimal;
- <u>or</u> consider each pair of points, assign each of the supermarkets to one of the two points, calculate the sum of distances and select the pair which minimizes this sum.

Because there are many more partitions than pairs it is much easier to use the second method. When the set of points must be divided into more than two clusters one must of course look at all possible k-sets (triples) of points and the amount of computation becomes too large. Fortunately algorithms from O.R. are available to solve this type of problem. When clusterings have been determined for several values of k it is interesting to see whether some of the clusters found can be isolated and considered particularly meaningful. Several ways have been suggested in the literature of which I will briefly outline one.

Suppose the following clusterings were found for several values of k:

k=2 : (1,2,3,4,5) (6,7,8,9,10)
k=3 : (1,2,3,4) (5) (6,7,8,9,10)
k=4 : (1,2,3,4) (5,6,7) (8,9) (10)
k=5 : (1,2) (3,4) (5,6,7) (8,9) (10)
k=6 : (1) (2) (3,4) (5,6,7) (8,9) (10)
k=7 : (1) (2) (3) (4) (5,6,7) (8,9) (10)

A cluster is defined as robust if its objects never join the objects of other clusters for larger values of k. For example cluster (1,2,3) for k=3 is robust but cluster (6,7,8,9,10) is not because one of its objects, number 5 joins objects 6 and 7 to form another cluster. MASLOC, a program based upon the K-Median algorithms and discussed in ref. 5 contains routines for determining robust clusters together with several other ways of evaluating the value of a clustering.

REFERENCES

Cluster analysis

1) Anderberg, *Cluster Analysis for Applications*, Academic Press, New York, 1973.
Introductory, contains computer programs (FORTRAN)

2) Bock, *Automatische Klassifikation*, Vandenhoeck and Dupprecht, Göttingen, 1974.
Extensive mathematical treatment.

3) Everitt, *Cluster Analysis*, Heinemann Educational Books, 1974, Introductory.

4) Kaufman and Rousseeuw, *Finding groups in data*, J. Wiley, 1984. To appear. Will contain listings of computer programs of selected methods.

5) Massart and Kaufman, *The interpretation of Analytical Chemical Data by the use of Cluster Analysis*, Wiley, New York, 1983.
Oriented towards chemical applications.

6) Sneath and Sokal, *Numerical Taxonomy*, Freeman, San Fransisco, 1973.
Oriented towards biology.

OPERATIONAL RESEARCH

Prof. Dr. L. Kaufman and Prof. D.L. Massart

Vrije Universiteit Brussel
Centrum voor Statistiek en Operationeel Onderzoek

INTRODUCTION

Operational research can be defined as the application of mathematical methods to solve problems involving the control of complex systems to provide solutions which best serve the organization, company or the decision maker.

This somewhat obscure definition was proposed over one hundred years after the first O.R.-models and methods were published.

The key word of the definition is best (solutions which best serve), and O.R. is strongly related to the optimization of mathematical functions as we will see in the next few examples.

Another important aspect is organization. In many fields and in particular in analytical chemistry more attention is paid to the detail of the work to be done than to the complex organization necessary to carry out that work. For example thousands of papers have been written on how to determine a biochemical parameter, but only a handful on how to design an optimal configuration for a clinical laboratory.

Many of the models used in O.R. concern problems of organization and therefore related to management science. Others are technical problems but which find their analog in organizational situations.

It may seem surprising to find so many management techniques in a field such as chemometrics. However, these techniques are certainly relevant for our purpose. The manager's job is to make the best possible use of the resources at his disposal in order to achieve a certain goal such as selling goods or manufacturing

B. R. Kowalski (ed.), Chemometrics. Mathematics and Statistics in Chemistry, 403–417.

a quantity of goods. The same purpose describes the work of most analytical chemists. Hence it is reasonable that the techniques used by modern managers can help them in their decisions.

Note that we have used the expression "help them in their decisions" and not make the decisions. Allthough O.R. methods are mathematical, they rarely offer ready-made solutions. The reason for this is that the models used in O.R. to describe a real situation cannot take into account all aspects of the problem or they would become mathematically too complex. Therefore in the most successfull models a compromise is found between the description of a real situation and the mathematical tractability.

This implies that the solutions obtained, should be understood more as a guide for evaluating different solutions than as the solution.

This is often not understood by chemists who assume that as O.R. techniques are mathematical they should lead to exact and unrefutable solutions. When they find that the solution obtained is not the ideal one because the model did not incorporate a certain aspect of the problem, they tend to conclude that O.R.-models are worthless.

We will now present a few small examples of O.R.-models. It is not our objective to be complete but rather to present a representative sample of models and problems which these models can help to solve. The models we will present are:

1. Linear programming;
2. Game theory;
3. Queuing theory;
4. Multicriteria analysis.

The first problem concerns an allocation situation solved by linear programming.

1. LINEAR PROGRAMMING

Linear programming is one of the oldest and most important methods of Operational Research. It has a considerable number of applications of which the best known is probably the diet or mixing problem. This problem can be formulated in the following simplfied form: given the prices of a number of ingredients p_1, p_2, p_3, p_4 , search for the composition of the cheapest mixture which satisfies a number of basic requirements. For example suppose:

ingredient 1 contains 5 calories per weight unit
ingredient 2 contains 10 calories per weight unit
ingredient 3 contains 25 calories per weight unit

and ingredient 4 contains 15 calories per weight unit.
Let us call x the number of weight units of ingredient 1 in the diet, y this number for ingredient 2, z for ingredient 3 and t for ingredient 4. Suppose the diet must contain at most 1500 calories. This can be expressed by the following inequality

$$5x + 10y + 25z + 15t \leq 1500.$$

At the same time in order to find the cheapest mixture one tries to

$$\text{minimize } p_1x + p_2y + p_3z + p_4t.$$

Let us now look at another example which might occur in an analytical laboratory.

A laboratory must carry out routine determinations of a substance P and uses two methods A and B to do this. With method A a technician can carry out 10 determinations per day, with method B 20 determinations per day. There is only room for 3 apparatuses for method B and there are 5 technicians working in the laboratory.

Method A, although it requires more man-hours is cheaper; it costs 100 units per determination, method B costs 250 units per determination. The total daily available budget is 10000 units. The question is: how should the technicians be allocated to the 2 methods, so that the number of possible daily determinations is as large as possible (is to be maximized).

Let us call
a the number of technicians working with method A;
and b the number of technicians working with method B.
The total daily number of determinations is given by

$$10a + 20b.$$

This function is to be maximized. It is called the <u>objective or economic function</u>. The number of technicians a and <u>b are called variables</u>. They must satisfy the following equations or restrictions (also called constraints):

$$a + b = 5 \text{ (personnel constraint)} \quad (1)$$
$$b \leq 3 \text{ (apparatus constraint)} \quad (2)$$
$$(10 \cdot 100)a + (20 \cdot 250)b \leq 10000 \text{ (budget constraint)} \quad (3)$$

and of course $a \geq 0$ and $b \geq 0$.

The problem is to find the values of a and b which maximize the economic function and which satisfy the constraints. These constraints can also be shown graphically:

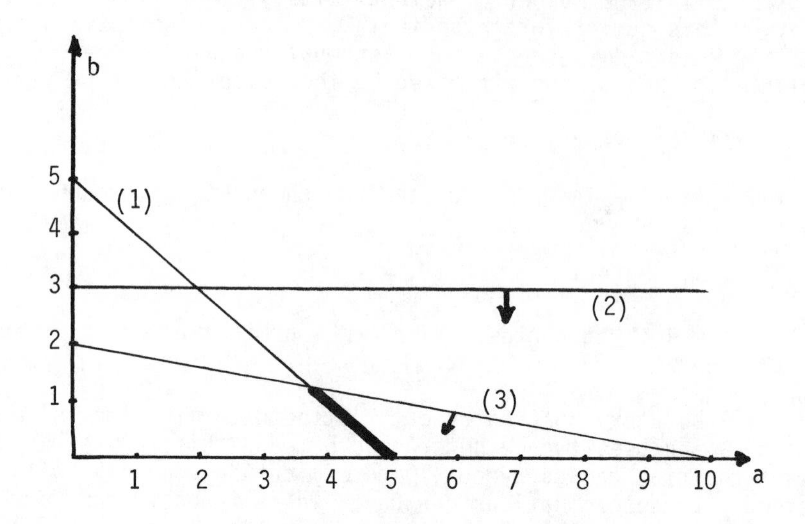

The values of a and b which satisfy all 3 constraints are given on the heavy line (they are called feasible solutions). To find the feasible solution which maximizes the economic function (the optimal solution) we can consider all solutions which have one particular value of the economic function: for example consider the straight line

$$10a + 20b = 30$$

it is given on the following drawing:

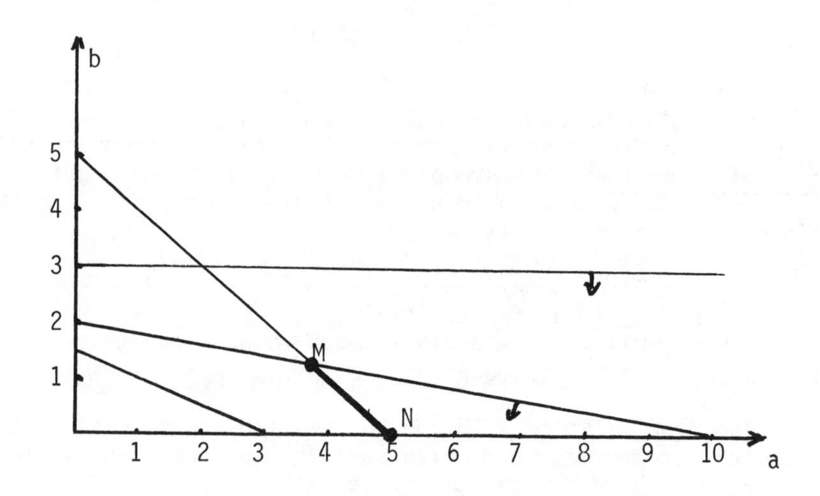

This line does not contain a single feasible solution (these are all on the line MN). By increasing the value of the economic function one obtains better and better sclutions. The new lines obtained in this way for example.

$$10a + 20b = 35, 40 \text{ etc.}$$

are parallel to the one drawn. Eventually one will reach the hard line at the point a = 5 and b = 0 but we can continue until we obtain point M which is the optimal solution. For this point it is easy to see that

$$a = 3.75$$
$$b = 1.25$$

$$10a + 20b = 62.5.$$

Clearly the method we used here is only practical for very small problems with two variables a and b. For problems with more than two variables an algorithm called the simplex method is used. It consists of moving from corner to corner in the region of feasible solutions until the best one is found.

Finally we observe that in the solution found 3.75 technicians should work with apparatus A. This can be solved by having 3 technicians work full time and one technician three-quarters of the time with this apparatus. If this is not feasible (for instance because working with apparatus A requires a different qualification than with apparatus B), then the solution obtained cannot be used. This is not a rare occurence and one then uses a method called integer programming. In this method some or all of the variables are only allowed to take integer values.

Another limitation of linear programming is that the objective and the constraints must be linear functions. In many problems this is not the case and one must resort to a more complex technique called nonlinear programming.

2. GAME THEORY

Consider the following example taken from ref. 3. An analyst has to decide which of 5 possible species is present in a solution. To do this he uses 3 spot tests. Suppose that for each test we know the probabilities of finding each of the 5 possible species. These probabilities are given in the following table:

		cationic species				
		1	2	3	4	5
Test	1	0.3	0.4	0.5	1	0
	2	0.2	0.3	0.6	0	1
	3	0.1	0.5	0.3	0.1	0

The probability of detecting species 1 with the first test is 0.3. The probability of detecting species 4 with the first test is 1 and species 5 cannot be detected with the first test. The question is: which test should be carried out (we suppose that the tests are destructive and only one test can be carried out). This situation is called a game against nature. In such a situation the selection of a test is a question of choosing a suitable criterion. Many solutions have been proposed. I will briefly outline one of these methods: the method of Laplace.

Suppose one has no information about the nature of the species present in the solution. If test 1 is selected the overall probability of identifying the species in the solution is given by:

$$\frac{1}{5}\,0.3 + \frac{1}{5}\,0.4 + \frac{1}{5}\,0.5 + \frac{1}{5}\,1.0 + \frac{1}{5}\,.0 = 0.44.$$

For test 2 the probability is 0.42 and for test 3 it is 0.20. Test 1 is selected because for this test the overall probability is the largest.

The method of Laplace can be improved if one has an idea of the probability of occurence of each of the 5 species. Instead of using the probability 1/5 one then considers probabilities which can be based on previous experience. For example if species 5 has a larger probability of occuring than the other species most probably test 2 will be preferred to test 1 as test 2 detects species 5 while test 1 does not.

There are other strategies for this type of situation. The best known ones are based on the minmax concepts introduced by Von Neumann and Morgenstern.

3. QUEUING THEORY

The time taken for an analysis is one of the possible performance characteristics of an analytical procedure. In a similar way the time between the arrival of a sample and the communication of the result, is a possible performance charactristic of an analytical laboratory. Usually a large part of this time is spent waiting. In particular two waiting times are significant:

a) the time a sample waits before processing, due to the apparatus or personnel being occupied and
b) the time lost because no samples are available.

If these times are too long there is clearly a lack of agreement between the work load or its distribution and the laboratory. Queuing theory studies the relationship between the work load or distribution and these waiting times. The following notations are used:

1) the mean rate of arrival of samples: λ (expressed for example in samples per hour or per minute).
2) the mean rate of analysis: μ (also called the mean <u>service rate</u> per channel).
3) the mean <u>service time</u>: $\bar{t}$.

$$\bar{t} = \frac{1}{\mu}$$

4) the number of apparatuses available: m (the number of channels).

These parameters can usually be observed in the laboratory and are used in a queuing model to determine the following values:

1) the mean waiting time before analysis: $\bar{w}$
2) the mean queue length: $\bar{n}_q$
3) the probability that no samples are available: p_0.

In queuing theory it is not sufficient to know the mean arrival time and the mean service time, one must also know the distribution of the arrivals and the distribution of the service time. In the simplest model called M|M|1 the following assumptions are made:

1) The number of arrivals during a given time interval t has a Poisson distribution. The probability of n arrivals during this interval is given by:

$$P_n(t) = \frac{e^{-\lambda t}\ (\lambda t)^n}{n!}$$

where (λt) is the average number of arrivals during the interval and λ is the average number of arrivals during a time unit.
2) The service time has an exponential distribution. This means that the probability that the service time equals t is given by

$$\mu\ e^{-\mu t}$$

where μ is the mean number of analyses per time unit.

3) The service time is independent of the arrivals.
4) There is only one channel or apparatus.

A queue with one channel or apparatus can be represented in the following way:

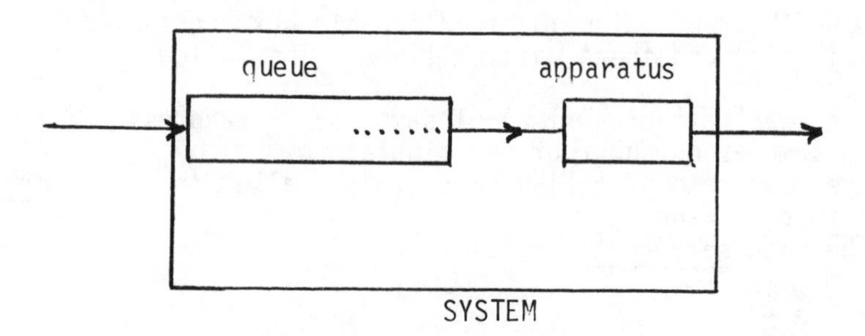

The reason for calling such a model M|M|1 is the following:
the first M refers to arrivals
the second M refers to service times
and the 1 is the number of channels.
The M stands for Markov. It can be shown that both the arrival times and service times described above, satisfy the so-called mathematical MARKOV principle which states that future events only depend upon the present state of the system and not upon past events.

There are more general models among which the much used M|M|m system in which there are m parallel apparatuses or channels. It can be represented in the following way:

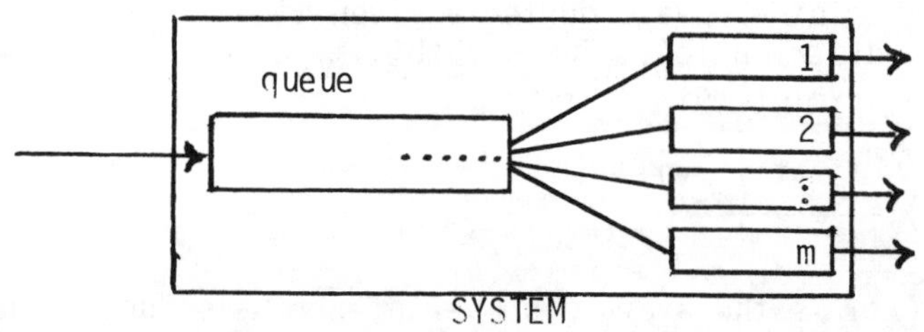

There are also much more general models which are obtained by generalizing the various hypotheses:

1) instead of m parallel channels, an entire network can be considered;
2) the Markov hypothesis for the arrivals and service times can be replaced by general distributions;
3) the classical priority system, FIFO (first in first out) can be replaced by different or more general orders of being

served: an example is the LIFO (last in first out) system.

Let us now look at some results for the basic M|M|1 model. We call the probability of n samples being present in the system p_n and the mean number of samples in the system $\bar{n}$. We then have the following results:

$$P_n = \rho^n(1-\rho): \ \rho = \frac{\lambda}{\mu} = \text{traffic intensity}$$

$$\bar{n} = \frac{\rho}{1-\rho}$$

$$\bar{n}_q = \frac{\rho^2}{1-\rho}$$

$$\bar{w} = \frac{\rho}{\mu(1-\rho)}$$

and the probability that the waiting time is at most T is given by

$$P(W \leq T) = 1. - \rho \ e^{-\mu T(1-\rho)}$$

All these results assume that the traffic intensity $\rho < 1$. This condition can also be written as

$$\lambda < \mu$$

and is necessary to ensure that the system reaches a state of equilibrium. Indeed, if λ is larger than μ the queue becomes longer and longer.

It should be observed that not all laboratories can be studied with queuing theory. For example in some laboratories the sample input is time dependent: in the early morning the lab is almost empty while in the evening there is a long queue of samples. This is the case in many clinical and industrial control laboratories. In terms of queuing theory such labs never reach a steady state or equilibrium. The solution of such queuing problems requires complex mathematics and usually only approximate solutions can be found.

Other laboratories, for instance analytical labs in research departments, process a more or less steady flow of samples. If the structure of the lab is quite simple, queuing theory can be applied. However, most labs are too complex and cannot be described by the simple models of queuing theory. If this is the case, digital simulation can be considered a reasonable alternative. In this approach the laboratory is described with a mathematical model.

The behavior of the laboratory is then described over a long period of time using random number generators for the events which affect the working of the laboratory such as arrival of samples and the end of analyses.

Another reason for using simulation is that it allows the model of the lab to be altered and to study the effect of the alterations on the important parameters which are being investigated. For instance in a spectroscopic laboratory the same person may be responsible for both the acquisition and the interpretation of the spectra or alternatively two persons can be involved.

In many simulation studies the following steps are carried out:

1) Formulation of the problem
   In this step the objective of the study must be stated. Examples are
   . what is the effect of adding a technician or an apparatus on the average throughput time?
   . should properties be given to certain samples?
   . what is the effect of automatic dataprocessing?

2) Collection and processing of laboratory data
   A detailed observation of the system is necessary before a model can be formulated: this should at least include
   . the distribution of the arrival of samples
   . the distribution of the service times
   . the presence of different groups or types of samples (for example samples which require a high priority)
   . the mean down-time of the instruments.
   If the objective is to reduce costs a detailed analysis of the cost structure is necessary.

3) Formulation of the mathematical model
   This is the most difficult part of the study because all the important factors must be included. On the other hand if too many factors are included the model cannot be solved.

4) Programming the model
   Several computer languages have been developed especially for simulation studies. Among these GPSS and GASP are often used for queuing problems and are therefore suitable for the simulation of laboratories. The basic idea is that with the use of a random number generator the important events in the queuing system are generated: among these the most important are arrivals of samples of different types and the end of analysis of a sample.

5) Estimation of the parameters
   The program is first run with the data obtained from the labo-

ratory. This makes it possible to obtain model values for the different parameters, such as mean waiting time, mean queue length, etc. If these parameters differ strongly from the observed values the model must be changed. If there are only slight differences the model is valid and can be used to answer the questions and hypotheses. (To check whether there are large differences between the observed and calculated parameters, statistical tests such as the chi-square test can be used).

6) Check the hypotheses

From all these points it is clear that simulation is a slow and complicated process which requires multidisciplinary cooperation.

4. MULTICRITERIA ANALYSIS

In many optimization and decision problems there is more than one optimization criteria. These criteria are usually conflicting and to make a decision, one is forced to compromise between the different criteria.

A simple example (from ref. (3)) is the choice of an instrument such as a spectrophotometer. One has to make a compromise between cost and quality of the apparatus. A more costly apparatus usually has a larger resolving power, which has a direct bearing on characteristics such as precision, accuracy and information content of the spectra.

There are several approaches to multicriteria problems of which we will outline one: the method of ROY. The object of the method of ROY is not to find a single optimal solution (such as for example in linear programming) but rather to find a set of solutions that are better than the other ones. These solutions are said to outrank the others.

Let us consider a simple example.

One has to select one of six procedures. Suppose there are four criteria described in the following table.

number	criterion	possible values
1	time	120,60,30,15 (minutes)
2	precision(relative)	±1, ±3, ±10 (%)
3	use of toxic reagents	yes, no
4	interference	yes, no

The method supposes that each criterion receives a weight. This weight represents the relative importance of the criterion. If the decision-maker cannot assign weights to the criteria then all weights are given the value one. The following weights have been given:

criterion	weight
1	$\lambda_1 = 5$
2	$\lambda_2 = 4$
3	$\lambda_3 = 3$
4	$\lambda_4 = 3$

From these values one sees that the speed and the precision are considered as more important than the two other criteria.

An important step then consists in determining the values of the criteria for each procedure (a procedure is also called a solution to the multicriteria problem). These values are given in the following table:

criterion	procedure 1	2	3	4	5	6
1	120	60	60	30	30	30
2	1	1	3	3	3	10
3	n	y	n	n	y	n
4	n	n	n	y	n	n

The procedures are then compared with each other in the following way. Consider procedures 1 and 2: procedure 1 is better than procedure 2 for criterion 3 (no toxic reagents are used) but worse for criterion 1 (it is much slower). One writes $N^+ = 3$ and $N^- = 1$. In the following table all procedures are compared.

	1	2	3	4	5	6
1	-	$N^+=3$ $N^-=1$	$N^+=2$ $N^-=1$	$N^+=2,4$ $N^-=1$	$N^+=2,3$ $N^-=1$	$N^+=2$ $N^-=1$
2		-	$N^+=2$ $N^-=3$	$N^+=2,4$ $N^-=1,3$	$N^+=2$ $N^-=1$	$N^+=2$ $N^-=1,3$
3			-	$N^+=4$ $N^-=1$	$N^+=3$ $N^-=1$	$N^+=2$ $N^-=1$
4				-	$N^+=3$ $N^-=4$	$N^+=2$ $N^-=3$
5					-	$N^+=2$ $N^-=3$
6						-

In the next step one obtains a numerical value for the relationship between two procedures. This is done using the weights of the criteria. Consider again procedures 1 and 2.
We just found $N^+=3$ and $N^-=1$. We then calculate the following expression:

$$P = \frac{\sum_{N^+} \lambda}{\sum_{N^-} \lambda} = \frac{3}{5} = 0.6$$

For procedures 1 and 4 it becomes: $\frac{\lambda_2+\lambda_4}{\lambda_1} = \frac{4+3}{5} = 1.4$. The following table contains the values of p for all pairs of procedures. Only the values larger than one have been retained because only these values correspond to a dominance of the first procedure over the second. It should also be observed that the values in the lower triangular part of the matrix are found by taking 1 divided by the corresponding element in the upper triangular part.

Until now we have only taken into account that one procedure is either better or worse than another for a criterion. It is possible that one procedure is so much worse than another for some criterion that one cannot call it better even if for all other criteria it is better: this is called a discrepancy relation. In our example we consider that a difference of more than 5% on criterion 2 is not allowed. This implies that when procedure 6 is compared with the other procedures it cannot outrank them. The values in the 6th row of the matrix are therefore deleted.

	1	2	3	4	5	6
1	-	-	-	1.40	1.40	-
2	1.67	-	1.33	-	-	-
3	1.25	-	-	-	-	-
4	-	1.14	1.67	-	1.00	1.33
5	-	1.25	1.67	-	-	1.33
6	1.25	2.00	1.25	-	-	-

At this stage a dominance threshold T is introduced with a value of at least one and usually higher. The idea is not to judge a procedure to be better than another when only a slight difference is observed. In this way one takes into account the uncertainty involved in choosing the weights λ. In this example T is given the value 1.33 and all p values smaller than 1.33 are eliminated. Then one can construct a dominance graph in which the vertices represent procedures and in which there is an edge from one vertex to another if the corresponding P-value is at least equal to T. In the example the graph is:

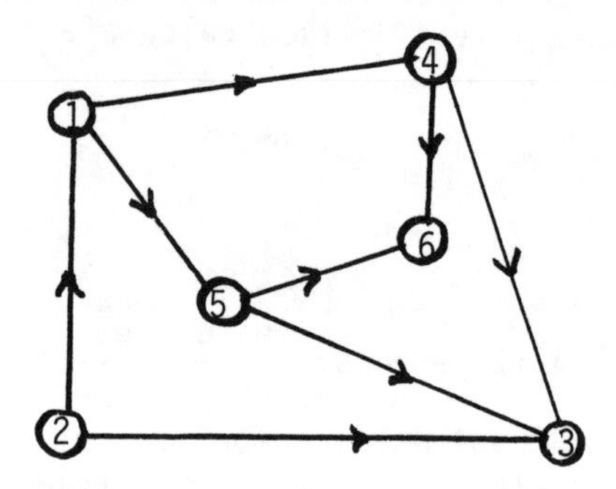

In this graph one determines the kernel which is defined as a set of nodes which satisfy the two following conditions:

a) no vertex of the kernel dominates another vertex of the kernel
b) all vertices outside the kernel are dominated by at least one of the vertices of the kernel.

In this example the kernel consists of the vertices 2,4 and 5 which means that these procedures are to be preferred over the others and the others can be eliminated from consideration.

REFERENCES

General

1) Wagner, *Principles of Operations Research*, Prentice Hall International editions, London, 1975.
   1000 pages.

2) Hillier and Lieberman, *Introduction to Operations Research*, Holden Day Inc., 1967.

3) Massart, Dijkstra and Kaufman, *The Evaluation and Optimization of Laboratory Methods and Analytical Procedures*, Elsevier, 1978.

4) Massart and Kaufman, *Operations Research in Analytical Chemistry*, Analytical Chemistry 47, pp. 1244A, 1975.

Linear and Integer Programming

5) Beale, *Mathematical Programming in Practice*, Pitman, 1968.

6) Zionts, *Linear and Integer Programming*, Prentice Hall, 1974.

Queuing Theory

7) Gross and Harris, *Fundamentals of Queuing Theory*, Wiley Interscience, 1974.

8) Cox and Smith, *Queues*, J. Wiley, 1961

9) Adeberg and Doerffel, *Euroanalysis*, Budapest, 1975.

Graph Theory and Networks

10) Ford and Fulkerson, *Flows in Networks*, Princeton University Press, 1962.

Multicriteria Analysis

11) B. Roy, Revue METRA, 11 (1), pp. 121, 1972.

12) M. Zeleny (Editor), *Multiple Criteria Decision Making*, Springer, 1976.

VARIATION IN SPACE: AN INTRODUCTION TO REGIONALIZED VARIABLES[1]

John C. Davis

Kansas Geological Survey, Univ. of Kansas, Lawrence, KS, USA 66044

ABSTRACT

Regionalized variable theory provides a way for statistically estimating variation in a property sampled over a spatial domain. Regionalized variables, which are intermediate in behavior between purely random variables and deterministic functions, commonly occur in nature. Examples include variations in composition within ore bodies or concentrations of pesticides in soils. The degree of spatial continuity of a regionalized variable is expressed in the semivariogram. Kriging is a procedure used to estimate values of regionalized variables at unsampled locations, and can be used to create a map of the expected values of a property. In addition, a map of the standard error of the estimates can be made.

INTRODUCTION

Chemometricians may regard the objects of their analytical endeavors as members of a homogeneous population, or in certain circumstances as a collection derived from two or more distinct but internally homogeneous populations. Such an operational philosophy is appropriate when analyzing batches from a manufacturing process or a collection of biopsy material from a socioeconomically defined group. However, this approach is unrealistic when applied to geochemical materials such as samples of rock from an ore body, or to environmental observations taken over an area. Each of these observations has an associated geographical (and sometimes temporal) attribute that cannot be ignored. Indeed, a major objective of most geochemical and

B. R. Kowalski (ed.), Chemometrics. Mathematics and Statistics in Chemistry, 419–437.

environmental studies is to identify and characterize the spatial component of variation contained within observations.

The term geostatistics is widely applied to a special branch of applied statistics developed by Georges Matheron of the Centre de Morphologie Mathématique in Fontainebleau, France. Geostatistics was devised to treat problems that arise when conventional statistical theory is used in estimating changes in a property that varies spatially. Because geostatistics is an abstract theory of statistical behavior, it is applicable to many circumstances in different areas of geology, ecology, and other natural sciences.

A key concept of geostatistics is that of the regionalized variable, which has properties intermediate between a truly random variable and one which is completely deterministic. Typical regionalized variables are functions describing natural phenomena which have geographic distributions, such as the elevation of the ground surface, changes in grade within an ore body, or the amount of pesticide in soil. Unlike random variables, regionalized variables have continuity from point to point, but the changes in the variable are so complex that they cannot be described by any tractable deterministic function.

Even though a regionalized variable is spatially continuous, it is not usually possible to know its value everywhere. Instead, its values are known only though samples, which are taken at specific locations. The size, shape, orientation, and spatial arrangement of these samples constitute the support of the regionalized variable, and the regionalized variable will have different characteristics if any of these are changed.

THE SEMIVARIANCE

A primary objective of geostatistics is to estimate the form of a regionalized variable in one, two, or three dimensions. Before this can be done, it is necessary to first consider one of the basic statistical measures of geostatistics, the semivariance, which is used to express the rate of change of a regionalized variable along a specific orientation. Estimating the semivariance involves procedures similar to those of time series analysis.

The semivariance is a measure of the degree of spatial dependence between samples along a specific support. For the sake of simplicity, we will assume the samples are point measurements of a property such as parts per million of selenium in soil samples. For computational tractability, we will further assume that the support is regular; that is, the samples are uniformly spaced along straight lines. If the spacing between samples along a line is some distance Δ, the semivariance can be estimated for distances which are multiples of Δ:

$$\gamma(h) = \sum_{i}^{n-h} (X_i - X_{i+h})^2 / 2n \qquad (1)$$

In this notation, X_i is a measurement of a regionalized variable taken at location i, and X_{i+h} is another measurement taken h intervals away. We are therefore finding the sum of the squared differences between pairs of points separated by the distance Δh. The number of points is n, so the number of comparisons between pairs of points is n-h.

If we calculate the semivariances for different values of h, we can plot the results in the form of a _semivariogram_, which is analogous to a correlogram. When the distance between sample points is zero, the value at each point is being compared with itself. Hence, all the differences are zero, and the semivariance for $\gamma(0)$ is zero. If Δh is a small distance, the points being compared tend to be very similar, and the semivariance will be a small value. As the distance Δh is increased, the points being compared are less and less closely related to each other and their differences become larger, resulting in larger values of $\gamma(h)$. At some distance the points being compared are so far apart that they are not related to each other, and their squared differences become equal in magnitude to the variance around the average value. The semivariance no longer increases and the semivariogram develops a flat region called a _sill_. The distance at which the semivariance approaches the variance is referred to as the _range_ or _span_ of the regionalized variable, and defines a neighborhood within which all locations are related to one another.

For some arbitrary point in space, we can imagine the neighborhood as a symmetrical interval (or area or volume, depending on the number of dimensions) about the point. If the regionalized variable is stationary, or has the same average value everywhere, any locations outside the interval are completely independent of the central point, and cannot provide information about the value of the regionalized variable at that location. Within the neighborhood, however, the regionalized variable at all observation points is related to the regionalized variable at the central location and hence can be used to estimate its value. If we use a number of measurements made at locations within the neighborhood to estimate the value of the regionalized variable at the central location, the semivariogram provides the proper weightings to be assigned to each of these measurements.

The semivariance is not only equal to the average of the squared differences between pairs of points spaced a distance Δh apart, it is also equal to the variance of these differences. That is, the semivariance can also be defined as:

$$\gamma(h) = \frac{\Sigma\{(X_i - X_{i+h}) - \frac{\Sigma(X_i - X_{i+h})}{n}\}^2}{2n} \quad (2)$$

Note that the mean of the regionalized variable X_i is also the mean of the regionalized variable X_{i+h}, because these are the same observations, merely taken in a different order. That is,

$$\frac{\Sigma X_i}{n} = \frac{\Sigma X_{i+h}}{n}$$

Therefore, their difference must be zero

$$\frac{\Sigma X_i}{n} - \frac{\Sigma X_{i+h}}{n} = 0$$

We can combine the summations

$$\frac{\Sigma X_i - \Sigma X_{i+h}}{n} = \frac{\Sigma(X_i - X_{i+h})}{n} = 0$$

Substituting into Eq. 2, we see that the second term in the numerator is zero, so the equation is equal to Eq. 1. Note that this relationship is strictly true only if the regionalized variable is stationary. If the data are not stationary, the mean of the sequence changes with h and Eq. 2 must be modified.

As you might expect, there are mathematical relationships between the semivariance and other statistics such as the autocovariance and the autocorrelation. If the regionalized variable is stationary, the semivariance for a distance Δh is equal to the difference between the variance and the spatial autocovariance for the same distance (Fig. 1).

Unfortunately, it often happens that regionalized variables are not stationary, but rather exhibit changes in their average value from place to place. If we attempt to compute a semivariogram for such a variable, we will discover that it may not have the properties we have described. However, if we reexamine the definition of semivariance given in Eq. 2, we note that it contains two parts, the first being the difference between pairs of points and the second being the average of these differences. If the regionalized variable is stationary, we have shown that the second part vanishes, but if it is not stationary, this average will have some value. In effect, the regionalized variable can be regarded as composed of two parts, a residual and a drift. The drift is the expected value of the regionalized variable at a point i, or computationally, a weighted average of all the points within the neighborhood around point i. The drift will have the form of a smooth approximation of the original regionalized variable. If the drift is subtracted from the regionalized variable, the residuals $R_i = X_i - \overline{X}_i$ will themselves be a regionalized variable and will have local mean values equal to zero. In other words, the residuals will be stationary and it will be possible to compute their semivariogram.

Here we come to an awkwardly circular problem. The drift could be estimated if we knew the size of the neighborhood and the weights to be assigned points within the neighborhood. However, the weights can be calculated only if we know the semivariances corresponding to the distances between point i, the center of the neighborhood, and the various other points. Having once calculated the drift, it could be subtracted from the observed values to yield stationary residuals, which in turn could be used to estimate the neighborhood size and form of the semivariogram.

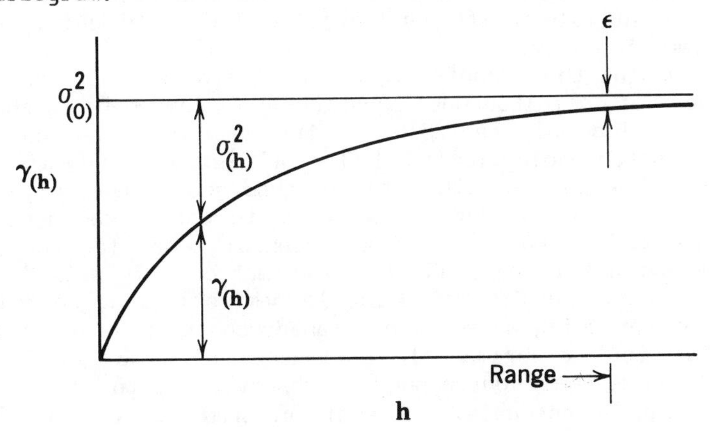

Figure 1. Relationship between semivariance γ and autocovariance σ^2 for a stationary regionalized variable. σ^2_0 is the variance of the observations, or the autocovariance at lag 0. For values of h beyond the range, $\gamma(h) = \sigma^2_0$.

At this stage we must relax our definitional rigor and resort to trial and error. We first concede that we cannot determine the neighborhood in the sense that we have been using the term. Instead, the neighborhood is defined as a convenient but arbitrary interval within which we are reasonably confident that all locations are related to one another. Within this arbitrary neighborhood, we assume that the drift can be approximated by a simple expression such as

$$\overline{X}_0 = \Sigma b_1 X_i$$

or

$$\overline{X}_0 = \Sigma(b_1 X_i + b_2 X_i^2)$$

where the first represents a linear drift and the second a quadratic drift. The calculations involve all of the points within the arbitrary neighborhood, so there is an interrelation between neighborhood size, drift, and semivariogram for the residuals. If the neighborhood is large, the drift calculations will involve many points and the drift itself will be very smooth and gentle. Consequently, the residuals will tend to be more variable and their semivariogram will be complicated in form. Conversely, specification of a small neighborhood size will result in a more variable drift estimate, smaller residuals, and a simpler semivariogram.

Determining the b coefficients for the drift requires solving a set of simultaneous equations of somewhat foreboding complexity. The only variables in the equations are the semivariances corresponding to the different distances between point i and the other points within the neighborhood. However, we do not yet have a semivariogram from which to obtain the necessary semivariances. We must assume a reasonable form for the semivariogram and use it as a first approximation. It will be much easier to guess the form of a simple semivariogram, so this is an argument for using as small a neighborhood size as possible.

Next, the experimental estimates of the drift are subtracted from their corresponding observations to yield a set of experimental residuals. A semivariogram can be calculated for these residuals and its form compared to that of the semivariogram which was first assumed. If the assumptions which have been made are appropriate, the two will coincide, and we have successfully deduced the form of the drift and the semivariogram. Most likely they will differ, and we must try again.

The process of attempting to simultaneously find satisfactory representations of the semivariogram and drift expression is a major part of "structural analysis." It is to a certain extent an art, requiring experience, patience, and sometimes luck. The process is not altogether satisfying, because the conclusions are not unique; many combinations of drift, neighborhood, and semivariogram model may yield approximately equivalent results. This is especially apt to be true if the regionalized variable is erratic or we have only a short sequence. In such circumstances it may be difficult to tell when (or if) we have arrived at the proper combination of estimates.

The semivariogram expresses the spatial behavior of the regionalized variable or its residual. A semivariogram which is tangent to the x-axis at the origin is described as parabolic and indicates that the regionalized variable is exceptionally continuous. A semivariogram which is linear in form indicates moderate continuity of the regionalized variable. A truly random variable will have no continuity and its semivariogram will be a horizontal line equal to the variance. In some circumstances the semivariogram will appear to not go through the origin but rather will assume some non-zero value. This is referred to as the "nugget effect." In theory, $\gamma(0)$ must equal zero; the nugget effect arises because the regionalized variable is so erratic over a very short distance that the semivariogram goes from zero to the level of the nugget effect in a distance less than the sampling interval.

Modelling the Semivariogram

In principal, the experimental semivariogram could be used directly to provide values for the estimation procedures we will discuss later. However, the semivariogram is known only at discrete points representing distances Δh; in practice, semivariances may be required for any distance, whether it is a multiple of Δ or not. For this reason, the discrete experimental semivariogram must be modelled by a continuous function that can be evaluated for any desired distance.

Fitting a model equation to an experimental semivariogram is a trial-and-error process, usually done by eye. Clark (1) describes and gives examples of the manual process, while Olea (2) provides a program which computes a linear semivariogram having the same slope at the origin as the experimental semivariogram.

Ideally, the model chosen to represent the semivariogram should begin at the origin, rise smoothly to some upper limit, then continue at a constant level. The _spherical_ model has these properties. It is defined as

$$\gamma(h) = \sigma_0^2\left(\frac{3h}{2a} - \frac{h^3}{2a^3}\right) \qquad (3)$$

for all distances up to the range, a, of the semivariogram. Beyond the range, $\gamma(h) = \sigma_0^2$. The spherical model is usually described as the ideal form of the semivariogram. Another model that is sometimes used is the exponential:

$$\gamma(h) = \sigma_0^2(1 - e^{(-h/a)}) \tag{4}$$

The exponential never quite reaches the limiting value of the sill, but approaches it asymptotically. Also, the semivariance of the exponential model is lower than the spherical for all values of h less than the range.

The linear model is simpler than either the spherical or exponential, as it has only one parameter, the slope. The model has the form:

$$\gamma(h) = \alpha h \tag{5}$$

and plots as a straight line through the origin. Obviously, this model cannot have a sill, as it rises without limit. Sometimes the linear model is arbitrarily modified by inserting a sharp break at the sill value so that

$$\begin{aligned} \gamma(h) &= \alpha h \qquad \text{for } h < a \\ \gamma(h) &= \sigma_0^2 \qquad \text{for } h \geq a \end{aligned} \tag{6}$$

Armstrong and Jabin (3) have criticized the use of such a model, because the kriging estimation procedure presumes the semivariogram is a continuous, smoothly varying function. However, for distances much less than the range, the linear model is a perfectly good approximation. Both the spherical and exponential models are almost coincident with a straight line near the origin. If the regionalized variable has been sampled at a sufficient density, relative to the range, there will be no significant differences between estimates made assuming a linear model and those obtained using a spherical or other model.

KRIGING

If measurements have been made of a regionalized variable at geographically scattered sampling points and the form of the semivariogram is known, it is possible to estimate the value of the regionalized variable at any unsampled location. The estimation procedure is called kriging, named after D.G. Krige, a South African mining engineer and pioneer in the application of statistical techniques to mine evaluation. Kriging can be used to make contour maps of a spatially distributed property, but unlike conventional contouring algorithms, it has certain statistically optimal properties. Perhaps most importantly, the method provides measures of the error or uncertainty of the contoured variable. Kriging uses the information from the semivariogram to find an optimal set of weights which are used in the estimation of the regionalized variable at unsampled locations. Since the semivariogram is a function of distance, the weights change according to the geographic arrangement of the samples.

Punctual kriging is the simplest form of kriging, in which the observations consist of measurements taken at dimensionless points, and the estimates are made at other locations which are themselves dimensionless points. Punctual kriging is used, for example, in contour mapping where the observations may be the concentration of a trace element, such as selenium, measured in a set of soil samples. Constructing a contour map requires that estimations of the concentration of selenium be made at closely spaced locations over the map area. Once made, contour lines can be drawn through these estimates to produce a map showing the continuous distribution of selenium in soils.

To simplify the problem, we may assume that the variable being mapped is statistically stationary, or free from drift. The value at an unsampled location may be estimated as a weighted average of the known observations. That is, the value at point p is based on a small set of nearby known control points:

$$\hat{X}_p = \Sigma\, W_i X_i$$

We expect that the estimate $\hat{X}_p$ will differ somewhat from the true (but unknown) value X_p by an amount which we may call the estimation error:

$$\varepsilon_p = (\hat{X}_p - X_p) \tag{7}$$

If the weights used in the estimation equation sum to one, the resulting estimates are unbiased provided there is no drift. This means that, over a great many estimations, the average error will be zero, as overestimates and underestimates will tend to cancel one another. However, even though the average estimation error may be zero, the estimates may scatter widely about the correct values. This scatter can be expressed as the error variance,

$$\sigma_\varepsilon^2 = \frac{\Sigma(\hat{X}_p - X_p)^2}{n} \tag{8}$$

or as its square root, the standard error of the estimate:

$$\sigma_\varepsilon = \sqrt{\sigma_\varepsilon^2} \tag{9}$$

It seems intuitively reasonable that nearby control points should be most influential in estimating the value at an unsampled location on a surface, and more distant control points should be less influential. It also seems reasonable to expect that the weights used in the estimation process, and the error in the estimate, should be related in some way to the semivariogram of the surface. In a simple example, Clark (1) demonstrates that this is so.

Suppose we wish to estimate the value of X at a point p from three nearby points, using as our estimator a weighted average of the three known values:

$$\hat{X}_p = W_1X_1 + W_2X_2 + W_3X_3$$

The weights are constrained to sum to one, so the estimate is unbiased if there is no trend. Suppose that weight W_1 is chosen to be equal to 1.0. Then, weights W_2 and W_3 must be zero and the estimate at p is

$$\hat{X}_p = 1.0\ X_1 + 0.0\ X_2 + 0.0\ X_3$$

or

$$\hat{X}_p = X_1$$

Obviously, the estimation error is simply $\varepsilon = X_p - X_1$, since X_1 is the estimate $\hat{X}_p$. If many other locations like X_p are estimated from points arranged in a manner spatially similar to X_1, the estimation variance can be calculated as the average squared difference between these pairs of points. For convenience, we may call these other estimated locations X_{pi} and the other estimating points X_{1i}. Then,

$$\sigma_\varepsilon^2 = \frac{1}{n} \sum_{i=1}^{n} (X_{pi} - X_{1i})^2$$

If this equation is compared to Eq. 1, you will see that the estimation variance is equal to twice the semivariance for a distance equal to the separation between points X_{pi} and X_{1i}.

We have chosen one particular combination of weights to arrive at an estimate of $\hat{X}_p$ and to determine the estimation error. There is an infinity of other possible combinations of weights that could be chosen, each of which will give a different estimate and a different estimation error. There is, however, only one combination which will give a minimum estimation error. It is this unique combination of weights that kriging attempts to find.

The derivation of the kriging equations are given in many sources; a simple discussion is contained in Clark (1) and a complete derivation is provided by Olea (4) for the case of punctual kriging. Optimum values for the weights can be found by solving a set of simultaneous equations, which includes values from a semivariogram of the variable being estimated. The weights are optimal in the sense that the resulting estimates are unbiased and have minimum estimation variance. No other linear combination of the observations can yield estimates which have a smaller scatter around their true values.

In the simplest possible situation, we may wish to make a kriged estimate of the value $\hat{X}$ at a point p from three known observations, X_1, X_2, and X_3. Three weights, W_1, W_2, and W_3 must be found for the kriging equation. To find these requires the solution to a system of three simultaneous equations:

$$W_1\gamma(h_{11}) + W_2\gamma(h_{12}) + W_3\gamma(h_{13}) = \gamma(h_{1p})$$

$$W_1\gamma(h_{12}) + W_2\gamma(h_{22}) + W_3\gamma(h_{23}) = \gamma(h_{2p})$$

$$W_1\gamma(h_{13}) + W_2\gamma(h_{23}) + W_3\gamma(h_{33}) = \gamma(h_{3p})$$

In this notation, $\gamma(h_{ij})$ is the semivariance over a distance h corresponding to the separation between control points i and j. For example, $\gamma(h_{13})$ is the semivariance for a distance equal to that between known points 1 and 3; $\gamma(h_{1p})$ is the semivariance for a distance equal to that between known point 1 and the location p where the estimate is to be made. The left-hand matrix is symmetrical because $h_{ij} = h_{ji}$. It has zeroes along the main diagonal because h_{ii} represents the distance from a point to itself, which is zero. Assuming the semivariogram goes through the origin, the semivariance for zero distance is zero. Values of the semivariance are taken from the semivariogram, which must be known (or estimated) prior to kriging.

However, a fourth equation is needed to insure that the solution is unbiased, by constraining the weights to sum to one. The fourth equation is

$$W_1 + W_2 + W_3 = 1.0$$

This gives a set of four equations but only three unknowns. The remaining degree of freedom is used to assure that the solution will have the minimum possible estimation error. A slack variable, called a Lagrange multiplier, λ, is added to the equation set. The complete set of simultaneous equations has the following appearance:

$$\begin{array}{l} W_1\gamma(h_{11}) + W_2\gamma(h_{12}) + W_3\gamma(h_{13}) + \lambda = \gamma(h_{1p}) \\ W_1\gamma(h_{12}) + W_2\gamma(h_{22}) + W_3\gamma(h_{23}) + \lambda = \gamma(h_{2p}) \\ W_1\gamma(h_{13}) + W_2\gamma(h_{23}) + W_3\gamma(h_{33}) + \lambda = \gamma(h_{3p}) \\ W_1 \quad + \quad W_2 \quad + \quad W_3 \quad + 0 = 1 \end{array} \tag{10}$$

Rearranging in matrix form,

$$\begin{bmatrix} \gamma(h_{11}) & \gamma(h_{12}) & \gamma(h_{13}) & 1 \\ \gamma(h_{12}) & \gamma(h_{22}) & \gamma(h_{23}) & 1 \\ \gamma(h_{13}) & \gamma(h_{23}) & \gamma(h_{33}) & 1 \\ 1 & 1 & 1 & 0 \end{bmatrix} \cdot \begin{bmatrix} W_1 \\ W_2 \\ W_3 \\ \lambda \end{bmatrix} = \begin{bmatrix} \gamma(h_{1p}) \\ \gamma(h_{2p}) \\ \gamma(h_{3p}) \\ 1 \end{bmatrix} \tag{11}$$

In general terms, we must solve the matrix equation

$$[A] \cdot [W] = [B]$$

for the vector of unknown coefficients, [W]. The terms in matrix [A] and vector [B] are taken directly from the semivariogram or from the mathematical function which describes its form. Once the unknown weights have been determined, the variable at location p is estimated by

$$\hat{X}_p = W_1 X_1 + W_2 X_2 + W_3 X_3 \tag{12}$$

The estimation variance is

$$\sigma_\varepsilon^2 = W_1\gamma(h_{1p}) + W_2\gamma(h_{2p}) + W_3\gamma(h_{3p}) + \lambda \tag{13}$$

That is, the variance of the estimate is essentially the weighted sum of the semivariances for the distances to the points used in the estimation, plus a contribution from the λ coefficient which is equivalent to a constant term. Kriging has two powerful advantages over conventional estimation procedures such as those used for contour mapping. Kriging produces estimates which, on the average, have the smallest possible error, and also produces an explicit statement of the magnitude of this error.

To construct a contour map of a regionalized variable, we must estimate the value of the variable at a regular grid of points covering the map area, and also determine the standard errors of these estimates. From these we can construct two maps; the first is based on the estimates themselves and is a "best guess" of the configuration of the mapped variable. The second is an error map showing the confidence envelope which surrounds this estimated surface; it expresses the relative reliability of the first map. In areas of poor control, the error map will show large values, indicating that the estimates are subject to high variability. In areas of dense control the error map will show low values, and at the control points themselves the estimation error will be zero.

The system of equations used to find the kriging weights must be solved for every estimated location, unless the samples are arranged in a regular pattern so the distances between points remain the same. If one of the regularly spaced estimation points coincides with a sample, the estimate will be identically equal to the sample value and the estimation variance will be zero. This is what is meant by the oft-heard statement that kriging is an "exact interpolator." Of course, we do not ordinarily produce estimates for locations which are already known, but this does occasionally occur when using punctual kriging for contouring. If any of the control points happen to coincide with grid nodes, kriging will produce the correct, error-free values. We also can be assured that the estimated surface must pass exactly through all control points, and that the confidence bands around the estimated surface go to zero at the control points.

In this example, we have assumed that an estimate is made using only three control points in order to simplify the mathematics as much as possible. In actual practice, we would expect to use more points, perhaps many more, in making each estimate. Every control point used in an estimate must be weighted, and finding each weight requires another equation. Most contouring routines use 16 or more control points to estimate every grid intersection, which means a set of at least 17 simultaneous equations must be solved for each location. When used in this manner as a contouring procedure, kriging is computationally expensive.

In theory, the number of points needed to estimate a location varies with the local density of control. All control points within the neighborhood around the location to be estimated provide information and should be considered. In practice, many of these points may be redundant, and their use will improve the estimate only slightly. Practical rules-of-thumb have been developed for contour mapping by kriging which limit the number of control points actually needed to a subset of the points within the zone of influence or neighborhood. The optimum number of control points is determined by the semivariogram and the spatial pattern of the points (5). The structural analysis thus plays a doubly critical role in kriging; it provides the semivariogram necessary to solve the kriging equations, and also determines the neighborhood size within which the control points are selected for each estimate.

An Example

You may now appreciate that kriging, even the highly simplified variations which we have considered, may be arithmetically tedious. The practical application of kriging to a real problem is only possible by using a computer, because the estimations must be made repeatedly for a large number of locations in order to characterize the changes in a regionalized variable throughout an area. As an example, consider the map shown in Figure 2, adapted from a study by Olea (5). Although this example does not involve the mapping of a geochemical variable, the steps in the analysis are the same. The regionalized variable is the elevation of the water table of the Equus Beds, an aquifer in south-central Kansas. A structural analysis, made over a much wider area, shows that a linear drift or trend must be removed in order to make the regionalized variable statistically stationary. The range of the semivariogram for the residuals from the drift is 28 miles. The semivariogram can be modelled as a linear function with a slope of 60 feet^2 per mile (Fig. 3).

In order to map the water table elevation by kriging, the contouring program generated estimates of the water table at locations spaced at equal intervals across the map area. The map in Figure 3 represents 31 × 61 = 1891 kriging estimates, each based on the eight nearest control wells selected by an octant search around the location being estimated. Each estimate in turn required the solution of a set of eleven simultaneous equations.

In addition to the map of the water table itself, kriging also was used to produce a map of the standard error of the estimates, shown in Figure 4. The standard error is zero at each of the 47 observation wells, but increases with distance away from the known control points. With 95% probability, the true surface of the ground water table lies within a confidence interval defined by plus or minus twice the indicated value. For example, at point A the water table is estimated to be at about 1480 feet. Because there are relatively few observation wells near this location, the map of standard error indicates a value of over 6 feet. Therefore, the true elevation at this point must be 1480 ±12 feet, or between 1468 and 1492 feet, with 95% probability.

The estimation of the value of a variable at a point from observations which are themselves points is only one application of kriging. The method can be extended to the estimation of the value of an area from samples which consist of areas, or for the estimation of the value contained within a volume from samples which are volumes. The latter application is especially important in mining, where the estimated quantities may be the grade of ore in a stope block, and the observations are the assay grades of diamond-drill core samples. The estimation procedure

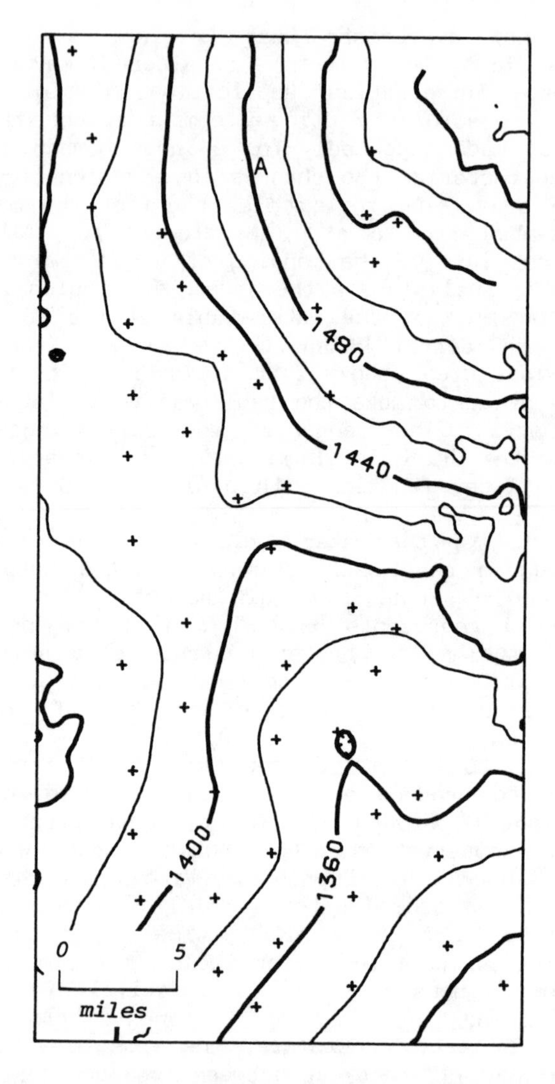

Figure 2. Map showing elevation of the water table in the Equus Beds, a major aquifer in south-central Kansas. Map produced by kriging, assuming a first-order drift. Contours are in feet above sea level. Crosses indicate observation wells.

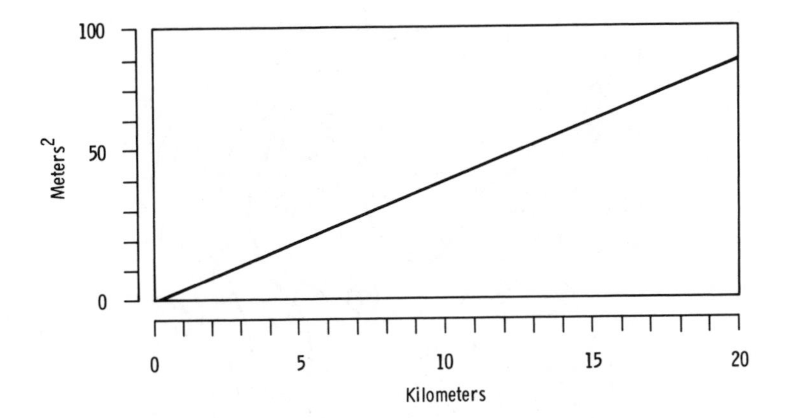

Figure 3. Linear semivariogram of water table elevations in an area which includes the map in Figure 2. Semivariogram has a slope of 60 feet^2/mile (3.5 meters^2/kilometer) within a 28-mile (45-kilometer) neighborhood.

is essentially the same as that presented here, but there are additional complexities that arise because of variation within the areas or volumes. An excellent introduction to the use of kriging for mine evaluation is given in the slim volume by Clark (1). More extensive discussions, including considerations of more advanced topics in geostatistics, are presented by David (6) and Journel and Huijbregts (7).

REFERENCES

1. Clark, I., 1979, Practical Geostatistics: Applied Science Publishers Ltd., London, 129 pp.
2. Olea, R. A., 1977, Measuring spatial dependence with semivariograms: Kansas Geological Survey Series on Spatial Analysis No. 3, Univ. Kansas, Lawrence, 29 pp.
3. Armstrong, M. and Jabin, R., 1981, Variogram models must be positive-definite: Mathematical Geology, v. 13, no. 5, pp. 455-459.
4. Olea, R. A., 1975, Optimum mapping techniques using regionalized variable theory: Kansas Geological Survey Series on Spatial Analysis No. 2, Univ. Kansas, Lawrence, 137 pp.
5. Olea, R. A., 1982, Optimization of the High Plains Aquifer observation network, Kansas: Kansas Geological Survey Groundwater Series 7, Univ. Kansas, Lawrence, 73 pp.
6. David, M., 1977, Geostatistical Ore Reserve Estimation: Elsevier Scientific Publishing Co., Amsterdam, 364 pp.

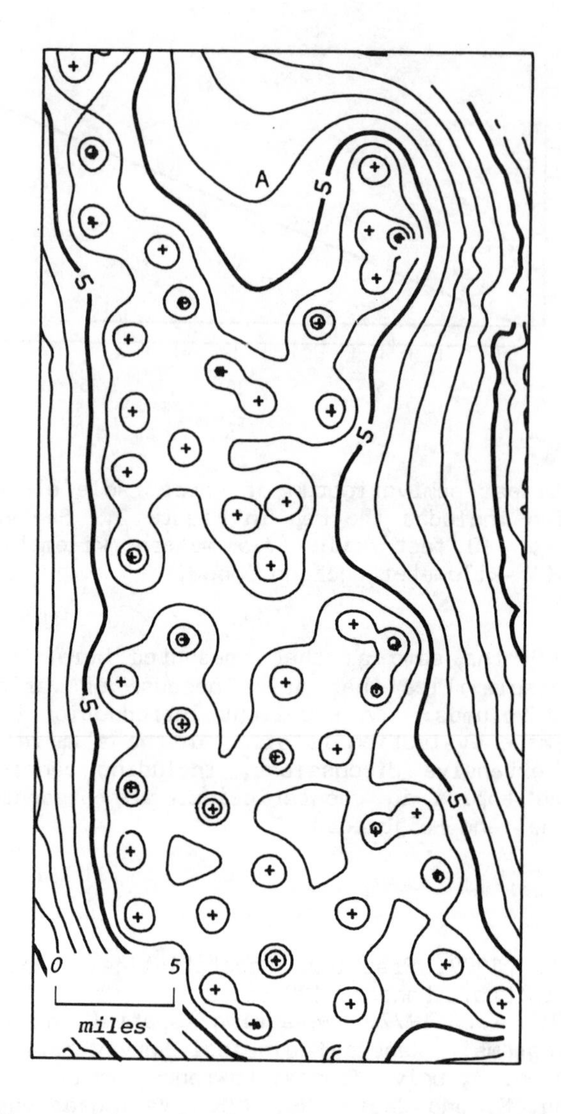

Figure 4. Map showing standard error of the estimates of water table elevations in the Equus Beds. Contour interval is 1 foot.

7. Journel, A. G. and Huijbregts, C. J., 1978, Mining Geostatistics: Academic Press, London, 600 pp.

Additional Reading

Henley, J., 1981, Nonparametric Geostatistics: Applied Science Publishers Ltd., London, 145 pp.

Matheron, G., 1965, Les Variables Régionaliseés et leur Estimation: Masson et Cie, Paris, 305 pp.
Matheron, G., 1971, The theory of regionalized variables and its applications: Les Cahiers du Centre de Morphologie Mathématique, Ecole Nationale Supérieure des Mines, Paris, 211 pp.

NOTE

[1]Some of the material in this article has been adopted from the manuscript for the second edition of "Statistics and Data Analysis in Geology," to be published by John Wiley & Sons, Inc., New York.

THREE-DIMENSIONAL GRAPHICS FOR SCIENTIFIC DATA DISPLAY AND ANALYSIS[1]

Stanley L. Grotch

Lawrence Livermore National Laboratory
Livermore, California 94550

ABSTRACT

Techniques for displaying scientific data in three dimensions are examined using a variety of real examples. Among the types of plots illustrated are: three-dimensional scatterplots, combined surface/contour plots, and combined surface/point displays. A variety of useful features which will enhance the utility of these plots is illustrated. Such displays have proved extremely valuable in providing insights into many problems of chemometrics.

INTRODUCTION

We are currently in the midst of a revolution of the most significant importance, particularly for those of us working in chemometrics. Within only the last few years the computer has brought us capabilities virtually inconceivable only a generation ago. We now have at our disposal computational and display capabilities of the most extraordinary power and versatility. This work focuses on one area of application which is likely to flourish in the future: three dimensional graphics for data display.

There is little doubt that the two-dimensional Cartesian plot is the most commonly used method of graphical data representation. In many real applications in chemometrics, however, three (or even more) dimensions are often encountered. For example, in chemistry one must often visualize the behavior of a dependent variable (commonly, a concentration or some

B. R. Kowalski (ed.), Chemometrics. Mathematics and Statistics in Chemistry, 439–466.

derived property) as a function of both time and temperature. Yet, in spite of the obvious need for 3D representation, few scientists presently use three dimensional graphics in their work. I strongly believe that as the hardware/software required to produce such graphics becomes more accessible, many more scientists will recognize the extraordinary potential of 3D data representation.

The 3D graphics shown here were all produced using one of two general purpose plotting programs developed by the author. The first program plots discrete points, and the second displays surfaces, contours and points in 3D. These programs, written in FORTRAN, may be run either interactively with the user responding to program queries, or more commonly, from a data file containing these responses. These plots may be seen by the user as generated, if desired. Virtually all of the features illustrated can be invoked by simple "yes/no" user responses.

DISSPLA, a proprietary graphics software package available for a variety of computers was used to perform the transformations necessary to produce the 3D graphics shown (1). Other similar software packages exist (2-3). These higher level programs greatly facilitate the programming effort needed to produce effective 3D graphics. Additionally, the capability to produce "device-independent" graphics provides the user with great flexibility in the choice of output devices.

THREE DIMENSIONAL SCATTERPLOTS

Three-dimensional point plotting will be illustrated using an example taken from a materials aging study. In such experiments, samples of material are subjected to differing time/temperature histories, and changes in a material property such as strength, weight, composition, etc. are measured as a function of time and temperature.

In this case, samples were aged under varying conditions, and a mathematical model was developed in which two parameters, A and B were derived. Changes in A and B from a baseline case (delta A and delta B) are to be related to changes in the load bearing capacity of the material (delta load). We wish to use 3D graphics with these data to find a suspected relationship between: delta load, delta A and delta B.

In Fig. 1 these data are shown as three, conventional two-dimensional plots (plotting all possible pairs of variables). With the possible exception of the delta load vs. delta B plot which is suggestive of a linear relationship, little insight is derived from these 2D plots.

In Fig. 2 we see four different 3D views of the same data using two-dimensional symbols (squares) to locate the points in space. From Fig. 2 we see:

1. Our viewpoint is of great importance in our perception of a scene and the conclusions we derive from it. If, for example, we had only the single view 2D, we might be misled into assuming that the points were colinear.
2. While these 2D symbols do show relative positions, they provide few depth cues to precisely locate the points in space.

   The plots can be made more realistic and hence more informative by incorporating several features as in Fig. 3:

   (1) Three-dimensional shaded symbols, scaled so that they decrease in size as their distance from the viewer increases.

   (2) A grid on the base plane.

   (3) Projections of the points on to the base plane.

   (4) Connections between the points in space and their projections.

In Fig. 3 we can quantitatively locate points in the lower plane. Labels can also be placed next to any points in space or their projections to facilitate identifying specific points.

However, one should be forewarned that there can be "too much of a good thing" for when these same features are applied to two other faces, a very cluttered Fig. 4 results. Nonetheless, judicious use of these features with more than one face can be extremely valuable in certain cases.

DRAWING A PLANE

Another feature found to be extremely useful in data analysis is drawing a plane through a set of data in three dimensions. In Fig. 5 we see four views of these aging data with a least squares planar fit also shown. By rotating the scene so as to look directly along the plane, we see in Fig. 6 that these data are in fact co-planar. A single suspect point is also clearly evident (the result of a typing error). Drawing a plane has often revealed an otherwise unexpected relationship or, conversely, has shown the inadequacy of a linear model in fitting a set of data.

Whenever a plane is drawn, the plotting program automatically produces a new data file containing the residuals of the fit. This file can then be used with the same plotting program to generate new plots from the same viewpoints, showing these residuals as a function of x and y. (Fig.7).

The user can also determine the plane by either: (1) specifying the coordinates of three non-colinear points or (2) providing the three coefficients determining a plane. To better differentiate points relative to the plane, connectors to the surface can be drawn, and only those points lying below the plane can be shaded. Alternatively, two plots can be drawn from the same viewpoint: in the first, only those points above the plane are shown, in the second, only those below. Similarly, different colors and/or shapes can be used to distinguish these points (see Fig 5).

THE IMPORTANCE OF VIEWPOINT--FISHER IRIS DATA

The utility of 3D scatterplots in cluster analysis and the importance of viewpoint are both well illustrated using the classical iris data set of R.A. Fisher (4). These data were used by Fisher in 1936 to establish the foundations of linear discriminant analysis. The original data set consisted of 50 measurements of four physical characteristics for each of three different species of iris. Using discriminant analysis, it can be determined that the three variables: sepal width, petal length ,and petal width are the "best" discriminators between these three classes.

In Fig. 8 we see these data plotted in the space of these three variables using different symbols (cubes, pyramids, and inverted pyramids) to differentiate between the varieties of iris. In the view of Fig. 8, the cubes seem to form a distinct cluster, but the separation between the two remaining groups is unclear. If the scene is tipped vertically as in Fig. 9, the visual separation of the cubes improves considerably, but the remaining classes still are confused.

However, if we merely rotate Fig. 8 through 90 degrees, Fig. 10 results. In this view the separation between all three classes is strikingly apparent. Particularly in cases of greater overlap, the use of color can significantly enhance the perceived separations and is of great value in cluster analysis.

A FEW PRELIMINARY CONCLUSIONS.

Extensive use of these methods suggests that:

1. The selection of an "optimal" viewpoint is often extremely important, but cannot usually be chosen, a priori. Therefore, real-time or near-real-time viewing of these images is of the utmost importance.
2. The application of any combination of the features illustrated should be under simple user control, typically

by a menu-like "yes/no" response. The "appropriate" choice of color, gridding, projections, symbol size, etc. is, in general, scene dependent and often subjective.

3. If these techniques are to be used routinely they must be: (1) simple to use, (2) rapid in execution, and (3) affordable, permitting experimentation and the generation of many mistakes.

THREE DIMENSIONAL SURFACES AND CONTOURS.

In chemometrics one often encounters problems where a function of two independent variables, $z = f(x,y)$, is either known or is fit to a set of data and is to be shown in three dimensions.

The contour plot has the great virtue of providing such a presentation using only two dimensions. Unfortunately, particularly in complex situations, contour plots do not give the viewer a good qualitative picture of the data. Generally in data interpretation, it is this "feel" that is the raison d'etre for the analysis.

To illustrate this point let us consider a well-known physical phenomenon. In Fig. 11 we see a plot of the contours of the hours of daylight over the earth as a function of: (1)latitude and (2) day of the year, for the latitude range ± 60 degrees (5). From these contours can the reader "visualize" the variation in daylight?

Among those representations providing a much better sense of the nature of the data, are a sheet or open mesh drawn in perspective in three-dimensions. Why not then get the best of both worlds by combining both types of displays on a single plot as is shown in Fig. 12? In Fig. 12 we clearly can see the inverse sinusoidal seasonal behavior between the two hemispheres. We can derive a broader insight and a better global understanding of a problem by visually and mentally jumping between these presentations in a series of different views (Fig. 13).

Another advantage of this combined plot is that one type of presentation will often suggest features that may not be evident in the other, and vice versa. For example, in Fig. 12 the presence of two days of constant total sunlight (the vernal and autumnal equinoxes) is apparent in the lower plot from the two parallel contours at 12 hours. This feature is certainly not obvious in the upper surface. However, the nature of the equinox can be made evident in the upper surface if vertical cutting planes are introduced, showing the constancy of the surface on these particular days (Fig. 14).

To delineate a given level, specific contours may also be plotted on the upper surface in contrasting colors, or the upper surface might be drawn using different colors above and below a specified horizontal plane. Truncating the function at user-specified levels can also provide very valuable visual insights. All of these features may be invoked by simple user responses.

COMBINATIONS OF POINTS AND SURFACES.

In chemometrics, experimental measurements are commonly used to develop mathematical models. In such cases, combinations of the features described above can be most informative. This will be illustrated using a real data example which will also show the dangers inherent in extrapolating empirically derived models.

In this example we are interested in modeling a set of experimental aging data which measure the mechanical behavior of a material after a series of different time/temperature exposures. As is often the case, here, our fundamental knowledge of the underlying physical processes is poor, and we must resort to empiricism. Two alternative models were developed and will be graphically compared both within the range of the experimental data and in extrapolation:

A. A quasi-linear model in time (t) and temperature (T). This model has an exponentially decaying initial phase from t=0 to t=5, followed by a linear response subsequently:

$$\text{Response} = a_1 + a_2 t + a_3 T \quad .$$

B. A full quadratic model in both t and T:

$$\text{Response} = b_1 + b_2 t + b_3 T + b_4 tT + b_5 t^2 + b_6 T^2 \quad .$$

In Fig. 15 we see the fit of the quasi-linear model to these data and in Fig. 16 a similar view showing the quadratic model. Cubes denote the experimental data and are shaded when they fall below the surface. The scatter of the data, the adequacy of the fit, and the presence of possible outliers are apparent in such plots.

We can directly compare the two models by merely overlaying the results through graphics software editing as is seen in Fig. 17. We see here that over the range of the experimental data both models yield approximately equivalent fits to these data. (In this instance, since the plots were to be superimposed, the lower contours were suppressed to avoid unnecessary confusion).

In accelerated aging experiments we are particularly interested in extrapolating model predictions beyond the range of the experimental measurements. In Fig. 18 we see this extrapolation in temperature for the two models shown side-by-side. The original high temperature measurements have been extended to room temperature. As this comparison shows, the two models are reasonably well behaved, and once again yield approximately the same predictions at lower temperatures.

Before we become overly complacent and accept the premise that these two models are equivalent, let us also compare them at longer times. In Fig. 19 we see the superimposed predictions of these models when extrapolated to approximately twice the original time span of the experimental data. Here, we see rather dramatically that their behavior is distinctly different at longer times.

This real data example, demonstrates rather vividly the dangers faced by the unwary in extrapolating models, particularly when they are empirically derived. In this instance although both models agreed quite well in their fit to the experimental data, and even in extrapolation in one dimension, they were quite different when extended in a second variable. The need for additional experimental data to better differentiate between the models is also apparent in this example.

Such 3D graphical presentations serve a number of functions:

1. The adequacy of model fits and the presence of suspect points or regions of poor fit are usually apparent in a series of rotated 3D views.
2. A better sense of the global nature of the model is obtained. The distribution of the experimental data and the selection of possible new experimental conditions becomes clearer. The degree of extrapolation is also better sensed in such views than by merely examining tables of numbers.
3. Using software editing, different alternative models can be compared or contrasted as in the superimposed views of Figs. 17 or 19, or in the side-by-side view of Fig. 18.

FINAL CONCLUSIONS

In chemometrics one commonly considers multivariate data. The computer revolution now permits us to graphically exhibit such data in a variety of ways using three dimensions. These graphical procedures have proved of great practical value in providing a better understanding of experimental data in a broad range of chemometrics applications.

[1]Work performed under the auspices of the U. S. Department of Energy by the Lawrence Livermore National Laboratory under contract number W-7405-ENG-48.

REFERENCES

1. DISSPLA is a proprietary graphics package developed by Integrated Software Systems Corp., 1186 Sorrento Valley Blvd., San Diego, CA 91121.

2. C. Machover, "A Guide to Sources of Information About Computer Graphics," IEEE Computer Graphics & Applications, Vol 1, No. 1, 73-85, Jan. 1981.

3. J. E. Scott, "Introduction to Interactive Computer Graphics," J. Wiley, 1982.

4. R. A. Fisher, "The Use of Multiple Measurements in Taxonomic Problems," Annals of Eugenics 7 pp. 179-188 (1936).

5. W. D. Sellers, "Physical Climatology," University of Chicago Press, Chicago (1965).

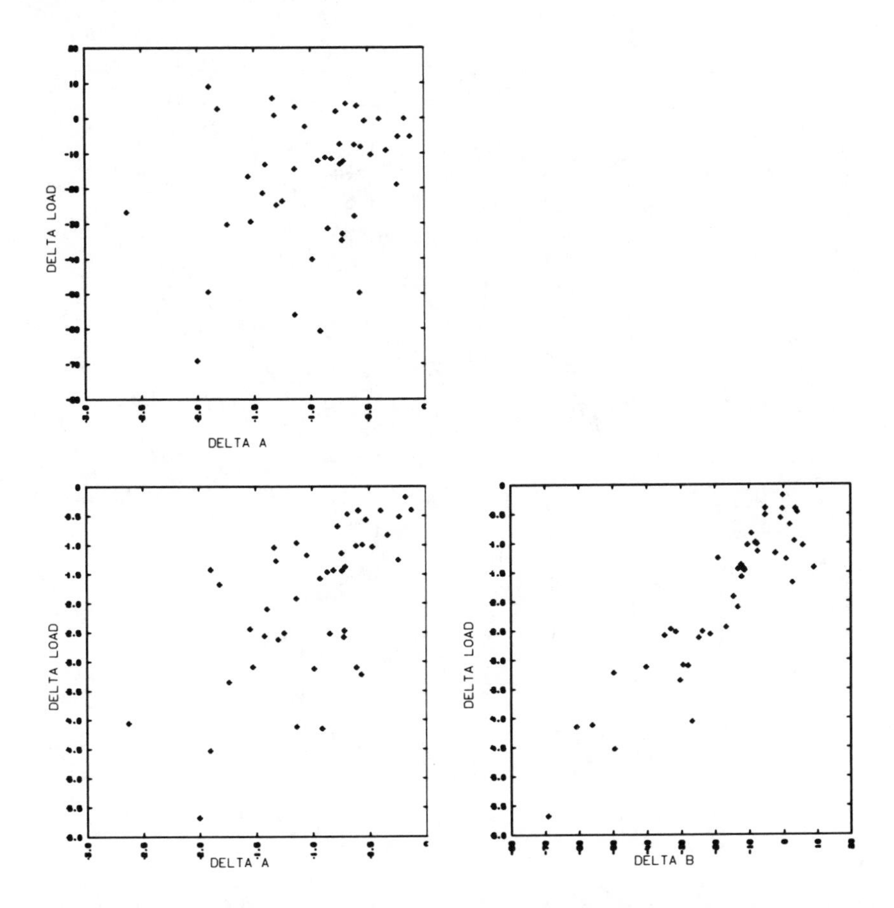

Fig. 1. Results of a material aging study in which all possible pairs of variables are plotted in 2D. The lower right subplot suggests a possible linear relationship.

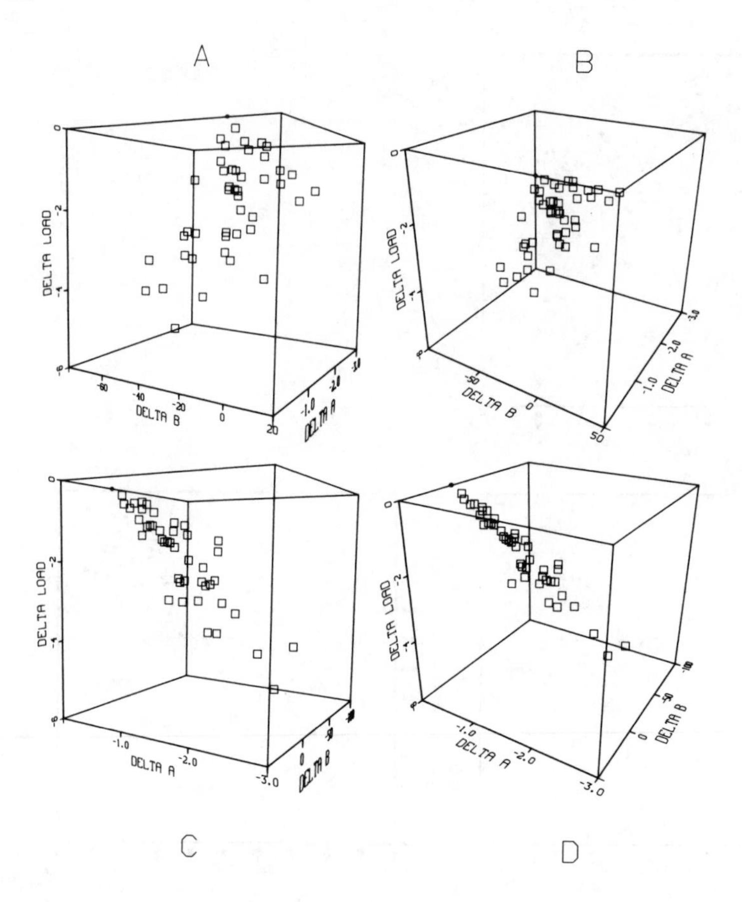

Fig. 2. Four rotated and tipped 3D views using 2D symbols for point location. Subplot D strongly suggests a linear relationship.

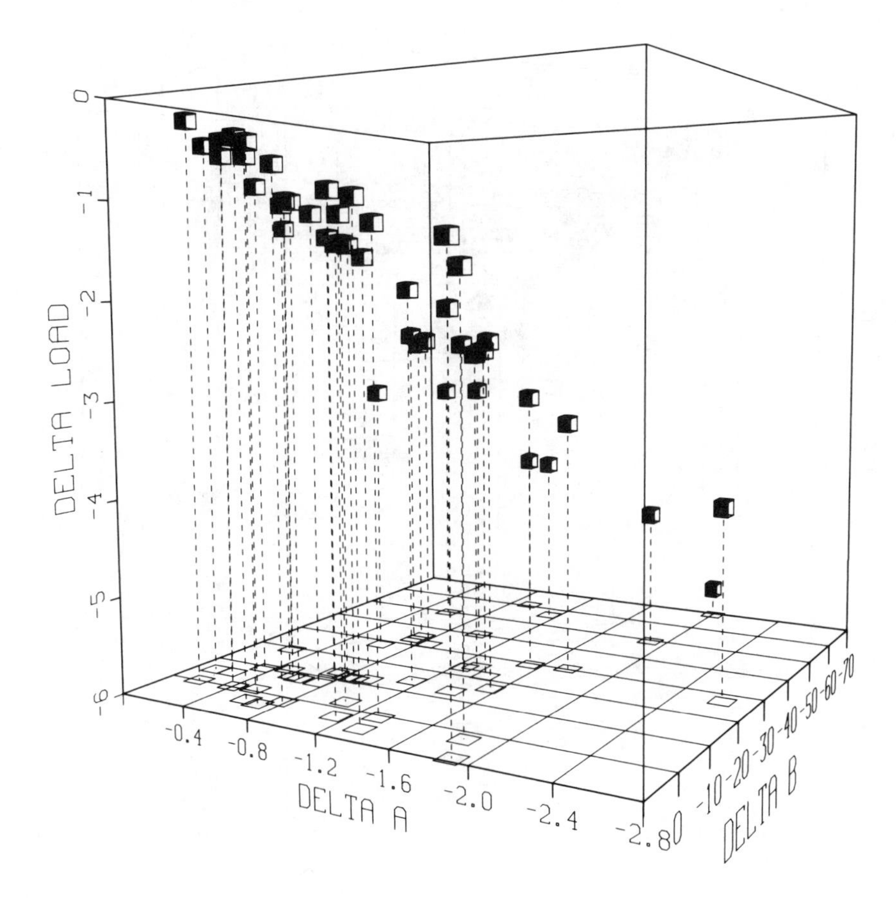

Fig. 3. Aging data in 3D including several enhancements to improve realism: 3D scaled shapes, lower surface gridding, projections, connectors to surface.

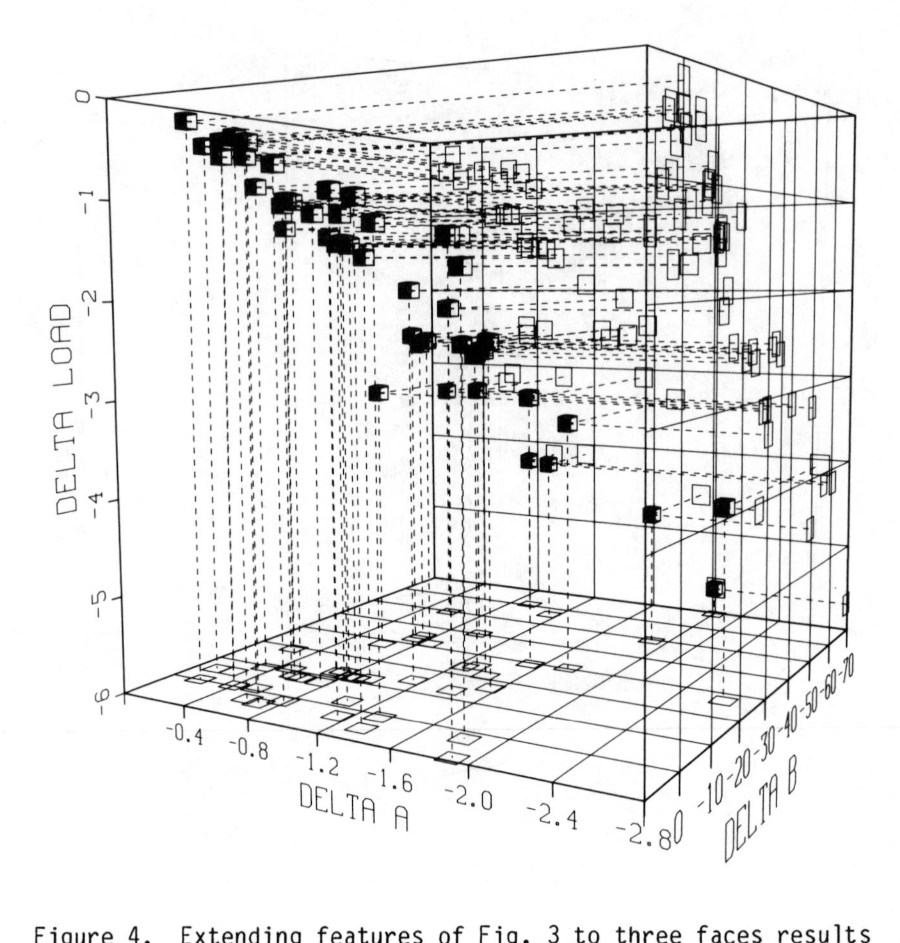

Figure 4. Extending features of Fig. 3 to three faces results in a confusing jumble of lines. Nonetheless, selective use of these enhancements to more than one face can be very valuable.

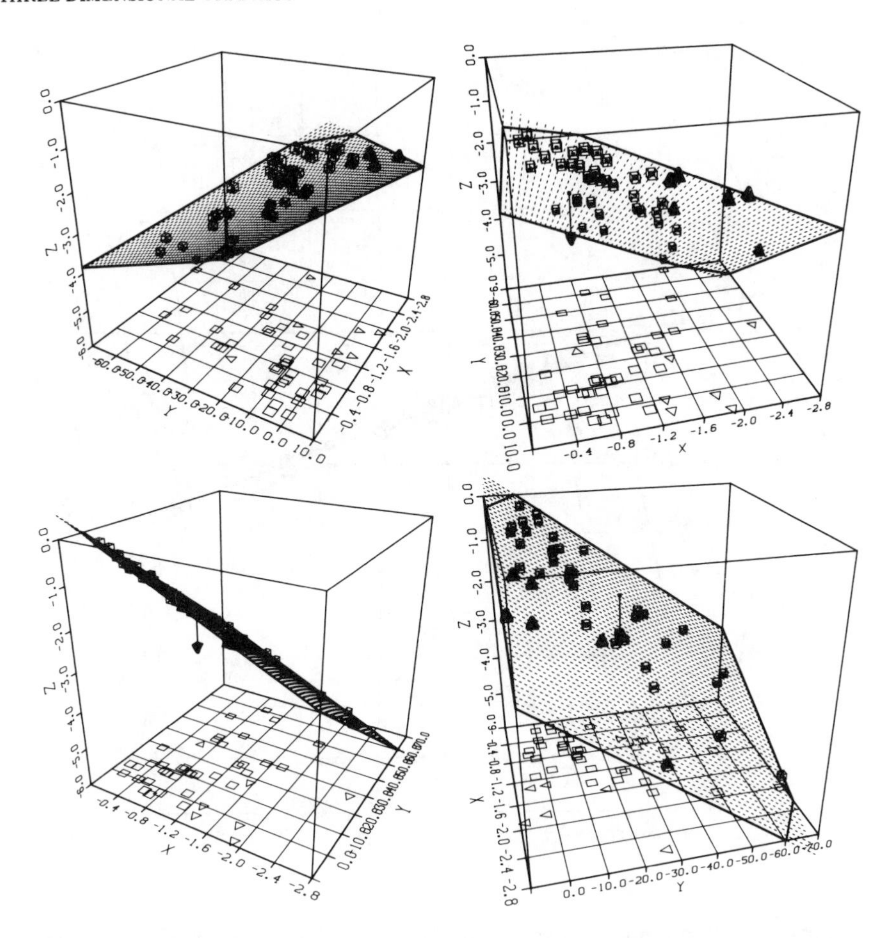

Figure 5. Four views of aging data with a least squares plane drawn. Points above the plane are shown as open cubes, those below as shaded pyramids.

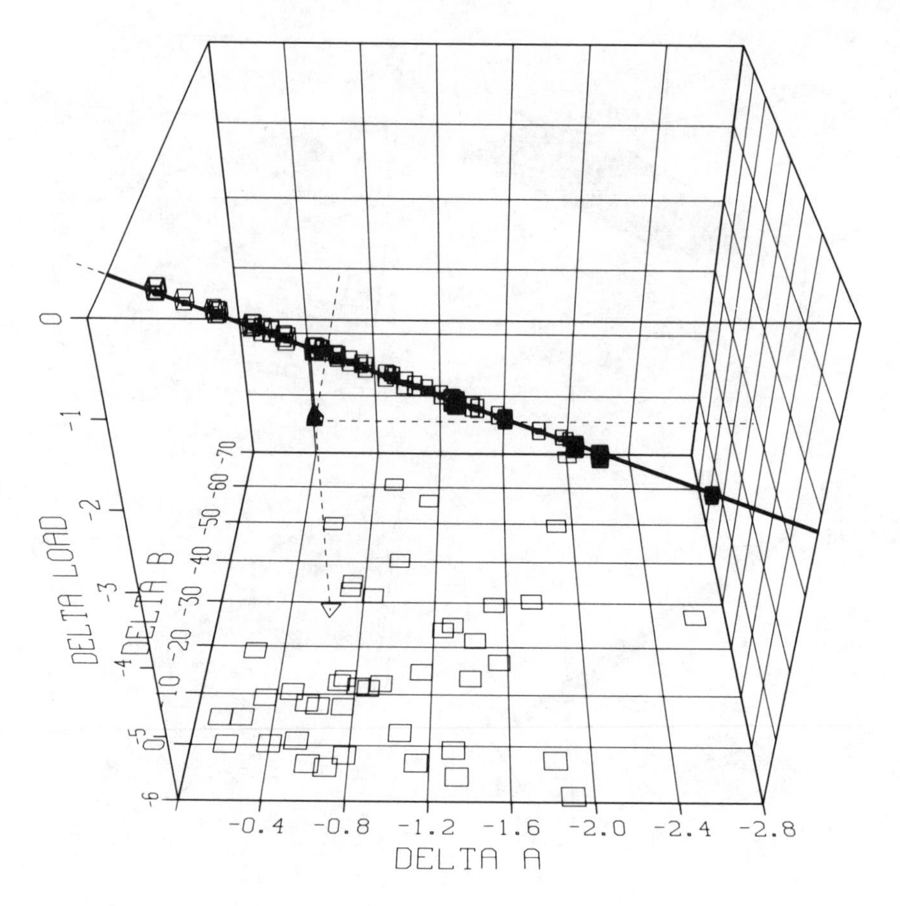

Figure 6. Viewing the aging data directly along the best fit plane clearly shows that the variables are linearly related. A single suspicious point is clearly evident and has been flagged.

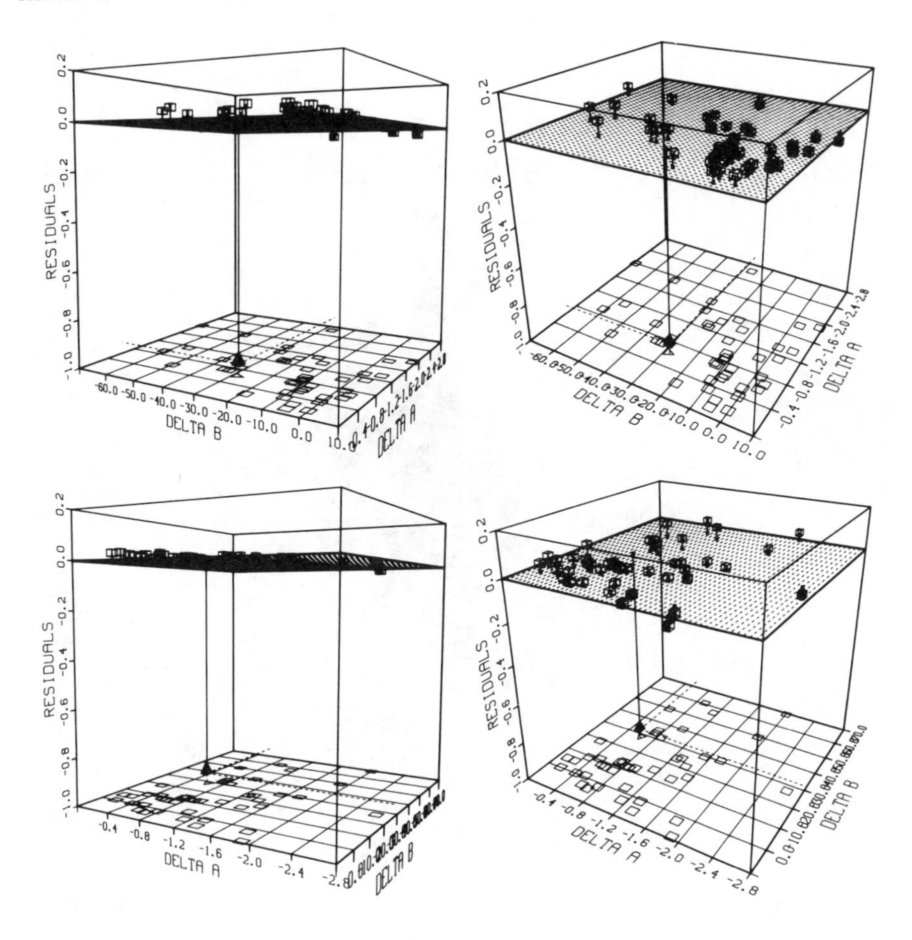

Figure 7. The residuals of the planar fit as seen from four different viewpoints. Once again, the single suspect point is clearly evident.

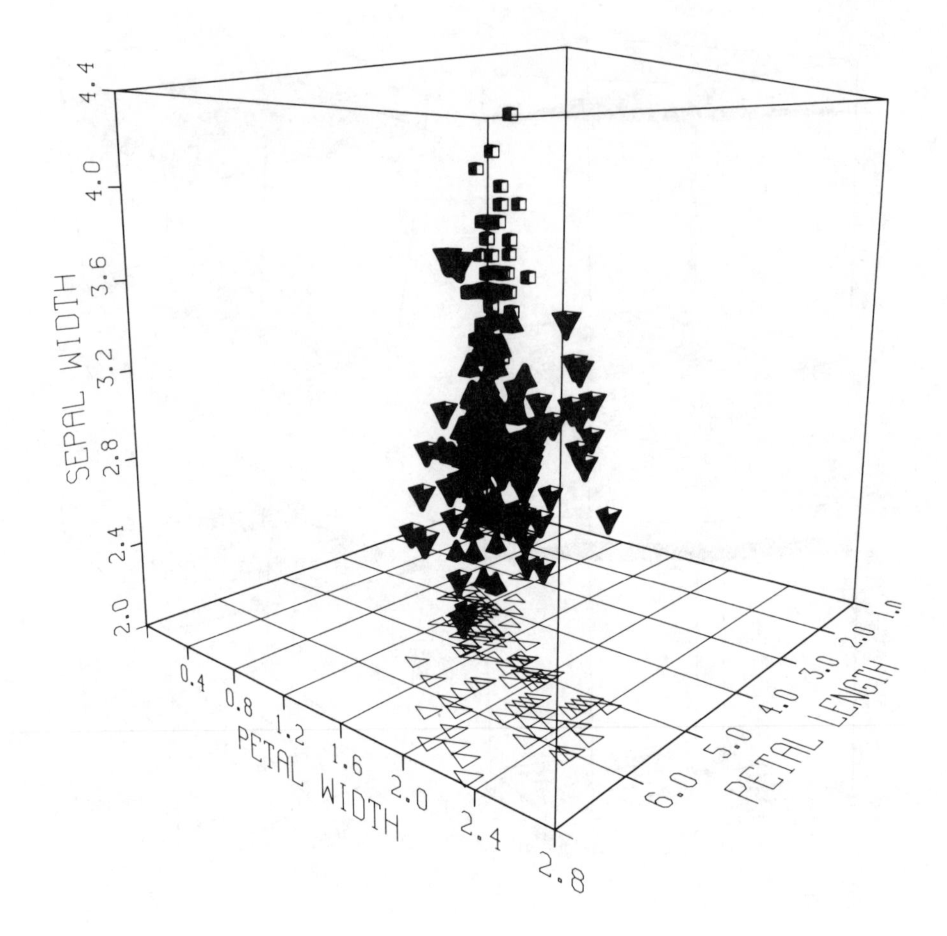

Figure 8. Fisher Iris data seen using "best" 3 dimensions. The cubes in the back appear distinct but the separation between two remainin classes is unclear.

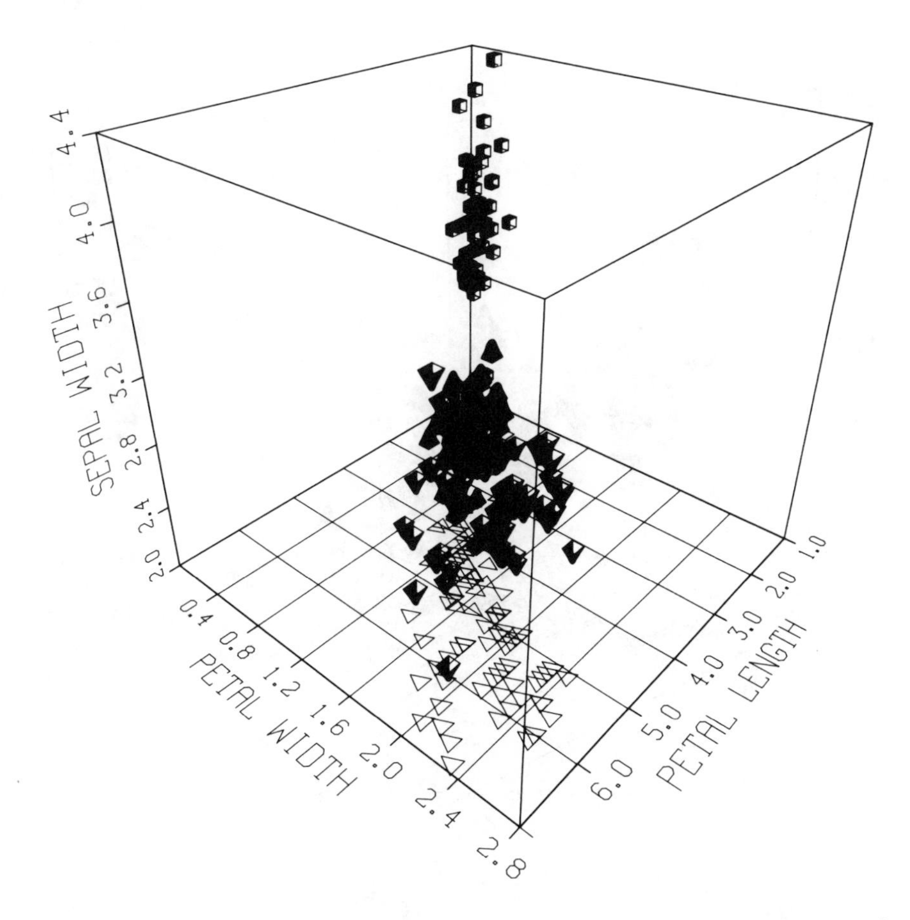

Figure 9. Tipping Fig. 8 vertically improves the separation of the class denoted by the cubes but does not improve our perception of the separation of the remaining classes.

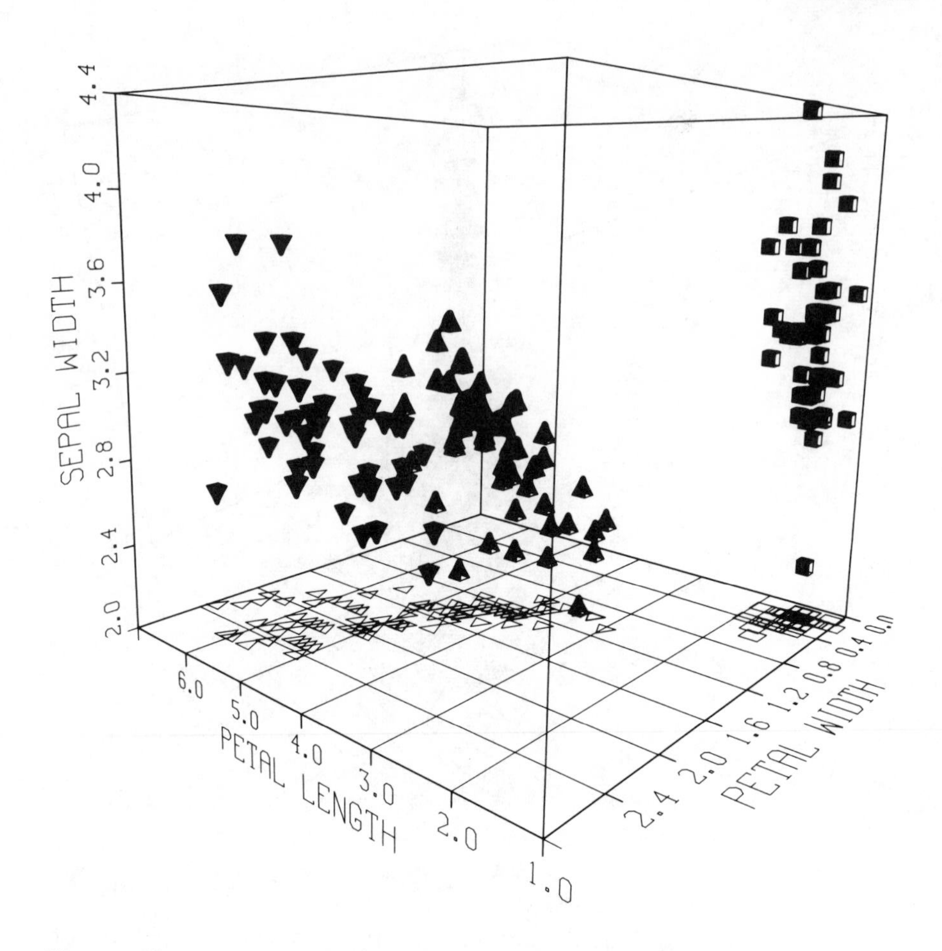

Figure 10. Rotating Fig. 8 through 90° clearly shows the separations of the three classes. The choice of viewpoint here is vital in "seeing" the separation.

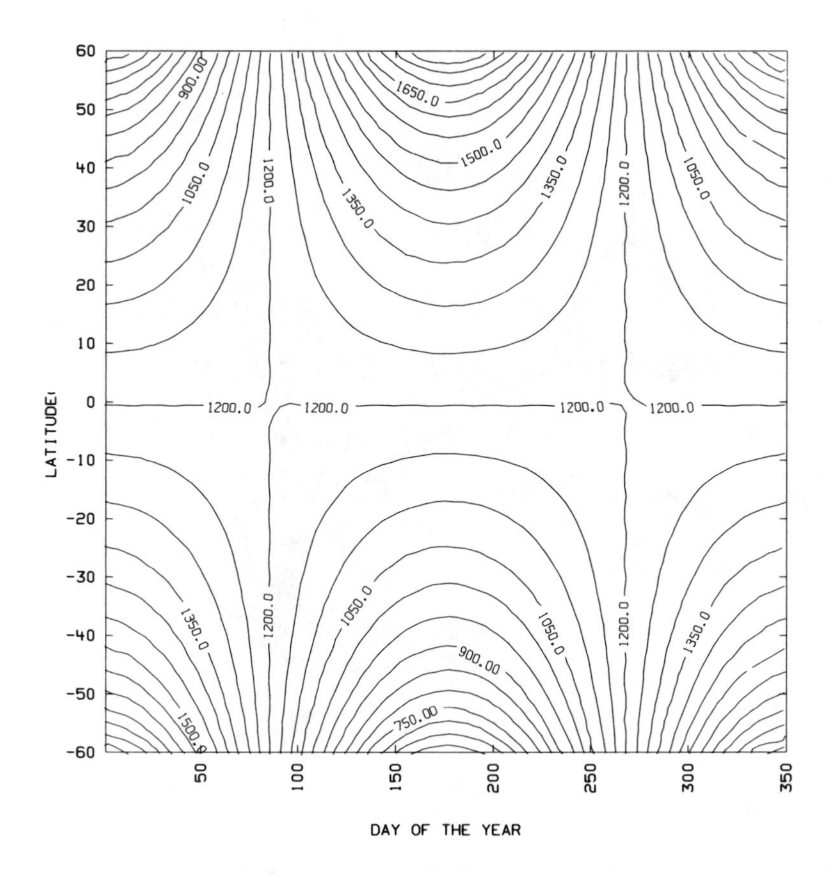

Figure 11. Contours of total hours of sunlight (multiplied by 100) over the latitude range ±60 as a function of day of the year. From this plot alone it is difficult for the novice to truly percieve the actual variation.

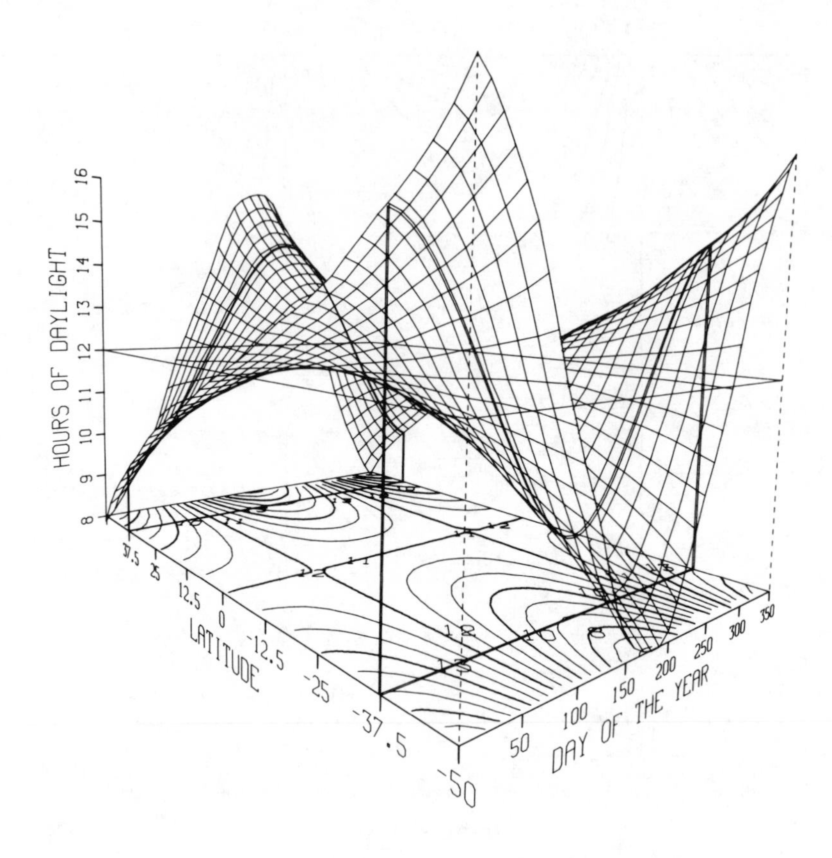

Figure 12. A combined surface/contour plot of the hours of daylight variation. The two cuts highlighted at ±37.5° latitude clearly show the universe seasonal variation between the two hemispheres.

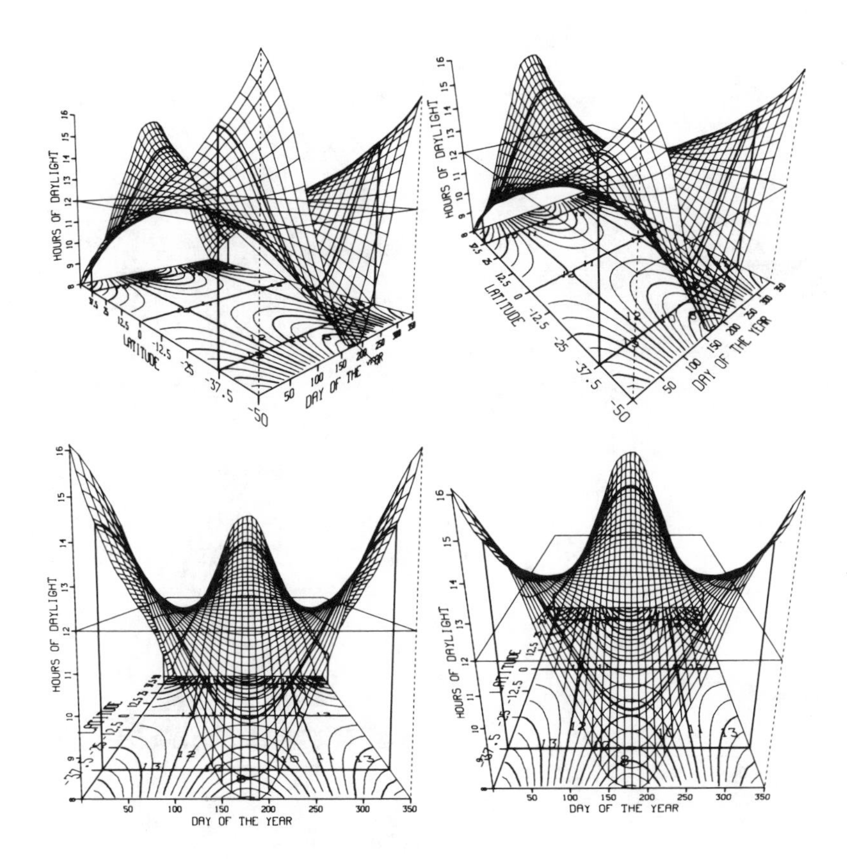

Figure 13. Four views of the hours of daylight variation. By examining a series of such views, the scientist gains a deeper understanding of the underlying variation.

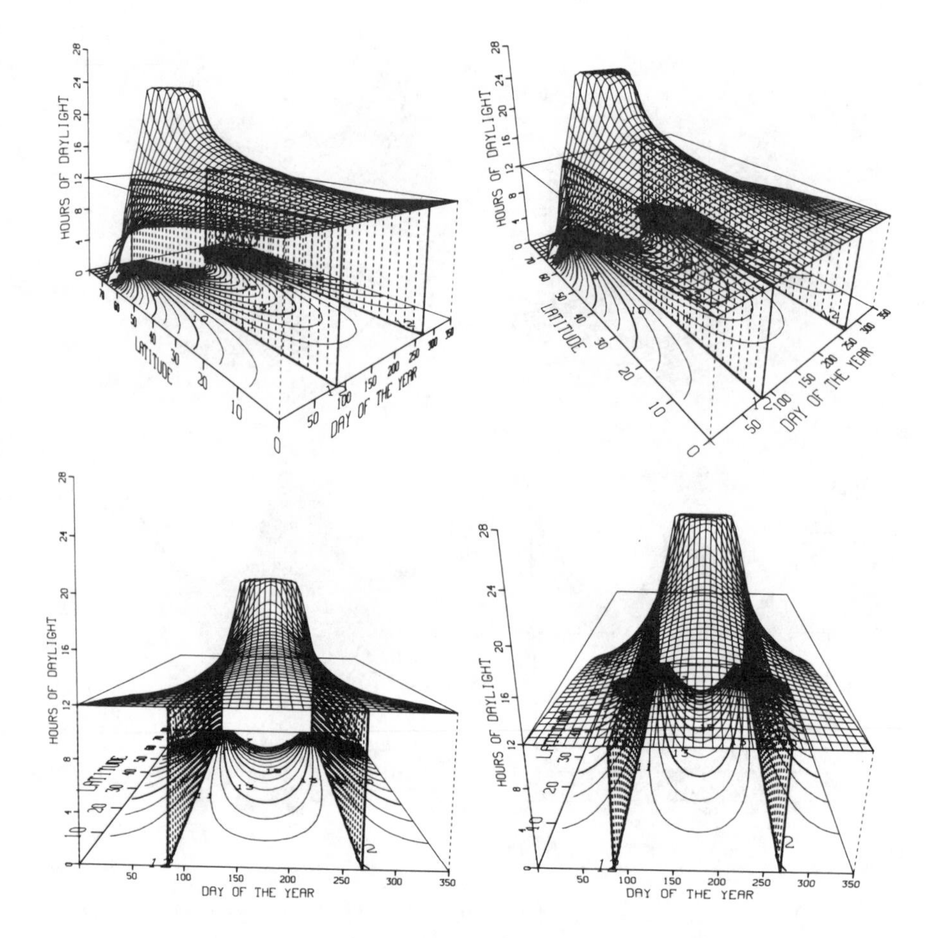

Figure 14. Hours of daylight in the northern hemisphere with the two equinoxes highlighted. Accenting such features can be of great value in calling attention to various aspects of data.

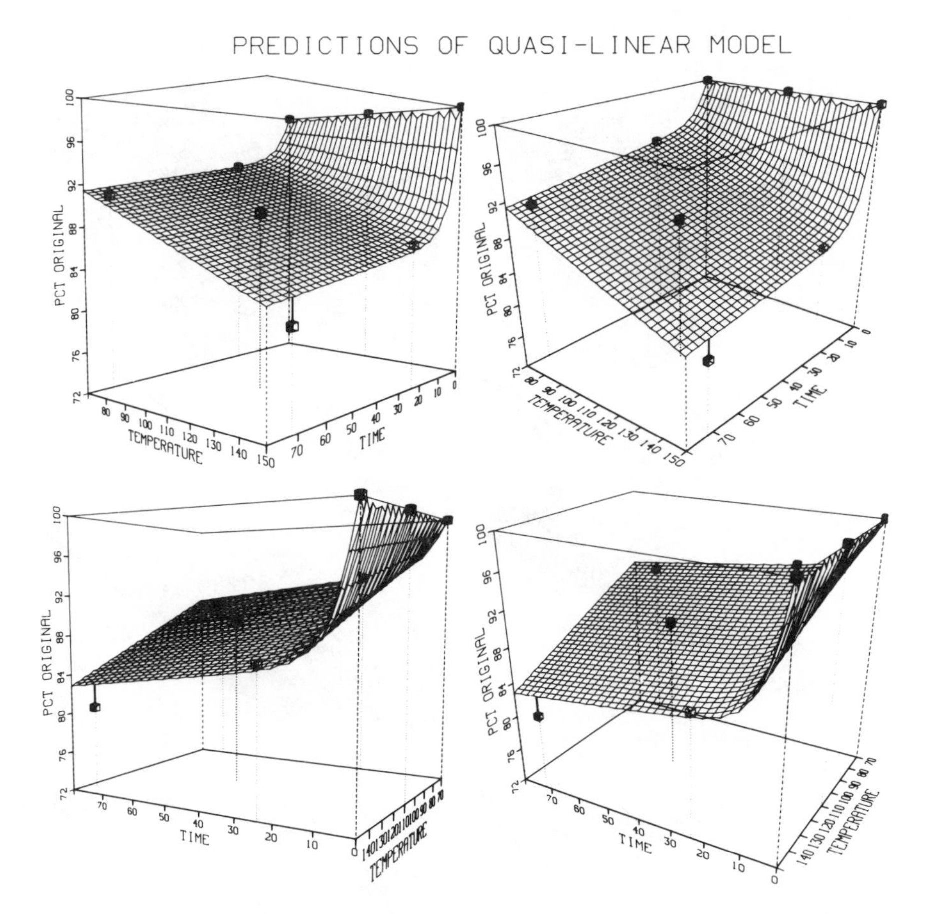

Figure 15. Four 3D views showing the predictions of a quasi-linear model fit to a set of aging data. In the range of the measurements the model fit is adequate.

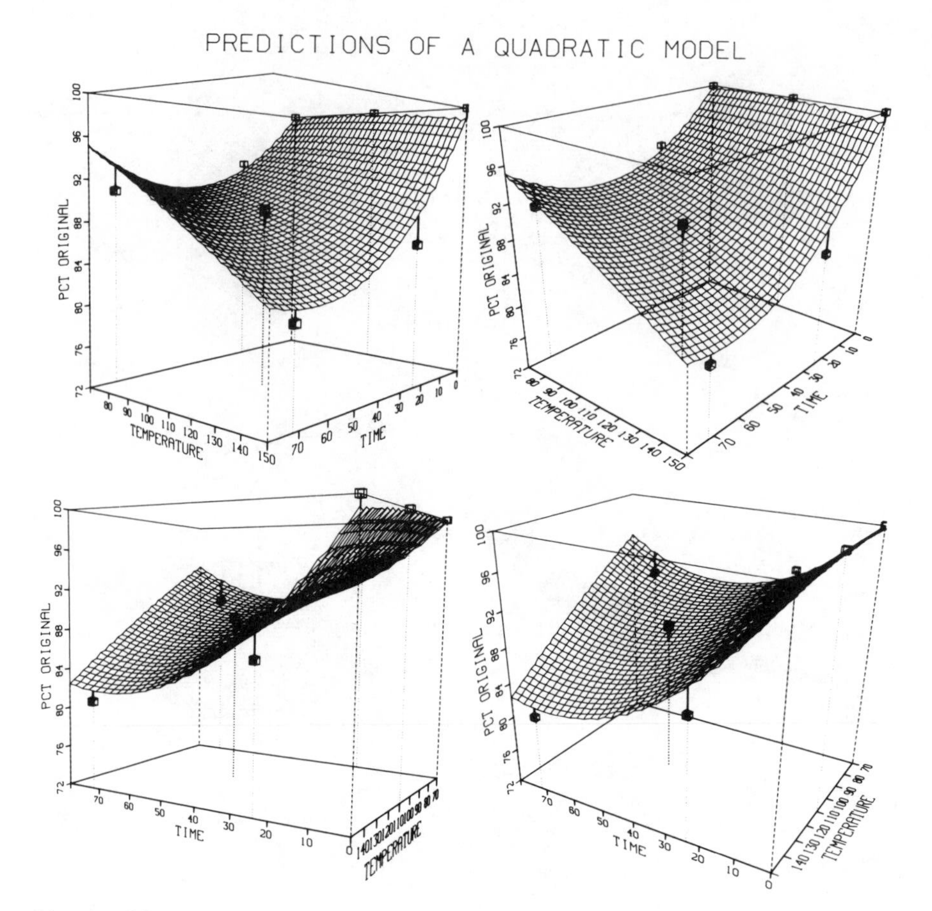

Figure 16. Analogous 3D views to Fig. 15 showing the fit to these data using a full quadratic model.

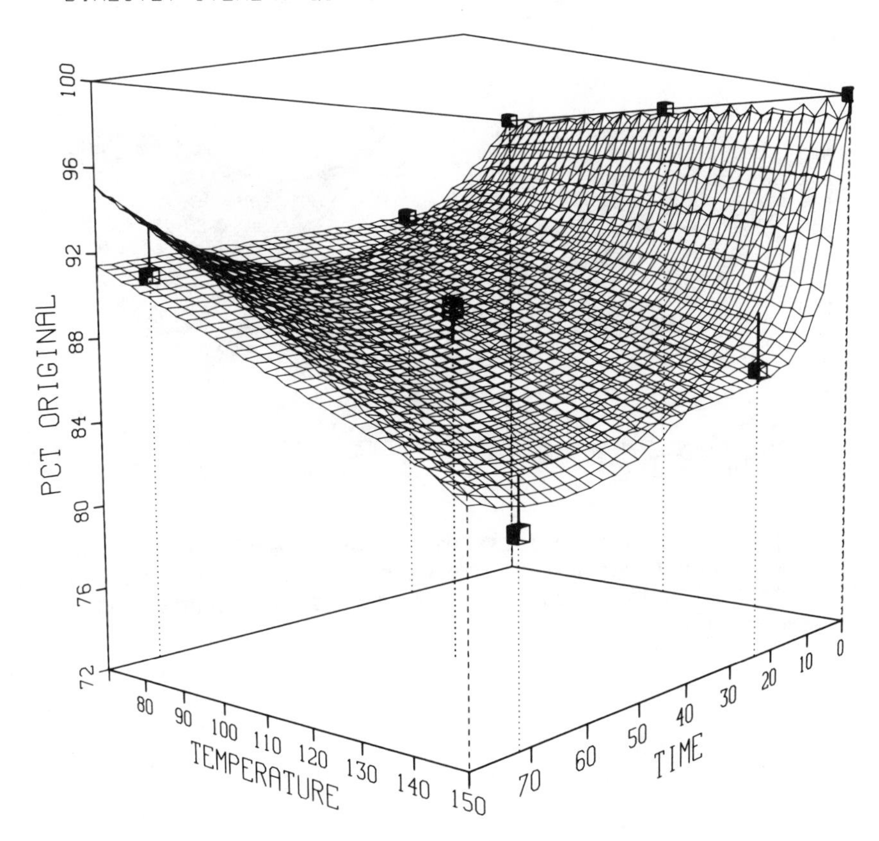

Figure 17. Direct comparison of the two models is performed by superimposing the two corresponding plots using software editing. When the initial plots are drawn to the same scale, and are mapped in the same manner, perfect registration will result.

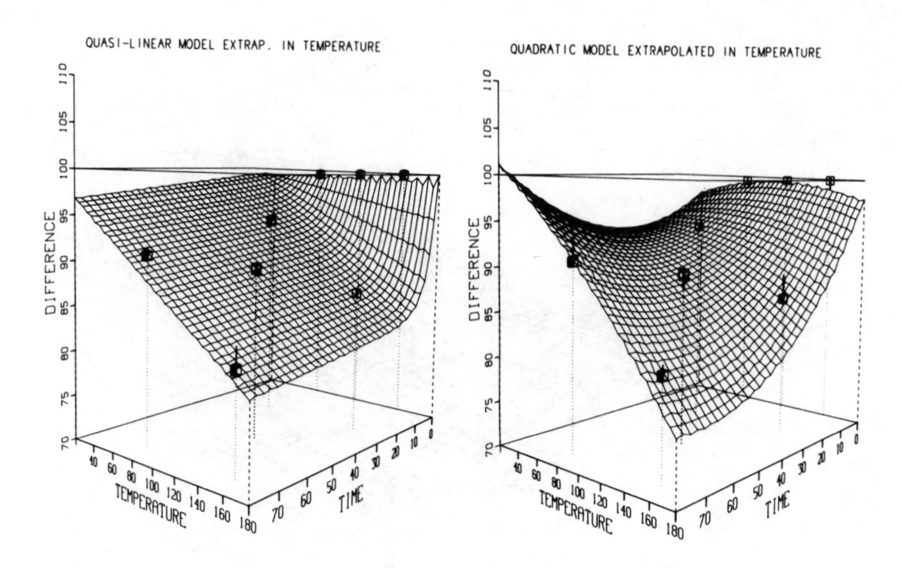

Figure 18. Side by side intercomparison of model predictions when the temperature variable is extrapolated beyond the range of the data. In this case, the two models give approximately the same results.

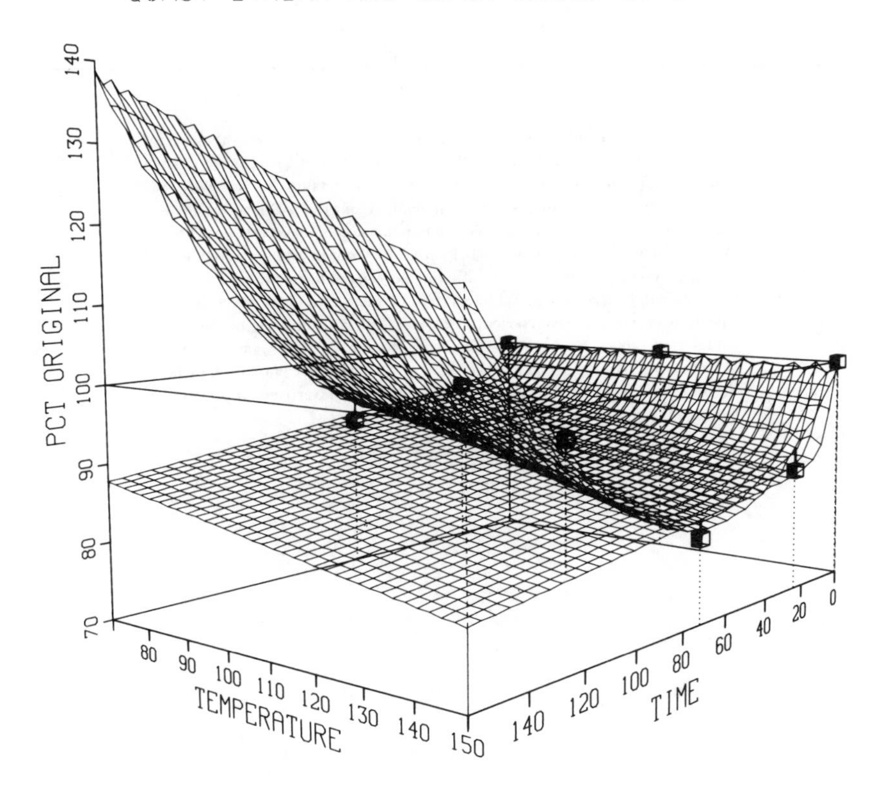

Figure 19. Comparison of two models when the time variable is extrapolated to about twice the range of the original data. The predictions of the two models are substantially different.

TEACHING CHEMOMETRICS

B.G.M. Vandeginste

Department of Analytical Chemistry
University of Nijmegen
Toernooiveld
6525 ED NIJMEGEN, The Netherlands

Two important developments took place during the past two decades, and have marked the face of analytical chemistry today: computers and instrumentation. Twenty years ago only a happy few dared the very user-unfriendly access to computers with: punched charts; paper tape; background processing without the possibility of interactive operation; long turn around times etc. Software packages being hardly available at that time, the development of algorithms was a very time-consuming activity. Likewise, at that time, analytical chemists were quite uninterested in the laboriously developed software because most of them lacked the necessary hardware for a meaningful application. Twenty years later, one can buy a microcomputer with a Random Access Memory capacity which is similar to the former academic computer center, for a fraction of the price. Consequently, complex data processing methods are now accessible for every respectable analytical laboratory. Sophisticated graphics capabilities allow information included in complex data structures in an ergonomic justified manner to be presented. There is no doubt that these developments will not slow down. On the contrary. There are good reasons to believe that the incredibly fast calculation capability of the computer center equipped with fast array processors will also be found on our desks over 10 to 20 years. Animated graphics on multi-dimensional data structures obtained with the analytical instruments of tomorrow, will become common property.

Todays general feeling that computers and microcomputers will govern the analytical chemistry practice triggered a big demand for academic and post-academic education in chemometrics. That big demand, of course, is also related to the fact that today the analytical chemist is capable of generating complex data

B. R. Kowalski (ed.), Chemometrics. Mathematics and Statistics in Chemistry, 467–479.

structures which were unthinkable twenty years ago. Major tools of today, HPLC, AAS, hyphenated methods like GC-MS, HPLC-UV-VIS, FIA,CFA, were not available at that time. Now, UV-VIS spectra can be recorded in less than 0.1 s. X-ray analyzers, in combination with an electron microscope are capable of recording X-ray spectra at the points of a grid over the microscope image in a few minutes. Data structures which were mainly one-dimensional, become now predominantly multi-dimensional.

Developments in the computer world also drastically changed the face of the analytical instrument. Switches and recorders are now replaced by keyboards and screens. A next step one can expect is the inclusion of intelligence in the instruments, which may lead to a generation of intelligent analyzers. Predecessors of such analyzers are the self-optimizing and self-calibrating systems which are now under development.

The tools in the pocket of the modern analytical chemists also expanded quite drastically. Jargon like RAM, ROM, IEEE, RS232, Pascal, Forth, LDA, LLM, KNN, ANOVA, PLS, SIMCA, ARTHUR, BMOP, IMSL, SAS, NAG, GSAM, UNEQ, SAMSON, SIMPLEX etc. is found in almost every present publication.

Although the number of mathematical and statistical tools is larger than ever, and the technical achievements are impressive, the aim of the chemical analysis has remained unchanged, namely the solution of the analytical problem at a lower price than the value of the generate analytical information.

Much of the conventional education in Analytical Chemistry is focused on the teaching of the fundamentals of instrumental methods. The measurement, however, is one step of the analytical process (figure 1). Of equal importance are the sampling strategies, the data processing, and the combination of analytical results into the requested information. The most perfect analytical method cannot neutralize the consequences of a bad sampling procedure. Well done measurements cannot cancel the consequences of a poor experimental design. A bad or wrong combination of analytical results distorts the informaiton which is hidden in the analytical data. Therefore, the teaching of Analytical Chemistry should embrace all steps of the analytical process, and should not be limited ot the instrumental methods. The right execution of the analytical determination requires a good

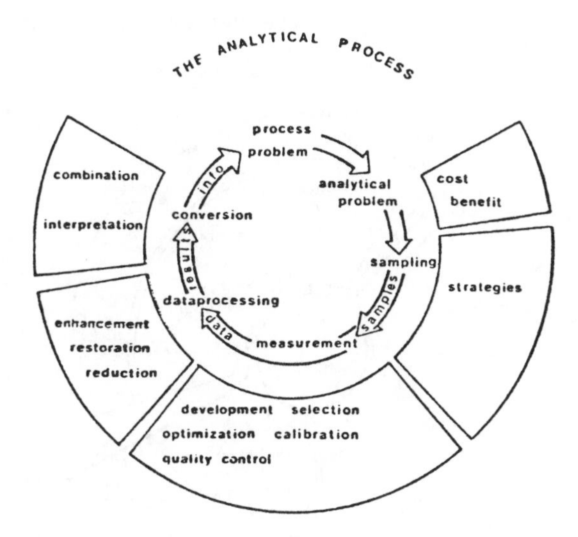

Figure 1 The analytical process

theoretical background in the method and a good skill for the manipulations. The optimization, quality control, calibration and design of the experiments require the application of advanced statistical and mathematical methods. So do the other steps in the analytical process (figure 2).

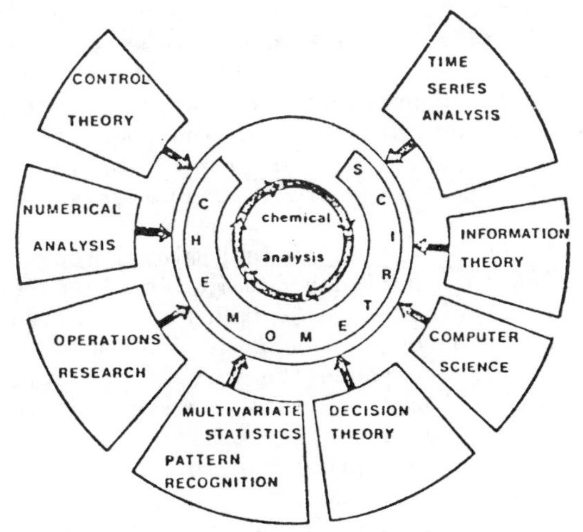

Figure 2

Chemometricians welded methods from operations reserach, contorl theory, multivariate statistics, information theory, numerical analysis and linear algebra into useful tools for the analytical chemist.

The chemistry curriculum in the Netherlands is schematically

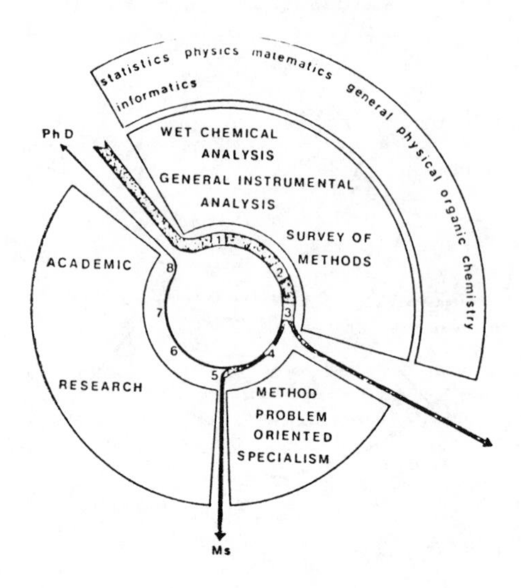

Figure 3

shown in figure 3. The first two years contain a uniform program for all students. Thereafter, a differentiated program is started where students have to choose a combination of one major topic (10 months) and two minor topics (6 months each). Such a major topic consists of full-time research and a number of classes. At the end of this curriculum, the masters certificate is obtained. Selected students get the opportunity to start a 4-year period research, in Nijmegen, the basics of some instrumental methods, linear algebra and some statistics are tought. During the practicals, the students get their first exposure to Chemometrics.

Chemometrics is one of the major (and minor) topics which can be chosen. During that period a combination is offered of a 30 hours of chemometrics class, and 30 hours of selected topics (e.g. workshops on pH in blood, asbestos in air, acid rain) (figure 4). The students are requested to write a review on a selected topic and each student presents a poster session every 6 weeks. The purpose of the posters is to train the students in presenting their results and to provide a forum where the practical applications of the theoretical concepts, which are presented during the class, are demonstrated and discussed (table 1).

Because of the wide variety and complexity of the statistical and mathematical tools, and our relatively short course time,

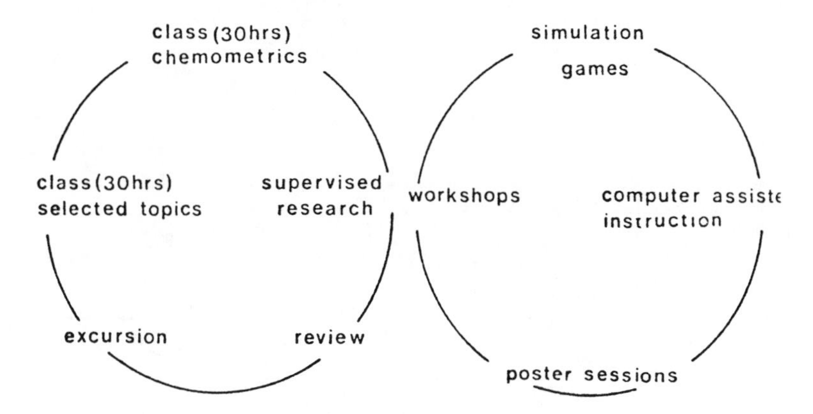

Figure 4

it is extremely important to design and employ very efficient teaching aids. The computer, being an indispensible tool for the research of the chemometrician, can also be exploited to the benefit of students' education.

Table 1

Analytical activity	topics
sampling for description; control; monitoring	statistics; autoregression; auto-/cross correlation; time series analysis
measurement calibration; selection; quality control; optimization	statistics; measurability; control charts; experimental design/Simplex; cost/benefit
data processing enhancement; restoration; reduction; reconstruction	filtering (time/frequency domain); curve fitting; deconvolution; regression; analysis of variance
result conversion combination; interpretation; reconstruction	multivariate statistics; pattern recognition; information theory; recursive estimation
organisation sample routing; delays; forecasting; laboratory description	Operations Research; queueing theory; digital simulation

Two efficient tools where the computer has a central part have been developed at our laboratory: simulation games (1) and a computer-aided instruction (2,3).

The structure of the simulation game is schematically shown in Table 2.

Table 2

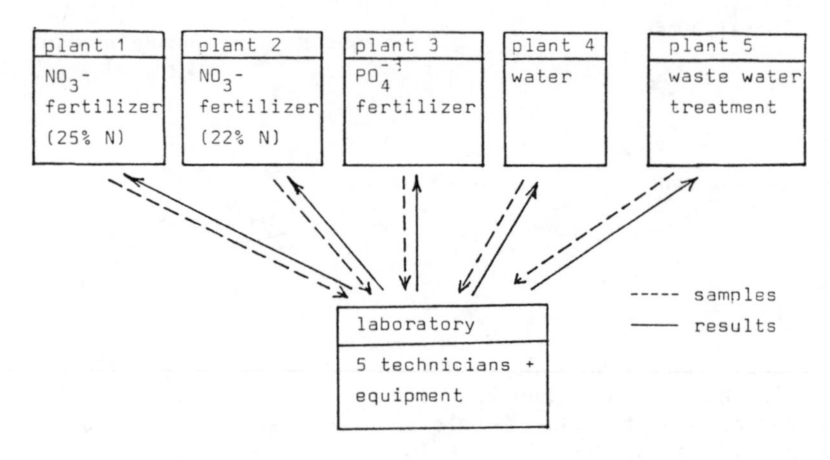

Five industrial plants are simulated, each having one fluctuating output variable of interest, e.g. the Nitrogen content in a continuous NO_3 - fertilizer stream, the hardness of the water used for steamboilers, the Ca^{+2} and SO_4^{2-} content in waste water, the Ca/SO_4 ratio in a phosphate fertilizer. Each of the processes is characterized by a required set point, the costs under uncontrolled operation, the standard deviation and Time **constant** (first order processes) of the fluctuations. For each of the industrial plants, a choice is available between 5 analytical methods (table 3) each with a specified (but unknown **for** the players) precision, analysis time, costs per analysis, and requirements for personnel and instrumentation. The analytical results are used to detect undesired deviations from the set point, whereafter a control action (feed forward hold loop) follows. The game can be played in two modes: as a competition game and as a mangement game. In the competition mode the same process is assigned to each student (maximum 10). The goal of the game is to obtain maximum profit.

Adjustable parameters for the players are: the analytical

Table 3

Analysis method	plants 1	2	3	4	5
	*				
1	1 Devarda N	6 Devarda N	8 grav. total $CaSO_4$	9 titration tot.hardness	10 electr. SO_4^{-2}
2	2 auto.anal. (NH_4^+)	2 auto.anal. NH_4^+	9 titration $Ca+SO_4^{-2}$	11 electr. Ca+Mg	10 electr. Ca^{+2}
3	3 electr. NO_3^-	7 electr. NO_3^-	10 electr.	12 foam test	14 Kjeldahl N
4	4 ν-act N	4 ν-act N	10 electr. Ca	13 titr. Ca+Mg	15 conduct
5	5 X-ray diffr. $CaCO_3/NH_4NO_3$	5 X-ray diffr. $CaCO_3/NH_4NO_3$	10 electr. SO_4		

*instrument number

method, the analysis frequency and a control factor. Futhermore the players have to process their collected data into process information.

In the competition mode, a group of 10 players forms the management of a laboratory with the task to analyze the five plants. Limitations on personnel and instrumentation prevent the separate optimization of the sampling and analysis of the plants. Consequently, the players have to optimize the laboratory activities as a whole. Because of the magnitude of the data stream, and the complexity of the decisions which have to be taken, the players are forced to organize their activites and to divide the tasks. The latter is the goal of the game. The hardware configuration consists of:

- a HP9845 which is the master console and give a continuous display of all process characteristics (stochastic and deterministic part), and all analytical activities. This display is not accessible for the players. At any time, the game operator may intervene in the course of the processes. At regular times an overview is displayed of the cost/benefit of each process;

- a terminal interfaced to the HP9845, for data entry by the players. Commands are: the process to be controlled, the analytical method, the batch size of the samples, the sampling interval, and a cancel command;

- a printer, which returns the analytical results (after a time equal to the analysis time), gives a review of the costs/benefit and prompts the players with messages (telex); this printer is the only source of information for the players;

- as an option an Apple II may be interfaced to the HP9845, equipped with a number of monitors in order to inform observers of the game. The observers receive continuous information on the analytical activities carried out on the processes. At any time the operator can produce a graphical display of the controlled and uncontrolled process. At regular intervals an overview is produced of the costs/benefit.

The software package (3) for the compter-aided instruction consists of a set of 14 modules, which are assembled in a database, and may be used for demonstrations during the classes, and for individual instruction and education as well. The driver of the package controls the whole session, without any need from the student to know any compuer-specific jargon. The package is easily updated with new modules, or obsolete modules can be replaced. Default input is provided as an option for a first inspection of the functioning of the algorithms. At the end of the run of a particular module, control is transfered to the driver, which displays again all available modules for a next choice. By typing the proper code, the requested modules is read from floppy disc, whereafter, execution is started. The majority of the modules operates in the graphics mode. Input requests are displayed without switching to the alphanumeric display. The powerful graphics capabilities of the used microcomputer (HP9845) allows to display several figures at a time, whereafter a selective replacement of parts of the screen is possible. Most modules request the student to define some parameters to generate the data which will be studied (e.g. a spectrum, a fluctuating process), whereafter the computer prompts for defining the adjustable parameters of the algorithm which will act on the data (e.g. type of filter in Fourier domain, cut-off frequency). A typical example of the lay-out of the graphics is given in figure 5. The two upper figures show the spectrum defined by the student and its power spectrum. The filter shape as defined by the student is given in the lower right hand corner and the result of the application of that filter on the signal is found in the left hand corner. In an interactive way the student can change the parameters of the chosen filter or he may decide to apply another filter type.

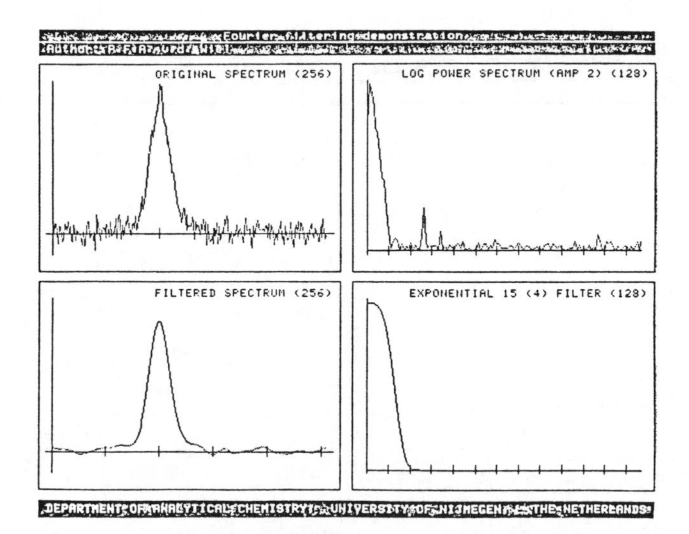

Figure 5

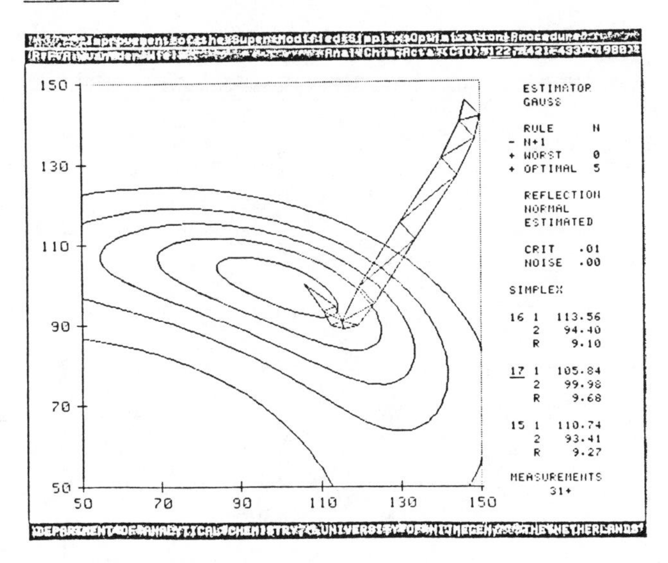

Figure 6

Figures 5 - 9 give an impression of the graphical displays of the program.

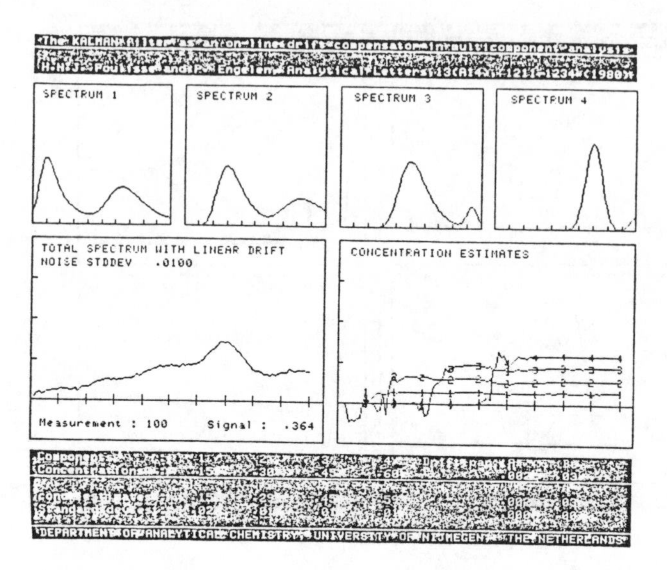

Figure 7

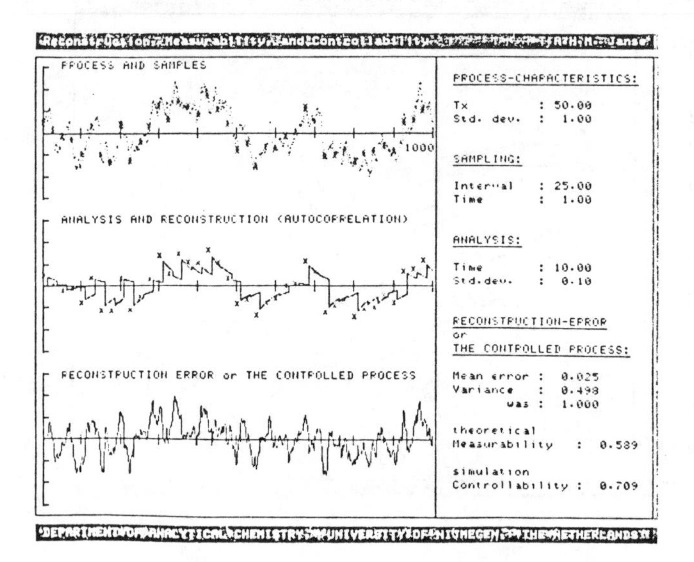

Figure 8

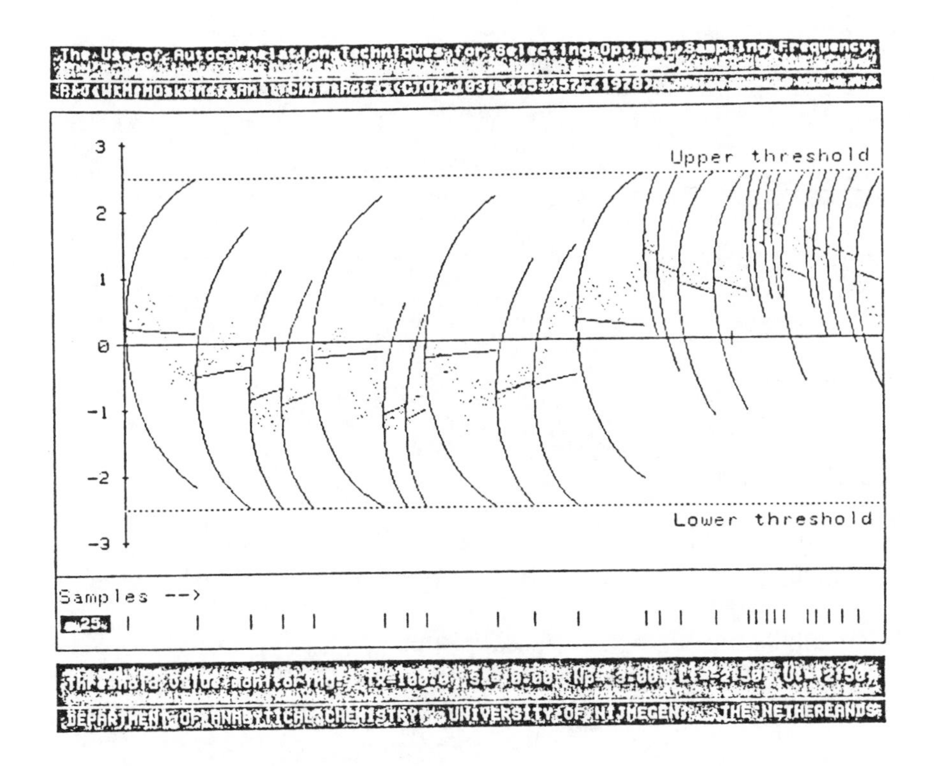

Figure 9

The table below, lists the modules, the options, and the adjustable parameters, description of the graphics and alphanumeric output.

Table 4

module	options	adjustable parameters of the algorithm	output
a1. sampling of lots	1-d lot: Time constant (Tx) of the process fluctuations in the lot	- grab size - number of grabs	
a2. process monitoring (threshold control)	1-st order process with a process time constant (Tx)	- lower threshold - upper threshold probability of undedected exceedings	figure 9
a3. process control	1-st order process with a process time constant (Tx)	- sampling interval - sampling time - analysis time - analysis precision	
b1. Simplex optimization	2- d response surface	- modified Simplex - supermodified Simplex	figure 6
b2. curve fitting	spectrum: Gaussian and/or Cauchy peak shape; number of peaks + peak parameters stand.dev. spectrum noise	- initial estimates	
b3. experimental design	number of variables (number of levels = 2) total number of experiments (single or in duplo	generators (number is defined by options)	- design lay-out - effect + confounding effects/inter-actions - test on residuals
b4. recursive parameter estimation	4-component spectrum: - concentrations of the 4 components - base line drift	filter parameters	figure 7
c1. autocorrelation and autoregression	time series: - time constant of stochastic part - deterministic part (drift, periodicity)	variables in the auto-regressive plots	figure 8
c2. detection limit	simulator of a spectrophoto-metric determination	- concentration - gain - base line - integration time - paired/unpaired measurements	
e1. sequential tests of significance	observations entered by user	- the risks α and β of errors of the first and second kind - the magnitude of the difference to detect	
e2. minimal spanning tree	data base from R.T.P.Jansen et al. Ann.Clin.Biochem. 18 (1981) 218-225.		
e3. Fourier filtering	spectrum: Gaussian peak + peak parameters, noise characteristic, number of data points	- filter type - filter parameter	figure 5
f1. laboratory simulation	simulator of two one-server systems in series	- mean interarrival times of samples - mean analysis time analyst 1 and 2 - type of distributions - simulation time	

Acknowledgement

The computer aided instruction program consists of a compliation of programs developed by many of the staff members of the department of analytical chemistry at the University of Nijmegen: T.A.H.M. Janse, P.F.A. van der Wiel, and prof. G. Kateman. The simulation games are based on the ideas of prof. G. Kateman and have been programmed by P.F.A. van der Wiel.

References

1. M.J.A. van den Akker, G. Kateman, Fres. Z. Anal. Chem. 282 (1976) 97-103.

2. B.G.M. Vandeginste, Anal. Chim. Acta 105 (1983) 199-206.

3. P.F.A. van der Wiel, G. Kateman, B.G.M. Vandeginste, T.A.H.M. Janse, presented at the CAC-II Conference, Petten, The Netherlands.

INDEX